MYP Mathematics

A concept-based approach

4 & 5
Extended

Rose Harrison • Clara Huizink
Aidan Sproat-Clements • Marlene Torres-Skoumal

OXFORD

How to use this book

Chapters

Chapters 1–3 each focus on one of the key concepts:

Form, Relationships and Logic.

Chapters 4–15 each focus on one of the twelve related concepts for Mathematics:

Representation, Simplification, Quantity, Measurement, Patterns, Space, Change, Equivalence, Generalization, Justification, Models, and Systems

Creating your own units

The suggested unit structure opposite shows just one way of grouping all the topics from different chapters, from both *Standard* and *Extended*, to create units. These units have been created following the official guidelines from the IB Building Quality Curriculum. Each unit is driven by a meaningful statement of inquiry and is set within a relevant global context.

- Each chapter focuses on one of the IB's twelve related concepts for Mathematics, and each topic focuses on one of the three key concepts. Hence, these units, combining topics from different chapters, connect the key and related concepts to help students both understand and remember them.

- Stand-alone topics in each chapter teach mathematical skills and how to apply them, through inquiry into factual, conceptual, and debatable questions related to a global context. This extends students' understanding and ability to apply mathematics in a range of situations.

- You can group topics as you choose, to create units driven by a contextualized statement of inquiry.

- This book covers the MYP Extended skills framework.
MYP Mathematics 4 & 5 Standard covers the Standard skills.

Using *MYP Mathematics 4 & 5 Extended* with an existing scheme of work

If your school has already established units, statements of inquiry and global contexts, you can easily integrate the concept-based topics in this book into your current scheme of work. The table of contents on page vii clearly shows the topics covered in each concept-based chapter. Your scheme's units may assign a different concept to a given topic than we have. In this case, you can simply add the concept from this book to your unit plan. Most topics include a Review in context, which may differ from the global context chosen in your scheme of work. In this case, you may wish to write some of your own review questions for your global context, and use the questions in the book for practice in applying mathematics in different scenarios.

Suggested plan

The units here have been put together by the authors as just one possible way to progress through the content.

Units 1, 5, 6, 9 and 10 are made up of topics solely from the *MYP Mathematics 4 & 5 Standard* book.

> Access your support website for more suggested plans for structuring units:
> www.oxfordsecondary.com/myp-mathematics

UNIT 2

Topics: Scatter graphs and linear regression, drawing reasonable conclusions, data inferences

Global context: **Identities and relationships**
Key concept: **Relationships**

UNIT 3

Topics: Equivalence transformations, inequalities, non-linear inequalities

Global context: **Identities and relationships**
Key concept: **Form**

UNIT 4

Topics: Rational and irrational numbers, direct and indirect proportion, fractional exponents

Global context: **Globalization and sustainability**
Key concept: **Form**

UNIT 7

Topics: Finding patterns in sequences, making generalizations from a given pattern, arithmetic and geometric sequences

Global context: **Scientific and technical innovation**
Key concept: **Form**

UNIT 8

Topics: Simple probability, probability systems, conditional probability

Global context: **Identities and relationships**
Key concept: **Logic**

UNIT 11

Topics: Circle segments and sectors, volumes of 3D shapes, 3D orientation

Global context: **Personal and cultural expression**
Key concept: **Relationships**

UNIT 12

Topics: Using circle theorems, intersecting chords, problems involving triangles

Global context: **Personal and cultural expression**
Key concept : **Logic**

UNIT 13

Topics: Algebraic fractions, equivalent methods, rational functions

Global context: **Scientific and technical innovation**
Key concept: **Form**

UNIT 14

Topics: Evaluating logarithms, transforming logarithmic functions, laws of logarithms

Global context: **Orientation in space and time**
Key concept: **Relationships**

UNIT 15

Topics: The unit circle and trigonometric functions, sine and cosine rules, simple trigonometric identities

Global context: **Orientation in space and time**
Key concept: **Relationships**

About the authors

Marlene Torres-Skoumal has taught DP and MYP Mathematics for several decades. In addition to being a former Deputy Chief Examiner for IB HL Mathematics, she is an IB workshop leader, and has been a member of various curriculum review teams. Marlene has authored both DP and MYP books, including Higher Level Mathematics for Oxford University Press.

Clara Huizink has taught MYP and DP Mathematics at international schools in the Philippines, Austria and Belgium. She has also been through the IB experience herself as a student and she is a graduate of the IB program.

Rose Harrison is the Lead Educator for MYP Mathematics at the IB Organization. She is an MYP and DP workshop leader, a senior reviewer for Building Quality Curriculum in the MYP, and has held many positions of responsibility in her 20 years' experience teaching Mathematics in international schools.

Aidan Sproat-Clements is Head of Mathematics at Wellington College in the UK, an IB World School. He has spent his career in British independent schools where he has promoted rigorous student-led approaches to the learning of Mathematics. He helped to develop the pilot MYP eAssessments for Mathematics.

Topic opening page

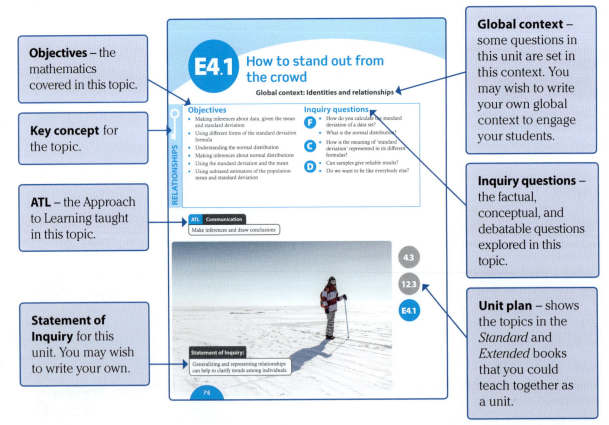

Objectives – the mathematics covered in this topic.

Key concept for the topic.

ATL – the Approach to Learning taught in this topic.

Statement of Inquiry for this unit. You may wish to write your own.

Global context – some questions in this unit are set in this context. You may wish to write your own global context to engage your students.

Inquiry questions – the factual, conceptual, and debatable questions explored in this topic.

Unit plan – shows the topics in the *Standard* and *Extended* books that you could teach together as a unit.

Learning features

Each topic has three sections, exploring:
- **factual**
- **conceptual**
- **debatable** inquiry questions

Problem solving – where the method of solution is not immediately obvious, these are highlighted in the Practices.

Explorations are inquiry-based learning activities for students working inidvidually, in pairs or in small groups to discover mathematical facts and concepts.

ATL highlights an opportunity to develop the ATL skill identified on the topic opening page.

Reflect and discuss – opportunities for small group or whole class reflection and discussion on their learning and the inquiry questions.

Worked examples show a clear solution and explain the method.

Practice questions written using IB command terms, to practice the skills taught and how to apply them to unfamiliar problems.

Objective boxes highlight an IB Assessment Objective and explain to students how to satisfy the objective in an Exploration or Practice.

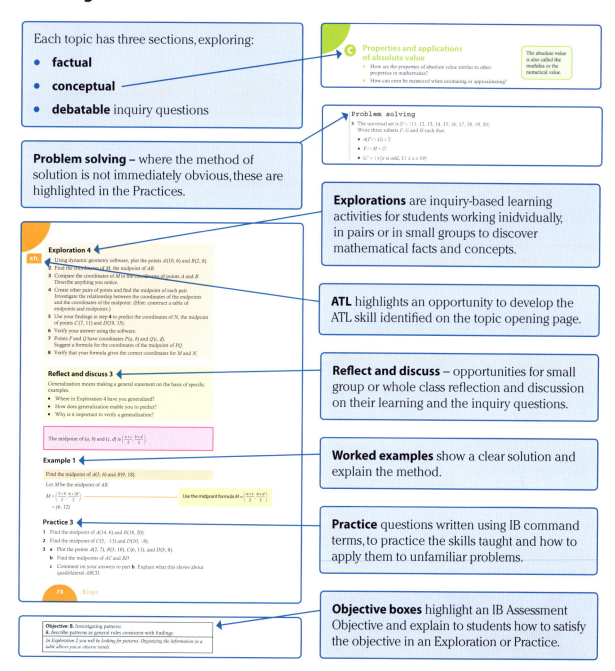

Technology icon

Using technology allows students to discover new ideas through examining a wider range of examples, or to access complex ideas without having to do lots of painstaking work by hand.

This icon shows where students could use Graphical Display Calculators (GDC), Dynamic Geometry Software (DGS) or Computer Algebra Systems (CAS). Places where students should not use technology are indicated with a crossed-out icon.

The notation used throughout this book is largely that required in the DP IB programs.

Each topic ends with:

Summary

The **standard deviation** is a measure of dispersion that gives an idea of how close the original data values are to the mean, and so how representative the mean is of the data. A small standard deviation shows that the data values are close to the mean. The units of standard deviation are the same as the units of the original data.

Σx means the sum of all the x values.

Using this notation, the mean of a set of values is $\frac{\Sigma x}{n}$, the sum of all the values divided by the number of values.

A formula to calculate the standard deviation

Using Σ notation, the mean of a set of discrete data values in a frequency table is $\frac{\Sigma x f}{\Sigma f}$, the sum of all the xf values divided by the total frequency.

A formula to calculate the standard deviation for discrete data presented in a frequency table is:

$$\sigma = \sqrt{\frac{\Sigma f(x-\mu)^2}{\Sigma f}}$$

The **normal distribution** is a symmetric distribution, with most values close to the mean and tailing off evenly in either direction. Its frequency graph is a bell-shaped curve.

Summary of the key points

Mixed practice

In questions **1** and **2** the data provided is for the population.

1 Find the mean and the standard deviation of each data set:

a 2, 3, 3, 4, 4, 5, 5, 6, 6, 6
b 21 kg, 21 kg, 24 kg, 25 kg, 27 kg, 29 kg
c

x	f
3	2
4	3
5	2

d

Interval	Frequency
1–5	2

Number of strawberries on each plant fed with special strawberry plant food:

| 14 | 15 | 17 | 17 | 19 | 19 | 12 | 14 | 15 | 15 |

Analyze the data to find whether there is an effect of using the special strawberry plant food.

3 Bernie recorded the weight of food in grams his hamster ate each day. Here are his results:

Day	1	2	3	4	5	6	7	8	9
Food (g)	55	64	45	54	60	50	59	61	49

a **Find** the mean and standard deviation of the weight of food the hamster ate over the 9 days.

b Bernie assumes that his hamster's food consumption is normally distributed.

Mixed practice – summative assessment of the facts and skills learned, including problem-solving questions, and questions in a range of contexts.

Review in context

The ancient art of origami has roots in Japanese culture and involves folding flat sheets of paper to create models. Some models are designed to resemble animals, flowers or buildings; others just celebrate the beauty of geometric design.

The resulting creases are illustrated in this diagram.

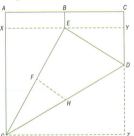

The artist thinks lengths FH and DY are equal.

b **Explain** why $EX = \frac{1}{2} EG$.

Review in context – summative assessment questions within the global context for the topic.

Reflect and discuss

How have you explored the statement of inquiry? Give specific examples.

Reflection on the statement of inquiry

Table of contents

Access your support website:
www.oxfordsecondary.com/myp-mathematics

Form

E1.1 What if we all had eight fingers? • Different bases — 2

Relationships

E2.1 Multiple doorways • Composite functions — 21

E2.2 Doing and undoing • Invserse functions — 35

Logic

E3.1 How can we get there? • Vectors and vector spaces — 54

Representation

E4.1 How to stand out from the crowd • Data inferences — 76

Simplification

E5.1 Super powers • Fractional exponents — 98

Quantity

E6.1 Ideal work for lumberjacks • Evaluating logs — 108

Measurement

E7.1 Slices of pi • The unit circle and trignometric functions — 124

Patterns

E8.1 Making it all add up • Arithmetic and geometric sequences — 145

Space

E9.1 Another dimension • 3D Orientation — 167

E9.2 Mapping the world • Oblique triangles — 183

Change

E10.1 Time for a change • Log functions — 202

E10.2 Meet the transformers • Rational functions — 216

Equivalence

E11.1 Unmistaken identities • Trigonometric identities — 233

Generalization

E12.1 Go ahead and log in • Laws of logarithms — 249

Justification

E13.1 Are we very similar? • Problems involving triangles — 264

Models

E14.1 A world of difference • Nonlinear inequalities — 282

Systems

E15.1 Branching out • Conditional probability — 299

Answers — 315

Index — 350

E1.1 What if we all had eight fingers?

Global context: Scientific and technical innovation

Objectives

- Understanding the concept of a number system
- Counting in different bases
- Converting numbers from one base to another
- Using operations in different bases

Inquiry questions

F
- How have numbers been written in history?
- What is a number base?
- How can you write numbers in other bases?

C
- How are mathematical operations in other bases similar to and different from operations in base 10?

D
- Would you be better off counting in base 2?
- How does form influence function?

| ATL | Communication |

Use intercultural understanding to interpret communication

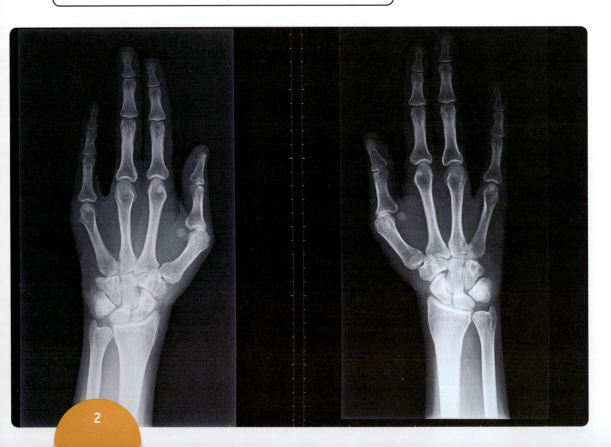

You should already know how to:

• use the operations of addition, subtraction and long multiplication in base 10, without a calculator	**1** Calculate these by hand: **a** 10 442 + 762 **b** 10 887 − 7891 **c** 27 × 43 **d** 14 078 × 71	
• understand place value	**2** Write down the value of the 5 in: **a** 351 **b** 511 002 **c** 15 **d** 1.5	

Numbers in different bases

- How have numbers been written in history?
- What is a number base?
- How can you write numbers in other bases?

Humans have used many different ways to record numbers. How would you write down the number of green bugs in this diagram?

You might have written a symbol 5, or the word "five". You could have used a word in a different language, or maybe even a tally: 卌. Each of these represents the number in a different way, but they all represent the same quantity.

Different cultures use different forms of notation to represent number. The ancient Egyptian hieroglyphic number system was an **additive system**. Each symbol has a different value and you find the total value of the number by adding the values of all the symbols together.

Egyptian numerals use these symbols:

stroke	heelbone	coiled rope
1	10	100

lotus flower	pointed finger	tadpole	scribe
1000	10 000	100 000	1 000 000

> Green shield bugs are sometimes called green stink bugs, as they produce a pungent odor if handled or disturbed.

The number 11 is written |∩, and 36 is written ||||||∩∩∩.

E1.1 What if we all had eight fingers?

Reflect and discuss 1

- Write each number in Egyptian numerals:

  5   32   126   99   100   10 240

- Write these numbers as ordinary (decimal) numbers:

- What are the advantages of the Egyptian number system?
- What are the disadvantages?

> Our number system, often called Arabic numerals, is a **place value system**. The position of a digit tells you its value.

The two numbers below have the same four digits, but digits do not always represent the same amount.

A digit in the furthest right column represents individual objects: *units*.
Moving left, the second column represents collections of ten units: *the tens column*.
The third column represents collections of ten tens: *the hundreds column*.
The pattern continues: each column is worth 10 times more than the column to its right.

The number 364 actually represents a collection of that many objects:

hundreds	tens	units
3	6	4

NUMBER

Reflect and discuss 2

- Does the symbol 0 have different values in the numbers 205 and 2051? Does it represent something different?
- The Roman numerals do not have a symbol to represent zero, so why does the place value system need a symbol for zero?

Our number system is known as decimal (also as base ten, or denary) because the value of each place value column is ten times the value of the column to its right.

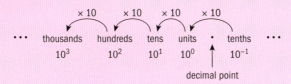

You can make a place value system with any natural number base. In binary (base two) each column is worth twice the column to its right.

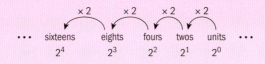

The binary number 10111_2 represents a set containing:

sixteens	eights	fours	twos	units
1	0	1	1	1

The subscript 2 means that the number 10111_2 is written in base 2.

Computer logical systems use base 2. This is why you see powers of 2 in lots of contexts relating to computers. For example, SD cards which store 8 GB (gigabytes), 16 GB, 32 GB, 64 GB and so on, rather than 10 GB, 20 GB, etc. Also, whereas the prefix *kilo* usually means 1000 (e.g. there are 1000 meters in a kilometer), in computing, the term *kilo* means 1024 (2^{10}), so there are 1024 bytes in a kilobyte.

Tip

Base 3 - ternary, or trinary
Base 4 - quaternary
Base 8 - octal
Base 12 - duodecimal
Base 16 - hexadecimal (widely used in computing)

E1.1 What if we all had eight fingers?

The base of a number system tells you how many unique symbols the number system has. Base 10 has ten unique number symbols, 0 through 9. Base 2 has two unique symbols, 0 and 1.

Example 1

Find the value of 12011_3 in base 10.

3^4	3^3	3^2	3^1	3^0
81	27	9	3	1
1	2	0	1	1

Write the powers of 3 above the digits, starting with $3^0 = 1$ at the right-hand side.

$81 + 2 \times 27 + 3 + 1 = 139$ — Add up the parts that make up the number.

$12011_3 = 139_{10}$ — Use subscripts to show the base.

Practice 1

1 Find the value of each binary number in base 10.

   a 10111_2 **b** 11001_2 **c** 1101101_2

2 Convert each number to base 10.

   a 21002_3 **b** 22101_3 **c** 412_5

   d 332_5 **e** 64_8 **f** 77_9

Problem solving

3 Write these numbers in ascending order:

   1010_4 100011_2 1011_3 1111_5

4 The number $1101011000_2 = 856_{10}$.

   a Describe the relationship between 1101011000_2 and 110101100_2.

   b Find the value of 1101011_2.

> Music is the pleasure the human mind experiences from counting, without being aware that it is counting.
>
> – Gottfried Leibniz

Exploration 1

1 Use this table to explore questions **a** to **g**.

3^7	3^6	3^5	3^4	3^3	3^2	3^1	3^0
2187	729	243	81	27	9	3	1

   a Explain how the value of 3^7 tells you that the number 1038_{10} will have 7 digits in base 3.

   b $1038_{10} = 729_{10} + 309_{10}$.

      Explain how this sum tells you that the first digit of 1038_{10} in base 3 will be 1.

▶ Continued on next page

c $309_{10} = 243_{10} + 66_{10}$.

Explain how this sum tells you that the second digit of 1038_{10} in base 3 will also be 1.

d $66_{10} < 81_{10}$.

Explain how this inequality tells you that the third digit of 1038_{10} in base 3 will be 0.

e $66_{10} = 2 \times 27_{10} + 12$.

Explain how this sum tells you that the fourth digit of 1038_{10} in base 3 will be 2.

f Continue the process to find the value of 1038_{10} in base 3.

g Check your solution by converting it back into base 10.

2 Use the method in step **1** to convert:

a 1000_{10} to base 2   **b** 513_{10} to base 3   **c** 673_{10} to base 4.

3 The following algorithm produces the digits of a number n in base b, starting with the units digit and then working to the left.

- Start with the number, n. Divide n by b and find the quotient q, and the remainder r.
- Write down the value of r.
- If $q > 0$, replace n with the value of q; otherwise stop.
- Repeat from the beginning with the new value of n, and record new remainders to the left of any you have already written down.

4 Here the algorithm is used to convert 1038_{10} into base 3.

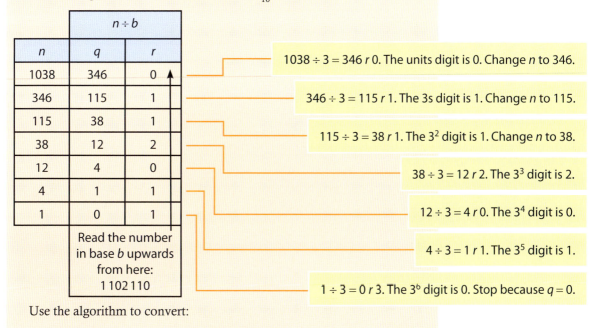

Use the algorithm to convert:

a 1000_{10} to base 2   **b** 513_{10} to base 3   **c** 673_{10} to base 4.

E1.1 What if we all had eight fingers?

Practice 2

1 Convert 999_{10} to:
   a base 2   b base 3   c base 4   d base 5

2 a Find the value of 472_8 in base 10.
   b Hence find the value of 472_8 in base 5.

3 By first converting to base 10, find the value of these numbers in the given base.
   a 223_5 in base 7   b 431_8 in base 2   c 214_6 in base 2
   d 1011_2 in base 6   e 110111_2 in base 8   f 110213_4 in base 9
   g 8868_9 in base 3   h 101101_2 in base 4   i 2468_9 in base 8

> Use the method you prefer to convert numbers to different bases.

Problem solving

4 Four students write the same number in different bases:

   Alberto: 1331_a Benito: 2061_b Claudio: 3213_c Donatello: 1000_d

   a Determine which of the four students used the largest base. Explain how you know.
   b Use your answer to part **a**, and any other information you can gain from the students' numbers, to list the numbers a, b, c and d in ascending order.
   c Determine the minimum possible value for b.
   d Find values of a, b, c and d such that $1331_a = 2061_b = 3213_c = 1000_d$.
   e Use your answer to **d** to find the value of the number in base 10.

Exploration 2

A bottle factory packs 12 bottles to a box. There are 12 boxes in a crate.

A shipping container will hold 12 crates.

1 Determine the number of bottles in each crate.
  Determine the number of bottles in each shipping container.

2 A customer orders 600 bottles. Find the number of crates and boxes to fulfil this order.

3 A customer orders 81 bottles. Determine the number of complete boxes and single bottles for this order.

4 The table on page 9 shows four different orders, with some information missing. Calculate appropriate values for the shaded cells.
  The basic price per bottle is €1.40.

▶ Continued on next page

E1 Form

Customer name	Total ordered	Notes	Containers	Crates	Boxes	Singles	Total cost
Mr Antinoro	a	n/a	0	5	0	0	b
Mr Drouhin	6000	10% discount on complete containers	c	d	e	f	g
Herr Müller	h	i (% discount on the whole order)	4	4	1	0	€8400
Mrs Symington	1500	n/a	j	k	l	m	n

5 An employee suggests that since all orders are made of a number of containers, crates, boxes and singles, the company does not need to repeat the headings every time, so an order of 3 crates, 7 boxes and 0 singles could be written as 370.

   a Write Mr Antinoro's, Mr Drouhin's and Herr Müller's orders using this convention.

   b Explain how notating the orders in this way relates to writing numbers in non-denary number bases.

   c Describe the problem in writing Mrs Symington's order in this way.

ATL Reflect and discuss 3

- Why do you think people most commonly use base 10 to count?
- When is the number 12 commonly used as a base? What makes the number 12 a convenient number to use?

Base 10 (decimal) uses ten different symbols to describe the whole numbers: 0, 1, 2, 3, 4, 5, 6, 7, 8 and 9. Similarly, base 2 uses two symbols: 0 and 1. Base 12 (duodecimal) requires twelve symbols, but you cannot use '10' or '11' because these both involve two digits.

> When you write numbers in bases greater than 10, letters are used for the extra symbols needed.
>
> In base 12, the symbols are 0, 1, 2, 3, 4, 5, 6, 7, 8, 9, A and B.

Sometimes lowercase letters a and b are used instead of uppercase A and B.

E1.1 What if we all had eight fingers?

Example 2

Find the value of $A3B_{12}$ in base 10.

12^2	12^1	12^0
144	12	1
A	3	B

The letter A represents 10, and B represents 11.

$A3B_{12} = (10 \times 144 + 3 \times 12 + 11 \times 1)_{10}$
$\phantom{A3B_{12}} = 1487_{10}$

Example 3

Find the value of 500 in base 16.

	$n \div b$	
n	q	r
500	31	4
31	1	F
1	0	1

$500_{10} = 1F4_{16}$

$500 \div 16 = 31\ r\ 4$. The units digit is 4. Change n to 31.

$31 \div 16 = 1\ r\ 15$. The symbol for 15 is F. Change n to 1.

$1 \div 16 = 0\ r\ 1$. The 16^2 digit is 1. Stop because $q = 0$.

Practice 3

1 Convert to base 10:

   a 210_{12} **b** 301_{16} **c** BAA_{12} **d** $G0_{20}$

2 Find the value in base 16:

   a 190_{10} **b** 2766_{10} **c** 47806_{10}

   d 48879_{10} **e** 51966_{10} **f** 64206_{10}

Problem solving

3 Hexadecimal (base 16) codes are often used as passwords for home Wi-Fi hubs and other digital services.

   a Explain why there are 256_{10} possible 2-character passwords in hexadecimal.

   b Find the number of possible 10-character hexadecimal passcodes.
       Give your answer to a suitable degree of accuracy.

NUMBER

Counting in 20s

US President Abraham Lincoln's "Gettysburg Address" began:

"Four score and seven years ago our fathers brought forth on this continent a new nation, conceived in liberty, and dedicated to the proposition that all men are created equal."

A 'score' means a group or set of 20, so 'four score and seven' is 87.

In the French number system, 80 is 'quatre-vingt' or 'four twenties' and 90 is 'quatre-vingt-dix' or 'four twenties and ten'.

Objective C: Communicating
iii. move between different forms of mathematical representation

Different number bases and symbols are different forms of representation. In this Exploration, show that you can move between different symbols and number bases.

ATL Exploration 3

The Mayans had a base 20 number system, but only three symbols:

They used the symbols ▬ and • in combination to create numbers up to 19. For example:

••• = 3   •̄ = 6   •̿ = 11   ≡̇ = 16

1 Write the numbers 4, 10, 14 and 19 in Mayan numerals.

The numbers from 0–19 fulfil the same purpose in Mayan numerals as the numbers from 0–9 do in base 10. Effectively, they are the Mayan digits. For place value, they wrote digits vertically in ascending order. It was important to leave enough space between the digits so that they could tell the difference between, say, a 10 (two horizontal bars) and 105 (two horizontal bars with one in the units place and one in the twenties place).

20^2s place	•	1	20^2s place	▬	5
20s place	•	1	20s place	🫘	0
1s place	•	1	1s place	≡̇	13

$(1 \times 20^2) + (1 \times 20) + 1 = 421$ $(5 \times 20^2) + (0 \times 20) + 13 = 2013$

▶ Continued on next page

E1.1 What if we all had eight fingers?

2 Find the value of these Mayan numerals:

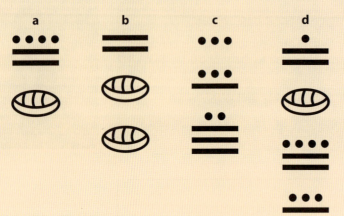

3 Write these numbers in Mayan:
   a 806   b 2005   c 10 145   d 16 125
   e 43 487   f 562 677   g 100 000   h 2 million

4 Find the value of this sum, showing all your steps:

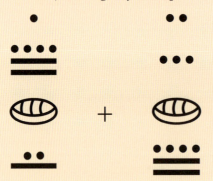

Express your answer in Mayan notation as well as in base 10.

5 Explain why the Mayan system is part additive and part place value.

6 Compare our base 10 number system with the Mayan system and write down any advantages or disadvantages that one has over the other.

The Mayan people kept stingless bees for honey. Technically, stingless bees do have stingers but they are so small that a sting from one is only about as painful as a mosquito bite.

C Performing operations

- How are mathematical operations in other bases similar to and different from operations in base 10?

In a calculation like this:

```
  8 9 2
+ 3 4 7
```

you add the units, then the tens, then the hundreds. When the addition gives you a number greater than or equal to 10 (the base) you write down the units digit and 'carry' the 10 to the next column.

NUMBER

You can add numbers in other bases in the same way as decimal numbers, 'carrying' when a number is greater than or equal to the base number.

Example 4

Find the value of $465_7 + 326_7$.

```
  4 6 5
+ 3 2 6
-------
     ¹4
```
Units column: $5 + 6 = 11_{10}$. In base 7 this is 1 seven and 4 units, or 14_7. Write 4 in the units column and carry a 1 into the sevens column.

```
  4 6 5
+ 3 2 6
-------
    ¹2 4
```
Sevens column: $6 + 2 + 1 = 9_{10} = 12_7$.

```
  4 6 5
+ 3 2 6
-------
  1124
```
7^2 column: $4 + 3 + 1 = 8_{10} = 11_7$.

So, $465_7 + 326_7 = 1124_7$.

Reflect and discuss 4

You could work out $465_7 + 326_7$ by adding, as in Example 4, or by:

- converting 465_7 and 326_7 into base 10
- adding
- converting the result back to base 7.

Which method would you prefer? Explain.

For subtraction, you can 'borrow' the base number, in the same way that you 'borrow' a 10 in base 10.

Example 5

Calculate $783_9 - 267_9$.

```
  7 8 3
- 2 6 7
-------
```
$7 > 3$, so you cannot subtract 7 from 3. Rewrite 7 8 3 as 7 7 ¹3, because 3 units, 8 nines and 7 eighty-ones is the same as 13 units, 7 nines and 7 eighty-ones.

```
  7 7 ¹3
- 2 6 7
-------
      5
```
Find $13_9 - 7_9$ by writing it in base 10:
$13_9 - 7_9 = 12_{10} - 7_{10} = 5_{10} = 5_9$

▶ Continued on next page

E1.1 What if we all had eight fingers?

$$\begin{array}{r} 7\,7^{1}3 \\ -\,2\,6\,7 \\ \hline 1\,5 \end{array}$$

7 − 6 = 1 in base 9 and in base 10.

$$\begin{array}{r} 7\,7^{1}3 \\ -\,2\,6\,7 \\ \hline 5\,1\,5 \end{array}$$

7 − 2 = 5

So, $783_9 - 267_9 = 515_9$.

Practice 4

1. Calculate:
   a. $124_8 + 321_8$
   b. $77_8 + 261_8$
   c. $563_8 + 241_8 + 757_8$

2. Calculate:
   a. $453_6 - 231_6$
   b. $341_6 - 153_6$
   c. $1231_6 - 402_6$

3. Calculate:
   a. $115_8 - 23_8 + 45_8$
   b. $2231_7 - 125_7 - 216_7$
   c. $8463_9 + 728_9 - 541_9 + 18_9$

4. Calculate:
   a. $92A1_{12} + 4436_{12}$
   b. $10000_{12} - 123_{12}$

Problem solving

5. Vorbelar the alien does not count in base 10, but in order to make things easy for us to understand, she uses our base 10 symbols in the usual order. She calculates 216 + 165 and obtains the answer 403.
   Find the value of 216 − 165, giving your answer using Vorbelar's base.

If you think dogs can't count, try putting three dog biscuits in your pocket and then giving Fido only two of them.

– Phil Pastoret

Reflect and discuss 5

When performing a calculation, is it important that the two numbers being added or subtracted are in the same base? Why or why not?

Exploration 4

There are many ways of setting out the calculation $742_{10} \times 36_{10}$.

Here is one approach.

$$\begin{array}{r} 7\,4\,2 \\ \times\,3\,6 \\ \hline {}_{1}2 \end{array}$$

First multiply the top row by 6, one column at a time, starting with the 2 in the units column.

6 × 2 = 12, so write 2 in the units column and 1 in the tens column.

▶ Continued on next page

7 4 2 × 3 6 ――― ²5 2	6 × 4 tens = 24 tens, so add 4 to the 1 in the tens column and write 2 in the hundreds column.
7 4 2 × 3 6 ――― 4 4 5 2	6 × 7 hundreds = 42 hundreds, so add 2 to the 2 in the hundreds column and write 4 in the thousands column. This completes 742 × 6.
7 4 2 × 3 6 ――― 4 4 5 2 ¹2 6 0	Now calculate 742 × 3 tens. 3 tens × 2 = 6 tens, so write 6 in the tens column. 3 tens × 4 tens = 12 hundreds, so write 2 in the hundreds column and 1 in the thousands.
7 4 2 × 3 6 ――― 4 4 5 2 2 2 2 6 0	3 tens × 7 hundreds = 21 thousands, so add 1 to the 1 in the thousands column and write 2 in the ten thousands column. This completes 742 × 3 tens.
7 4 2 × 3 6 ――― 4 4 5 2 + 2 2 2 6 0 ――― 2 6 7 1 2	Add together the two parts already found.

1 Use the method above to calculate $727_{10} \times 736_{10}$. Explain how you have used times tables.

2 Try to use the same method to calculate $727_9 \times 736_9$. Describe any difficulties.

3 Here is a multiplication table in base 9.
Use the table to help you calculate $727_9 \times 736_9$.

×	1	2	3	4	5	6	7	8	10
1	1	2	3	4	5	6	7	8	10
2	2	4	6	8	11	13	15	17	20
3	3	6	10	13	16	20	23	26	30
4	4	8	13	17	22	26	31	35	40
5	5	11	16	22	27	33	38	44	50
6	6	13	20	26	33	40	46	53	60
7	7	15	23	31	38	46	54	62	70
8	8	17	26	35	44	53	62	71	80
10	10	20	30	40	50	60	70	80	100

4 Explain why you need this table for long multiplication in base 9.

E1.1 What if we all had eight fingers?

Practice 5

1 Here is a multiplication table in base 5:

×	1	2	3	4	10
1	1	2	3	4	10
2	2	4	11	13	20
3	3	11	14	22	30
4	4	13	22	31	40
10	10	20	30	40	100

Use it to complete these multiplications:

a $123_5 \times 24_5$ **b** $401_5 \times 133_5$ **c** $213_5 \times 1310_5$

2 Copy and complete this multiplication table for base 6.

×	1	2	3	4	5	10
1	1	2		4	5	10
2	2	4		12		20
3						
4	4	12				40
5	5					
10	10	20		40		

Hence complete these multiplications:

a $155_6 \times 23_6$ **b** $212_6 \times 315_6$

Problem solving

3 Find two numbers (both greater than 1) whose product is 1215_6.

Because long multiplication relies on knowing multiplication table facts, multiplication in different bases is quite tricky. You can write out a multiplication table for the base you need, but in most cases it is easier to convert the numbers to base 10, perform the multiplication, and then convert the numbers back to the required base.

Division also relies on knowing multiplication table facts, so it is usually easiest to convert numbers to base 10 and then divide them.

A peek at binary numbers

- Would you be better off counting in base 2?
- How does form influence function?

Computers use base 2 (binary) for most of their operations. One reason for this is that the electrical signals that pass through the computer chips are either 'on' or 'off'. As the computer's memory exists in two states (on or off) it uses just two symbols for its counting system: 0 and 1.

Reflect and discuss 6

- Create a multiplication table in binary. Can you 'learn your times tables' in the new base?
- Try using it to perform some multiplications in base 2.
- Can you work out $1\,011\,100\,110 \times 110\,110\,100$ in base 2?

You may find multiplication slow in base 2 because there are a lot of digits, and long multiplication involves a lot of steps. There are fewer symbols in numbers with small bases, and thus numbers have more digits. In numbers with larger bases, there are more symbols, and so numbers have fewer digits.

> Hexadecimal is used in computers because each symbol in hexadecimal represents a string of 4 bits, or 4 binary digits. For example, 3F in hexadecimal converts to 0011 1111 in binary. That is why built-in passwords (in modems and routers for example) often use digits 0-9 and letters A-F.

Exploration 5

These hand diagrams show a way of counting in binary using your fingers. An extended finger/thumb represents 1 and a curled one represents 0.

This diagram represents zero.   This diagram represents one.

1 Find the number in base 10 represented by the diagrams below. It may help you to find the number in binary first.

2 Draw finger diagrams to show each of these base 10 numbers in binary:

a 32_{10} **b** 68_{10} **c** 101_{10} **d** 250_{10}

E1.1 What if we all had eight fingers?

Reflect and discuss 7

- Is counting in base 2 on your fingers a more efficient way of counting?
- How is counting in base 2 on your fingers more difficult than counting in base 10?
- When might it be useful to try to count in base 2 on your fingers?

Summary

Our number system, often called Arabic numerals, is a **place value system.** The position of a digit tells you its value.

Our number system is also known as decimal, or base ten, or denary, because the value of each place value column is 10 times the value of the column to its right.

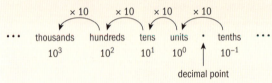

You can make a place value system with any natural number base. In binary, or base two, each column is worth twice the column to its right.

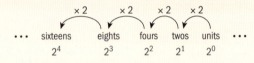

The subscript 2 means that the number $10\,111_2$ is written in base 2.

The base of a number system tells you how many unique symbols the number system uses. Base 10 has ten unique number symbols: 0 through 9. Base 2 has two unique symbols: 0 and 1.

When you write numbers in bases greater than 10, letters are used for the extra symbols needed.

In base 12 the symbols are 0, 1, 2, 3, 4, 5, 6, 7, 8, 9, A and B (or sometimes lowercase letters a and b).

You can add numbers in other bases in the same way as decimal numbers, 'carrying' when a number is greater than or equal to the base number.

For division and multiplication in a different base, it is usually most efficient to convert the numbers to base 10, do the calculations, then convert the answer back to the base in question.

Mixed practice

1 Find the value in base 10 of:

a 1101_2
b $110\,110_2$
c $1\,000\,101_2$
d 1221_3
e $12\,000_3$
f 4523_8
g $34\,337_8$
h $18A9_{11}$
i $4EA8_{16}$

2 Find the value in the base specified of:

a 100_{10} in base 2
b 100_{10} in base 3
c 100_{10} in base 4
d 255_{10} in base 2
e 1008_{10} in base 3
f 8162_{10} in base 5
g 271_{10} in base 12
h 456_{10} in base 12
i 1689_{10} in base 12
j 2785_{10} in base 12
k 1049_{10} in base 16
l 15342_{10} in base 16

3 You can measure angles in degrees, minutes and seconds. One degree = 60 minutes or 60′, and one minute = 60 seconds or 60″. 45° 15′ 30″ = 45 degrees, 15 minutes and 30 seconds.

Convert these angle measures:

a 42.8° into degrees, minutes and seconds
b 90.15° into degrees, minutes and seconds
c 2.52° into degrees, minutes and seconds
d 25°15′ into base 10 (including a decimal)
e 62° 30′ 30″ into base 10 (including a decimal)
f 58° 12′ 20″ into base 10 (including a decimal)

4 Calculate, giving your answers in the base used in the question:

a $1101_2 + 1001_2$
b $11\,101_2 + 101\,101_2$
c $1011_2 + 10\,111_2 + 1\,101\,101_2$

d $122_3 + 2101_3$ **e** $2122_3 + 20\,012_3$
f $3012_4 + 10\,332_4$
g $33\,010_4 + 312\,212_4 + 101_4$
h $565_{12} + 718_{12}$ **i** $1A91_{12} + 3302_{12}$
j $AB0_{12} + 10B8_{12}$ **k** $6E_{16} + AC_{16}$
l $F00D_{16} + FACE_{16}$

5 **Calculate**, giving your answers in the base used in the question:

a $11011_2 - 101_2$ **b** $111011_2 - 11101_2$
c $10110_2 - 111_2$ **d** $2210_3 - 102_3$
e $3011_4 - 231_4$ **f** $44125_6 - 5150_6$
g $50112_6 - 1045_6$ **h** $B791_{12} - 89A0_{12}$
i $FEED_{16} - BEEF_{16}$

6 If the current time was 10:45, **use** addition and subtraction to **find** the time that is:

a 35 minutes from now
b $2\frac{1}{2}$ hours from now
c 3 hours and 40 minutes from now
d 1 hour and 19 minutes ago
e 6 hours and 50 minutes ago

7 **Calculate**, giving your answers in the base used in the question.

> Either create multiplication tables like the ones in Practice 5, or convert the numbers to base 10 and then convert back.

a $110_2 \times 10_2$ **b** $1011_2 \times 11_2$ **c** $1101_2 \times 101_2$
d $2101_3 \times 21_3$ **e** $14_5 \times 23_5$ **f** $24_5 \times 131_5$

8 A set of weights contains weights in grams that are powers of 4. There are three of each size weight in the set. The lightest weight is 1 g, and the heaviest is 16 384 g.

a **Find** the total weight of these six weights: two of the smallest weights, three of the next size and one of the size above that.

b **Write down** the weights you would use to create a total weight of 26 g. Remember that there are only three of each type of weight.

c **Find** the value of 2313_4 in base 10.
Explain why this would be relevant if you wanted to find the total weight of three 1 g weights, one of the next size, three of the next size and two of the size after that.

Problem solving

d Ronnie has $331\,213_4$ grams in weights. Donnie has $231\,232_4$ grams in weights. Connie wishes to weigh out the same amount as Ronnie and Donnie combined, using the weights from her sets. Remembering that she has only three of each type of weight, **determine** the weights Connie should use.

e Ted weighs out 1330_4 grams. Ed weighs out twice that amount. Ned weighs out twice as much as Ed and Ted combined. Remembering that he has only three of each type of weight, **determine** the weights Ned should use.

9 A congressional committee decides to streamline the US monetary system. All coins will be phased out except for 1 cent, 5 cents and 25 cents. All existing bills (bank notes) will be phased out and replaced with bills valued $1, $5, $25, $125 and $625.

a A mathematician suggests that the dollar should be revalued to be worth 125 cents, not 100 cents as it is currently.

  i **Suggest** reasons why the mathematician thinks this might be a good idea.

  ii **Suggest** reasons why it might not be a good idea.

b **Find** 540_{10} in base 5.
Hence find the smallest number of notes that you would need to give somebody $540 under the new system.

c **Find** 732_{10} in base 5 and 246_{10} in base 5.

d Rania owes $732 to her credit card company. She makes a payment of $246.
Find the total amount outstanding (still owed), and the least number of notes she would need to repay the amount.

e In real life, monetary systems do not use regular bases, but have combinations of notes that follow irregular patterns. For example, in the UK, there are £1 and £2 coins, then notes valued £5, £10, £20 and £50.
Suggest reasons why it would not be convenient to use a perfectly regular base as suggested above.

Review in context

ATL

In the 1960s and 1970s, NASA's Pioneer and Voyager missions launched spacecraft to travel beyond the solar system. Each contained messages in case the craft was intercepted by extra-terrestrial life forms. Scientists tried to choose things that they hoped could be universally communicated, or which could be measured by aliens.

1. If an alien culture has a number system, there is a good chance that they will have some sort of place value system. They will probably have an idea of whole numbers, and maybe an awareness of prime numbers. One suggestion for a message to include might be to communicate the first prime numbers in binary. The first five numbers in binary are: 10, 11, 101, 111 and 1011.

   a **Write down** the values of these numbers in denary (decimal).

   b **List** the next ten prime numbers in denary.

   c **Hence list** the first 15 prime numbers in binary.

Problem solving

2. For the messages, the scientists did not use 1s and 0s; these symbols are known only to humans. Instead, they used horizontal and vertical lines: — and |.
   Here are some sequences given in binary using the symbols — and |.
   Determine whether — or | represents 1. Justify your answer.
   Describe each sequence and **predict**, using — and |, the next three terms.

   a |, |——, |——|, |————, ||——|

   b |, ||, |—, ||, |—|, |———, ||—|, |—|—|

   c |, ||, |——|, ||—||, |—|————|

3. The scientists also encoded some information about our solar system. They used information about quantities that can be counted, rather than measured, because measurements require units whereas counting does not. One quantity that can be counted is the number of 'days' in each planet's year. Earth has (roughly) 365 Earth days per year because that is how many times it rotates around its own axis while orbiting the sun. The table gives some information about the planets in our solar system:

Planet	Year length	Day length
Mercury	87.96 Earth days	1408 Earth hours
Venus	224.68 Earth days	5832 Earth hours
Earth	365.26 Earth days	24 Earth hours
Mars	686.98 Earth days	25 Earth hours
Jupiter	11.862 Earth years	10 Earth hours
Saturn	29.456 Earth years	11 Earth hours
Uranus	84.07 Earth years	17 Earth hours
Neptune	164.81 Earth years	16 Earth hours

   a **Calculate** the number of Earth hours in a Mars year. Give your answer in denary.

   b **Hence find** the number of Mars days in a Mars year. Give your answer in denary.

   c **Find** (to the nearest whole number) the number of Mars days in a Mars year in binary.

   d **Find**, in binary, the number of days in a year for each of the other outer planets (Jupiter to Neptune).

   e **Describe** any problems you would encounter when trying to perform a similar calculation for Venus.

E2.1 Multiple doorways

Global context: Scientific and technical innovation

Objectives

- Performing operations with functions, including function composition
- Using function composition to solve real-world problems
- Decomposing functions

Inquiry questions

- How do you add and subtract functions?
- What is function composition?

- How can you use function composition to model real-world situations?

- Does a composite function have a unique decomposition?

RELATIONSHIPS

ATL Critical-thinking

Draw reasonable conclusions and generalizations

You should already know how to:

• draw mapping diagrams for functions	**1** Draw a mapping diagram to represent the function shown on the graph. 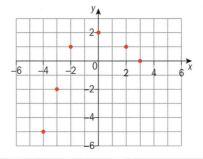
• find the natural domain and range of a function	**2** If $f:\mathbb{R} \to \mathbb{R}$, find the natural domain and range of: **a** $f(x) = 2x + 3$ **b** $f(x) = \sqrt{x}$ **c** $f(x) = \dfrac{1}{x}$ **d** $f(x) = \dfrac{2}{x-2}$ **e** $f(x) = \sqrt{4 - x^2}$
• evaluate functions	**3** $f(x) = 2x - 3$. Find: **a** $f(4)$ **b** $f(-1)$ **c** $f(3x)$

Function operations

- How do you add and subtract functions?
- What is function composition?

> The function $(f + g)(x)$ is equivalent to $f(x) + g(x)$.
> The function $(f - g)(x)$ is equivalent to $f(x) - g(x)$.

Example 1

$f(x) = 2x + 4$ and $g(x) = 4x - 1$.

Write the function given by:

a $(f + g)(x)$ **b** $(f - g)(x)$

a $(f + g)(x) = f(x) + g(x)$
$= (2x + 4) + (4x - 1)$ — Substitute $f(x)$ and $g(x)$ and simplify.
$= 6x + 3$

b $(f - g)(x) = f(x) - g(x)$
$= (2x + 4) - (4x - 1)$
$= -2x + 5$

ALGEBRA

ATL

Exploration 1

$f(x) = \sqrt{x}$ has domain $\{x : x \geq 0\}$.

$g(x) = x + 2$ has domain $x \in \mathbb{R}$.

$h(x) = 1 - x^2$ has domain $x \in \mathbb{R}$.

$k(x) = \dfrac{2}{x-1}$ has domain $\{x : x \neq 1\}$.

1 Write each function and state its domain.

   a $(f+g)(x)$ **b** $(f-g)(x)$ **c** $(h-f)(x)$ **d** $(g+k)(x)$

2 Find a rule for the domain of a function formed by adding or subtracting two functions. Justify your rule with a mathematical explanation or a mapping diagram.

Example 2

$f(x) = 2x^2 + 1$, $g(x) = \dfrac{1}{3x-2}$, and $h(x) = \sqrt{1-3x}$.

a State the domains of f, g and h.

b Write the function $k(x) = (g-f)(x)$ and state its domain.

a f has domain $x \in \mathbb{R}$.

For g, $3x - 2 \neq 0$,

so $3x \neq 2 \Rightarrow x \neq \dfrac{2}{3}$ *Exclude values of x that make the denominator zero.*

g has domain $\left\{x : x \neq \dfrac{2}{3}\right\}$

For h, $1 - 3x \geq 0$, so $1 \geq 3x \Rightarrow \dfrac{1}{3} \geq x$.   *Exclude values of x that make the radicand negative.*

h has domain $\left\{x : x \leq \dfrac{1}{3}\right\}$

b $k(x) = (g-f)(x)$

$= \dfrac{1}{3x-2} - (2x^2 + 1)$ *Write with a common denominator.*

$= \dfrac{1 - (2x^2+1)(3x-2)}{3x-2}$

$\dfrac{1 - (6x^3 - 4x^2 + 3x - 2)}{3x-2}$

$\dfrac{-6x^3 + 4x^2 - 3x + 3}{3x-2}$

k has domain $\left\{x : x \neq \dfrac{2}{3}\right\}$

For two functions $f(x)$ and $g(x)$, the domain of $(f+g)(x)$ and the domain of $(f-g)(x)$ is the intersection of the domains of $f(x)$ and $g(x)$.

Practice 1

1. $h(a) = 2a^2 + 1$ and $k(a) = -(a+3)$.
   State the domains of h and k.
   Find $(h+k)(a)$ and state its domain.

2. $h(x) = x^3 - 5x^2$ and $g(x) = 5 - x$. Find $(g-h)(x)$ and state its domain.

3. $h(x) = 2x^2 - x$ and $k(x) = \sqrt{3x+1}$.
   State the domains of h and k.
   Find $(h+k)(x)$ and state its domain.

4. $h(c) = 2c - 5$ and $k(c) = \sqrt{2-3c}$.
   Find $(k-h)(c)$ and state its domain.

5. $g(x) = 6x - 8$, $h(x) = \dfrac{1}{2x}$ and $k(x) = \sqrt{x+4}$. Find $(g+h-k)(x)$ and $(k-h-g)(x)$, and state their domains.

6. $f(x) = \dfrac{1}{x}$ and $g(x) = \dfrac{1}{x-2}$.

   a State the domain of f and the domain of g.

   b Write down the domain of $f + g$.

Problem solving

7. a Write a function f with domain $\{x : x \neq 3\}$.

   b Write a function g with domain $\{x : x \neq 0\}$.

   c Write a function with domain $\{x : x \neq 0 \text{ and } x \neq 3\}$.

Function composition is a way of combining functions.

The mapping diagram shows three sets A, B and C and two functions f and g.

f maps the elements of A onto B, and g maps the elements of B onto C:

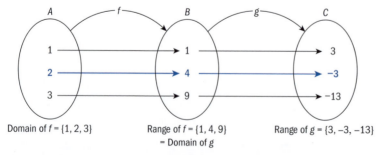

B is the range of f.
C is the range of g.

From the diagram, f maps 2 onto 4, and then g maps 4 onto -3.

So $f(2) = 4$ and $g(4) = -3$

$f(1) = 1$ and $g(1) = 3$

$f(3) = 9$ and $g(9) = -13$

Applying f to the elements of A and then applying g to the elements of B gives a mapping from A to C. The function that maps A to C is called the **composite function** of f and g. A composite function is a combination of two functions.

> The composite function of f and g is the function f followed by function g.
> You write this as $g \circ f(x)$ and say 'g of f of x'. You can also write $g(f(x))$.

From the diagram, $\quad g \circ f(1) = 3$

$$g \circ f(2) = -3$$
$$g \circ f(3) = -13$$

The domain of $g \circ f$ is the domain of f: $\{1, 2, 3\}$.

The range of $g \circ f$ is the range of g: $\{3, -3, -13\}$.

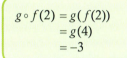

> $g \circ f(x) = g(f(x))$ means f followed by g.
> $f \circ g(x) = f(g(x))$ means g followed by f.

Example 3

> g is a function from A to B and f is a function from B to C.
> $g(1) = 2$, $g(3) = 4$, $g(5) = 6$ and $f(4) = 5$, $f(2) = 3$, $f(6) = 7$.
> **a** Draw a mapping diagram for the composite function $f \circ g$.
> **b** List the elements of A, B and C.
> **c** Find $f \circ g(1)$, $f \circ g(3)$ and $f \circ g(5)$.
> **d** State the domain and range of $f \circ g(x)$.

a

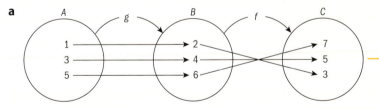

$f \circ g$ means g followed by f.

b $A = \{1, 3, 5\}$, $B = \{2, 4, 6\}$, $C = \{3, 5, 7\}$

c $f \circ g(1) = 3$, $f \circ g(3) = 5$, $f \circ g(5) = 7$

d Domain of $f \circ g(x)$ is $\{1, 3, 5\}$.

Range of $f \circ g(x)$ is $\{3, 5, 7\}$.

Reflect and discuss 1

ATL
- Explain why it makes sense to write 'function f followed by g' as: $g \circ f$.
- For the function $g \circ f(x)$, explain why $f(x)$, the image of x, must belong to the domain of g.
- If the range of f is the entire domain of g, what is the domain of $g \circ f$?

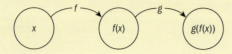

Practice 2

1 f is a function from A to B and g is a function from B to C.

$f(1) = 5$, $f(2) = 9$, $f(3) = 13$, $g(5) = -2$, $g(9) = 2$, $g(13) = 6$.

a Draw a mapping diagram for the composite function $g \circ f$.

b List the elements of A, B and C.

c Find $g \circ f(1)$, $g \circ f(2)$ and $g \circ f(3)$.

d State the domain and range of $g \circ f(x)$.

2 For each composite function, draw a mapping diagram, then state the domain and range.

a $h \circ g$, where $g(-5) = 1$, $g(-2) = 4$, $g(0) = 2$, $h(2) = -1$, $h(4) = -3$, $h(1) = 5$.

b $f \circ h$, where $h(1) = 2$, $h(3) = 5$, $h(4) = 7$, $f(2) = 3$, $f(5) = 4$, $f(7) = 6$.

c $g \circ f$, where $f(a) = b$, $f(b) = c$, $f(c) = d$, $g(d) = x$, $g(c) = y$, $g(b) = z$.

3 The mapping diagram here represents two functions f and h.

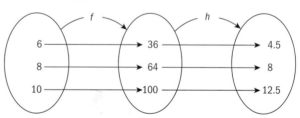

a Find $h \circ f(6)$, $h \circ f(8)$ and $(h \circ f)(10)$.

b State the domain and range of $h \circ f(x)$.

Problem solving

4 $f(x) = 2x + 3$ and $g(x) = x - 5$. The domain of f is $\{0, 1, 3, 5\}$.

a Draw a mapping diagram to represent $g \circ f(x)$.

b State the range of $g \circ f(x)$.

5 For two functions f and g:
- the domain of f is $A = \{1, 2, 3\}$
- the range of f and domain of g is $B = \{4, 5, 6\}$
- the range of g is $C = \{7, 8, 9\}$

Determine the number of different composite functions $g \circ f$.

ALGEBRA

The functions $f(x) = 2x$ and $g(x) = x^2 - 1$ both have domain and range $x \in \mathbb{R}$, the set of real numbers. Applying function f and then function g to input value 3 gives:

$$x \xrightarrow{f(x)=2x} 2 \times 3 \to 6 \xrightarrow{g(x)=x^2-1} 6^2 - 1 \to 35$$

$f(3) = 6$ and $g(6) = 35$ or $g \circ f(3) = 35$ or $g(f(3)) = 35$

Applying function f and then function g to input value x gives the composite function $g \circ f(x)$:

$$x \xrightarrow{f(x)} 2x \xrightarrow{g(x)} (2x)^2 - 1 \xrightarrow{g \circ f(x)} 4x^2 - 1$$

$g \circ f(x) = (2x)^2 - 1 = 4x^2 - 1$

ATL Reflect and discuss 2

- For the functions $f(x) = 2x$ and $g(x) = x^2 - 1$, find $f \circ g(x)$.
- Is $f \circ g$ equal to $g \circ f$? Explain.

Example 4

$f(x) = 5 - 3x$ and $g(x) = x^2 + 4$. Find:

a $g(f(1))$ **b** $f \circ g(2)$ **c** $f(g(x))$ **d** $g \circ f(x)$

a $f(1) = 5 - 3 \times 1 = 2$ — Start with the inside function f.
$g(2) = 2^2 + 4 = 8$
$\Rightarrow g(f(1)) = 8$

b $g(2) = 2^2 + 4 = 8$ — Start with the inside function g.
$f(8) = 5 - 3 \times 8 = -19$
$\Rightarrow f \circ g(2) = -19$

c $f(g(x)) = 5 - 3(g(x))$ — $g(x)$ goes in here.
$= 5 - 3(x^2 + 4)$
$= 5 - 3x^2 - 12$
$= -3x^2 - 7$

d $g \circ f(x) = g(5 - 3x)$ — $f(x)$ goes in here.
$= (5 - 3x)^2 + 4$
$= 25 - 30x + 9x^2 + 4$
$= 9x^2 - 30x + 29$

Practice 3

1 $f(x) = 3x$, $g(x) = x+1$ and $h(x) = x^2+2$. Find:

- **a** $f(g(3))$
- **b** $g(h(-2))$
- **c** $f \circ h(2)$
- **d** $h \circ g(-0.5)$
- **e** $g(f(0))$
- **f** $f(f(-4))$
- **g** $h \circ h(0)$
- **h** $h(f(-1))$
- **i** $f \circ h(x)$
- **j** $g(f(x))$
- **k** $g \circ g(x)$
- **l** $h(g(x))$

Problem solving

2 A composite function $f(g(x)) = 2(x^2 - 5) + 1$ and $f(x) = 2x + 1$. Find:

- **a** a suitable function $g(x)$
- **b** the composite function $g(f(x))$.

3 A composite function $g(f(x)) = \dfrac{1}{x^2 - 1}$, $x \neq \pm 1$ and $g(x) = \dfrac{1}{x-1}$, $x \neq 1$. Find:

- **a** a suitable function f and state its domain and range
- **b** the composite function $f \circ g(x)$.

4 $f(x) = 7x - 2$ and $g(f(x)) = 3x^2 - 70$. Find $g(33)$.

> In **4**, first find x.

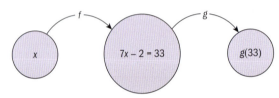

5 The graphs show the functions f and g. Both have domain $\{x : 0 \leq x \leq 6\}$.

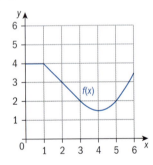 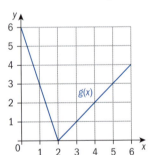

Using the graphs of f and g, evaluate:

- **a** $f \circ g(0)$
- **b** $f \circ f(2)$
- **c** $g \circ f(3)$
- **d** $g \circ g(4)$
- **e** $f \circ g(1)$
- **f** $g \circ f(5)$
- **g** $f(g(f(4)))$

6 Emily works 30 hours per week at a travel agency. She is paid a weekly salary of $500, plus a 2% commission on sales over $5000. One week she sells enough to get the commission.

- **a** Explain what x and the functions $f(x) = 0.02x$ and $g(x) = x - 5000$ represent in this context.
- **b** Determine whether $f \circ g(x)$ or $g \circ f(x)$ represents her commission.

7 $g(x) = 4x + 4$ and $f(x) = x^2 - 64$. Find x such that $f \circ g(x) = 0$.

ALGEBRA

C Modelling using compositions of functions

- How can you use function composition to model real-world situations?

Objective D: Applying mathematics in real-life contexts
i. identify relevant elements of authentic real-life situations

In Exploration 2, you will need to identify the meaning of the given functions in light of the problem.

Exploration 2

As a diver descends deeper into the ocean, the mass of water above the diver increases and so the pressure on the diver's body increases.

This table shows the diver's depth at different times.

Time t (seconds)	Depth d (meters)
0	0
20	15
40	32
60	44
80	65
100	80
120	90

The next table, below, shows the pressure exerted on a diver at different depths. The pressure is measured in atmospheres (atm), where 1 atm is the atmospheric pressure at sea level.

Depth d (meters)	0	10	20	30	40	50	60	70	80	90
Pressure p (atm)	1	2	3	4	5	6	7	8	9	10

1 Let $f(t)$ be the function connecting time t and depth d.

   $f(0) = 0$ and $f(20) = 15$. Find $f(60)$. Explain what $f(60)$ means.

2 Let g be the function connecting the depth d and pressure p.

   Explain what $g(60) = 7$ means.

3 Use the tables to find $g \circ f(120)$. Explain what $g \circ f(120)$ means.

4 Explain how to find the time before the diver is at a pressure of 9 atm.

E2.1 Multiple doorways

Practice 4

Problem solving

1. The price p of an orange, in cents, is a function f of its weight w, in grams.

Weight w (grams)	100	125	150	200	250	300	350	400
Price p (cents)	12	15	20	26	33	36	45	50

> Oranges are domesticated so you are unlikely to find them growing naturally in the wild.

The weight w of an orange, in grams, is a function g of its radius r, in cm.

Radius r (cm)	3.8	4.2	4.3	4.8	5.1	5.3
Weight w (grams)	190	210	215	240	250	265

a Find $f(125)$ and $f(350)$.

b Find $g(4.2)$ and $g(5.3)$.

c Evaluate $f \circ g(5.1)$ and explain what this means.

d Determine the function composition that describes the relationship between the radius of an orange and its price in cents.

2. A particular computer screen saver simulates an expanding circle. The circle starts as a dot in the middle of the screen and expands outward so its radius increases at a constant rate.

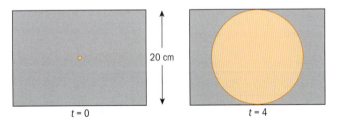

The screen is a rectangle with a height of 20 cm. The circle takes 4 seconds to expand to touch the top and bottom of the screen.

a Copy and complete this table for the radius of the circle at time t seconds:

Time t (seconds)	0	1	2	3	4
Radius r (cm)	0	2.5			10

b The radius r is given by the function $f(t)$. Use the values in your table to find the function $f(t)$.

c The area A of the circle is a function $g(r)$ of the radius r. Write the function $g(r)$.

d Hence express the area A of the circle as a function of time t.

3 A cylindrical container has a radius of 50 cm and a capacity of 500 cm³. It is being filled with water at a constant rate of 10 cm³ per second.

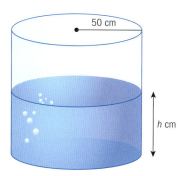

  a Write down a function for the volume $v(t)$ of water in the container after t seconds.

  b Write down a formula for the height h of water in the container in terms of v.

  c Find $h \circ v(t)$ and explain what it means.

  d Determine how long it takes for the height of the water in the container to reach 50 cm.

4 A pebble is thrown into still water and creates a circular ripple. The radius r of the water ripple increases at the rate of 2 cm per second.

  a Write down the radius r in terms of time t after the pebble was thrown.

  b $A(r) = \pi r^2$ is the area of the water ripple in terms of its radius. Explain the meaning of $A \circ r(t)$.

  c Find the area A of the water ripple after 60 seconds.

5 For each pair of functions f and g in the table below:

  a write down the meaning of $g \circ f$

  b describe a possible real life use for the composite function.

f	g
i Fuel consumption is a function of vehicle velocity.	Fuel cost is a function of fuel consumption.
ii Height of water in a tank is a function of time.	Volume of the tank is a function of height.
iii Radius of a circle is a function of time.	Volume is a function of the radius of a circle.
iv Sum of the angles of a regular polygon is a function of the number of sides of the polygon.	Size of each angle of a regular polygon is a function of the sum of the angles of the polygon.
v Bank savings is a function of time.	Interest earned is a function of amount of savings.

You can graph composite functions on any GDC or graphics software. For example, for functions $f_1(x) = x^2 + 2x + 1$ and $f_2(x) = 3x + 5$, the two composite functions $f_3(x) = f_1(f_2(x))$ and $f_4(x) = f_2(f_1(x))$ are shown in the graphs below.

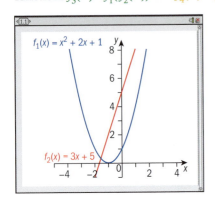 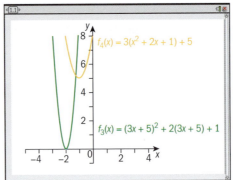

D Decomposing functions

• Does a composite function have a unique decomposition?

Exploration 3

1. Find two functions f and g such that their composition is:
   $g \circ f(x) = (x+2)^2 - 3$.

2. Expand and simplify $g \circ f(x)$ and explain how you might find the functions when the composite function is simplified.

3. Find two functions h and k such that their composition is:
   $k \circ h(x) = (x+2)^2 + (x+2) - 3$.

4. Simplify $k \circ h(x)$, and find functions h and k. Are the functions the same as you found in step **3**? Explain.

5. Determine whether or not it is better to simplify a function before finding its decomposition. Explain.

The word *composition* has meanings in photography, art, literature, music, chemistry, cooking, flower arranging, design, metallurgy, printing, and of course mathematics.

Reflect and discuss 3

Does every composite function have a unique decomposition?

Practice 5

Problem solving

For each question, find two functions f and g whose composition is the function $h(x) = f \circ g(x)$.

1. $h(x) = (2x-1)^3$
2. $h(x) = 2x^3 - 1$
3. $h(x) = \sqrt{6x-5}$
4. $h(x) = 6\sqrt{x} - 5$
5. $h(x) = \sqrt[3]{2x-1}$
6. $h(x) = 2^{3x-2}$

ALGEBRA

Reflect and discuss 4

- Compare your answers to Practice 5 with others in your class. Did you all find the same functions g and f?
- Write down a non-linear function that models a real life situation, and decompose the function. Explain what the decomposition of the function means in the context of the problem.

Summary

The function $(f+g)(x)$ is the same as $f(x)+g(x)$.

The function $(f-g)(x)$ is the same as $f(x)-g(x)$.

For two functions $f(x)$ and $g(x)$, the domain of $(f+g)(x)$ and the domain of $(f-g)(x)$ is the intersection of the domains of $f(x)$ and $g(x)$.

The composite function of f and g is the function f followed by the function g.

You can write $g \circ f(x)$ or $g(f(x))$, and say 'g of f of x.'

$g \circ f(x) = g(f(x))$ means f followed by g.
$f \circ g(x) = f(g(x))$ means g followed by f.

Mixed practice

1. $f(x) = 2 - 5x$, $g(x) = x^2 - 4x$, $h(x) = \frac{x-2}{4}$ and $k(x) = -\frac{2}{x}$, $x \neq 0$.

   Write down these functions in simplest form:

   a $(f+g)(x)$ **b** $(g-f)(x)$ **c** $(f+h)(x)$
   d $(g-h)(x)$ **e** $(h+k)(x)$ **f** $(k-h)(x)$
   g $(f+k)(x)$

2. **State** the domains of functions f and g below. **Find** $f \circ g(x)$ and $g \circ f(x)$, and **state** their domains.

   a $f(x) = x+1$, $g(x) = \sqrt{2x-1}$
   b $f(x) = x^2$, $g(x) = \sqrt{x-1}$
   c $f(x) = x^2 + x$, $g(x) = 2 - 3x$

3. Given the functions $p(t) = 4t+1$, $q(t) = t^2 - 4$ and $r(t) = \frac{1}{t}$, $t \neq 0$, **find** these functions and **state** their domains:

   a $p(q(t))$ **b** $q(r(t))$ **c** $p(r(t))$ **d** $r(q(t))$

4. Given the functions $f(x) = \frac{1}{x}$, $g(x) = 2x+4$ and $h(x) = x^2 - 2$, **find** in terms of f, g and h the functions:

   a $\frac{2}{x} + 4$ **b** $4x^2 + 16x + 14$

5. Given that $f(x) = \frac{x-2}{3}$ and $g(x) = 2x+1$, **find**:

   a $g(f(a))$ **b** $f(g(-b))$

Problem solving

6. If $f(x) = 3x+k$ and $g(x) = \frac{x-4}{3}$, **find** the value of k such that the composite functions $f \circ g$ and $g \circ f$ are equal.

7. If $f(x) = 2x-3$ and $g(x) = 2x^2$, **find**:

   a $f(g(x-1))$ **b** $g(f(-3x))$

8. If $f(x) = x-1$ and $g(x) = 4x+3$, **find** $g(f(3x^2-1))$.

9. The notation $[x]$ means the greatest integer not exceeding the value of x. If $f(x) = [x]$, $g(x) = 15x$, and $h(x) = \frac{8}{x}$, **find**:

   a $f\left(g\left(\frac{1}{2}\right)\right)$ **b** $f(h(-5))$ **c** $h(f(\pi))$

10. The number of bacteria B in food refrigerated at a certain temperature can be modelled by the function $B(T) = 20T^2 - 80T + 500$, where $1 < T < 15$, and T is in degrees Celsius. When the food is taken out of the refrigerator, the food temperature T can be modelled by the function $T(t) = 4t+2$, $0 \leq t \leq 4$, where t is the time in hours.

    a **Find** the composite function $B \circ T(t)$ and **explain** what it means in the context of the problem.

    b **Determine** how long it takes for the bacteria count to be 5000.

E2.1 Multiple doorways 33

11 A store owner adds 50% to the wholesale price to decide the selling price of items in the store. If items do not sell in 4 weeks, he sells them at a 20% discount.

  a **Write down** a function that represents the selling price s of an item in terms of its wholesale price w.

  b **Write down** a function that represents the discounted price d of an item in terms of the selling price.

  c **Write down** a composite function to represent the discounted price in terms of the wholesale price.

  d **Find** the discount price of an item whose wholesale price was $72.

12 From an online shopping store you receive a $5 discount coupon. You go online to spend it, and see that your favorite shoe store is having a 25% sale.

  a If x represents the original cost of a pair of shoes:

   i **write down** a function $f(x)$ that represents the cost effect of using your discount coupon

   ii **write down** a function $g(x)$ that represents the cost effect of a 25% sale

   iii **write down** a composite function that represents how much you would pay for the shoes if you first use the discount coupon and then apply the 25% sale

   iv **write down** a composite function that represents how much you would pay for the shoes if you first apply the 25% sale and then use the discount coupon.

  b You find a pair of shoes you like for $50. **Evaluate** the price of the shoes using the two composite functions you found in **a iii** and **iv**, and **state** the difference in price between them.

13 For each question, **find** two functions f and g whose composition is the function $h(x) = f \circ g(x)$.

  a $h(x) = -(x+1)^4 + 7$ **b** $h(x) = \sqrt[5]{4 - 3x}$

  c $h(x) = -5x^2 - 8$ **d** $h(x) = \dfrac{3}{x-4}$

  e $h(x) = 10^{\sqrt{x}}$

Review in context

1 An oil spill is modelled as a circle of radius r meters. At time $t = 0$ the radius is 70 m. The radius increases at a rate of 0.8 meters per minute.

  a Express the radius r as a function of time.

  b **Find** the composite function $A \circ r(t)$ where $A(r)$ is the area of a circle of radius r. **Explain** what it means.

  c **Determine** how long it takes for the area of the oil spill to exceed $100\,000$ m².

2 The formula $K(C) = C + 273$ converts Celsius temperature into Kelvin temperature. The formula $C(F) = \dfrac{5}{9}(F - 32)$ converts Fahrenheit temperature into Celsius temperature.

  a **Write down** a composite function to convert Fahrenheit temperature into Kelvin temperature.

  b The boiling and freezing temperatures in Fahrenheit are 212° and 32° respectively. **Find** the boiling and freezing temperatures in Kelvin.

3 The speed of a car, v (in km/h) at a time t hours, can be modelled using the function $v(t) = 40 + 3t + t^2$. The rate of fuel consumption, r (in liters/km) at a speed of v km/h can be modelled by the function $r(v) = \left(\dfrac{v}{500} - 0.1\right)^2 + 0.2$.

  a **Write down** the function $r(v(t))$, the rate of fuel consumption as a function of time.

  b **Find** the rate of fuel consumption for a trip that lasts 4 hours.

4 Composite functions can also be used to create and/or explain number puzzles. Ask a friend to think of any number, then:

  • multiply the number by 4 and then add 6
  • divide the result by 2 and then add 3.

  a **Write down** the steps as functions, and the composite function that shows the order in which the steps are executed.

  b Simplify the function, and **explain** how to find your friend's number.

5 Make up your own number puzzles which you can solve using composite functions.

E2.2 Doing and undoing

Global context: Scientific and technical innovation

Objectives
- Understanding onto and one-to-one functions
- Finding the inverse of a function algebraically
- Finding the inverse of a function graphically
- Restricting the domain of a function, if necessary, so that its inverse is also a function
- Justifying that two functions are inverses of each other
- Justifying that a function is self-inverse

Inquiry questions

F
- What are onto and one-to-one functions?
- How do you find the inverse of a function?

C
- How do you know when two functions are inverses of each other?
- Is the inverse of a function always a function?

D
- Are there functions that are self-inverses?
- Can everything be undone?

RELATIONSHIPS

ATL Critical-thinking

Test generalizations and conclusions

You should already know how to:

• decide whether a mapping diagram or a set of ordered pairs represents a function	**1** Decide which diagram represents a function. If it does not represent a function, state the reason why. **a** 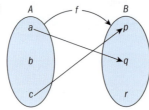 **b** 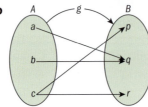 **c** 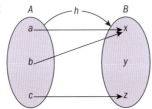 **2** $M = \{1, 2, 3, 4\}$. Determine whether or not each relation is a function from M to M. If it is not a function, state the reason why. **a** $R_1 = \{(2, 3), (1, 4), (2, 1), (3, 2), (4, 4)\}$ **b** $R_2 = \{(2, 1), (3, 4), (1, 4), (4, 4)\}$
• use the vertical line test to decide if a relation is a function	**3** Use the vertical line test to determine if each relation here is a function. **a** 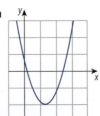 **b** 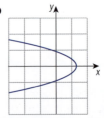
• form the composition of two functions	**4** For the functions $f(x) = 2x + 1$ and $g(x) = x^2 + 1$, find: **a** $f \circ g(x)$ **b** $g \circ f(x)$

F Types of functions

- What are onto and one-to-one functions?
- How do you find the inverse of a function?

For a relation between two sets to be a function, each element from the domain must map to one *and only one* element in the range.

> For a function $f: A \to B$:
> - Each element in the domain A must have an image in the range B.
> - Each element in A can have only one image in B. So if a function contains two ordered pairs (a, b_1) and (a, b_2), then $b_1 = b_2$.

Onto functions

This mapping diagram shows a function $f: A \to B$.

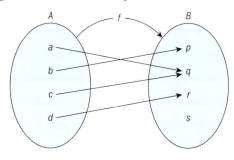

The element s in B is not the image of any element in A.

Each element in the domain A maps to one, and only one, element in B. The element s in B, however, is not the image of any element in A. Therefore, $f: A \to B$ is not an **onto** function.

> A function $f: A \to B$ is **onto** if every element in B is an image of an element in A.

In this graph of $y = \frac{1}{2}(x-1)$, if you draw horizontal lines through every possible y-value then these lines will all intersect the graph line, because every element in the range (y-values) is the image of an element in the domain (x-values).

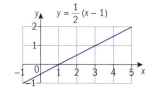

> The **horizontal line test for onto functions**: a function f is onto, if and only if its graph intersects any horizontal line drawn on it at least once.

Reflect and discuss 1

- Do all graphs of straight lines represent **onto** functions? Explain.
- Draw three different graphs that represent onto functions, and specify the domain and range of the functions.

Top right is the graph of the function $f(x): x^2$, $f: \mathbb{R} \to \mathbb{R}$. As x can assume any value, the domain of the function is the set of real numbers. The range is the set of non-negative real numbers, hence this function is not onto.

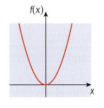

If, however, we define $f(x) = x^2$ such that $f: \mathbb{R} \to \mathbb{R}^+ \cup \{0\}$, then $f(x)$ is an onto function, and the graph for these y-values passes the horizontal line test.

One-to-one functions

This mapping diagram shows the function $f: X \to Y$, $X = \{1, 2, 3, 4\}$ and $Y = \{A, B, C, D\}$.

f is a one-to-one function.

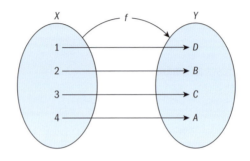

> A function $f: A \to B$ is a one-to-one function if every element in the range is the image of exactly one element in the domain. In other words, a function f is one-to-one if $f(a_1) = f(a_2)$ implies that $a_1 = a_2$.

Looking again at the graph of $y = \frac{1}{2}(x-1)$, every y-value is the image of exactly one x-value, so y is a one-to-one function. If you draw a horizontal line through every possible y-value, these lines will all intersect the graph line once only.

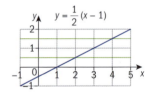

> The **horizontal line test for one-to-one functions**: a function f is one-to-one, if and only if its graph intersects any horizontal line drawn on it at exactly one point.

Reflect and discuss 2

- How does the horizontal line test for onto functions differ from the horizontal line test for one-to-one functions?
- Is every straight line a one-to-one function? Explain.
- Is a quadratic function one-to-one? Explain.
- Draw three different graphs that represent one-to-one functions.
- How can a function be onto but not one-to-one?
- How can a function be one-to-one but not onto?
- Can a function be neither onto nor one-to-one? Explain.

Example 1

Determine which of these functions are onto, one-to-one, both, or neither.

a 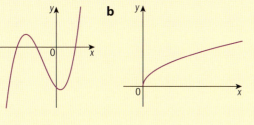 b c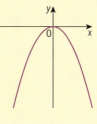

$f: \mathbb{R} \to \mathbb{R}$ $\qquad$ $f: \mathbb{R}^+ \to \mathbb{R}$ $\qquad$ $f: \mathbb{R} \to \mathbb{R}^- \cup \{0\}$

a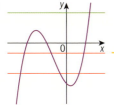

Use the horizontal line tests.

Horizontal lines intersect the graph in at least one point, so the function is onto. Some horizontal lines intersect at more than one point, so the function is not one-to-one.

b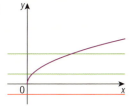

Some horizontal lines do not intersect the graph, so the function is not onto. Horizontal lines do not intersect the graph in more than one point, so the function is one-to-one.

c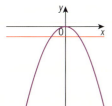

The range is the set of non-positive real numbers, hence it is onto. It is not one-to-one since horizontal lines intersect at more than one point.

Practice 1

1 Determine which of these functions are onto, one-to-one, neither or both.

a $f: \mathbb{R} \to \mathbb{R}$

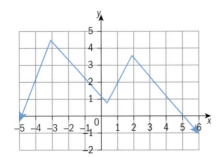

b $f: A \to B; A = \{x : -5 \leq x < 5\}, B = \{y : -3 \leq y < 4\}$

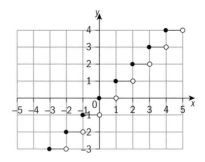

c $f: \{x : -2 \leq x \leq 8\} \to \{y : -1 \leq y \leq 1\}$

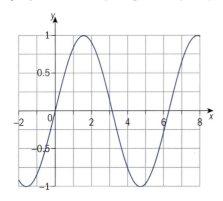

d $f: \mathbb{R} \to \mathbb{R}$

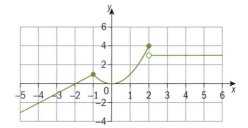

e $f: \mathbb{R} \to \mathbb{R}$

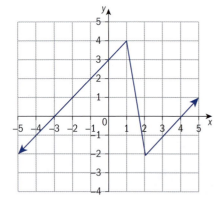

f $f: \mathbb{R} \to \{y : y \geq -3\}$

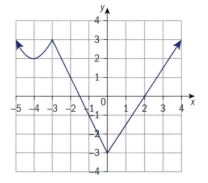

g $f : \mathbb{R} \to \mathbb{R}$

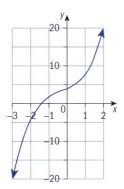

h $f : \{x : x \in \mathbb{R}, -3 \leq x \leq 5\} \to \mathbb{R}$

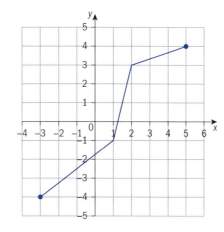

Problem solving

2 Create your own functions that are:

a onto, but not one-to-one

b one-to-one, but not onto

c neither onto nor one-to-one

d both onto and one-to-one.

> If you change the elements of the ordered pairs in a relation R by reversing them, you obtain the inverse relation $R^{-1} = \{(b, a) : b \in B \text{ and } a \in A\}$. Since a relation is a set of ordered pairs, every relation has an inverse which is also a relation.

Exploration 1

1 A friend gives you these instructions to get to her house:

- From your home, go north on route B21 for 3 km.
- Turn right on Kanuga Drive.
- At the 4th traffic light, turn left on Georgia Avenue.
- My house is the 5th one on the right.

Write down the directions to get back home from your friend's house. Explain how you worked them out.

2 a Write down a function f that models the steps in the algorithm below. State its largest possible domain and range.

- Start with a real number x.
- Add 4.
- Divide the result by 3.
- Subtract 5 from this result.
- The final answer is y.

▶ Continued on next page

E2.2 Doing and undoing

b Write down the inverse algorithm that starts with *y* and leads to *x*. Write down a function *g* that models the steps in this inverse algorithm. Look at your functions *f* and *g*. Explain how you can get from *f* to *g* algebraically.

c Rewrite *g*, interchanging the *x* and *y*, so that $g: x \to y$. Justify that *g* is a function, and state its largest possible domain and range.

d State whether *f* and *g* are:

  i one-to-one functions   **ii** onto functions.

e Draw the graphs of $f(x)$ and $g(x)$ on the same set of axes. Write down the equation of the line of symmetry.

> Simplify the functions before graphing them.

Step **2** of Exploration 1 goes through the algebraic process of finding the inverse of a function: reverse the process by making *x* the subject, then interchange the *x* and *y* variables.

> The notation for the inverse of a function *f* is f^{-1}.

> **Inverse functions**:
>
> **1** The inverse of a function $f: A \to B$ is the relation formed when the domain and range of the function are interchanged.
>
> **2** The graph of the inverse of a function is the reflection of the graph of the original function in the line $y = x$.
>
> **3** If the inverse relation is also a function, then it is the inverse function of the original function, $f^{-1}: B \to A$. We say that the function is *invertible*.

Example 2

Find the inverse of the function $f(x) = \frac{1}{5}(6x - 5)$.

$y = \frac{1}{5}(6x - 5)$	Write the function using *y* instead of $f(x)$.
$5y = 6x - 5$	Make *x* the subject.
$6x = 5y + 5 = 5(y+1)$	
$x = \frac{5}{6}(y+1)$	
$y = \frac{5}{6}(x+1)$	Interchange *x* and *y*.
$f^{-1}(x) = \frac{5}{6}(x+1)$	Write in inverse notation.

> You can also interchange the *x* and *y* variables first and then isolate *y*.

ALGEBRA

ATL **Reflect and discuss 3**
- Are all linear relations also linear functions? Explain.
- Is the inverse of any linear function also a function? Explain.

Example 3

a Find the inverse of the function $y = \dfrac{1}{x-2} + 3$, $x \neq 2$, algebraically.

b Confirm your answer to **a** graphically.

c Determine if the inverse is also a function.

a
$$y = \dfrac{1}{x-2} + 3$$

$$y - 3 = \dfrac{1}{x-2}$$

$$(y-3)(x-2) = 1$$

$$x - 2 = \dfrac{1}{y-3}$$

$$x = \dfrac{1}{y-3} + 2,\ y \neq 3 \quad \text{— State the excluded value.}$$

$$y = \dfrac{1}{x-3} + 2,\ x \neq 3 \quad \text{— Interchange } x \text{ and } y.$$

b

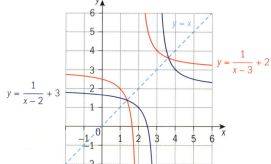

Graph the function and its inverse. Add in the graph of the line $y = x$ to check that one graph is a reflection of the other in $y = x$.

$f(x) = \dfrac{1}{x-3} + 2$ is the inverse of $f(x) = \dfrac{1}{x-2} + 3$ because it is a reflection in the line $y = x$.

c $f(x) = \dfrac{1}{x-3} + 2$, $x \neq 3$ is a function because it passes the vertical line test.

ATL **Practice 2**

1 Write down the inverse of each function, and state whether or not the inverse is also a function.

a $f(x) = \{(3, 1), (-1, 1), (-3, 1), (1, 1)\}$

b $g(x) = \{(2, 4), (-5, 3), (-2, -3), (1, 1), (-4, -4), (-1, -2), (3, -1)\}$

c $h(x) = \{(0, 0), (-2, 4), (-1, -4), (2, 2), (-3, -3), (-4, 4), (1, -1)\}$

E2.2 Doing and undoing

Problem solving

2 Copy each graph and draw the inverse on the same set of axes. Determine if the graph of the inverse of each function also represents a function.

> Use the vertical line test.

a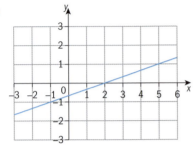

b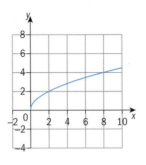

c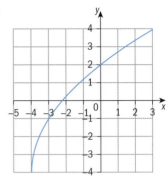

d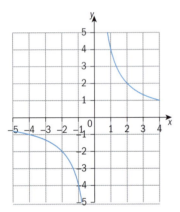

e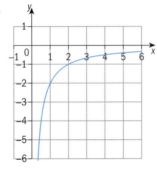

f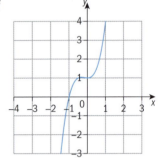

3 Find the inverse of each function algebraically. Determine if the inverse of the function is also a function.

a $y = -4x + 1$

b $y = 4 - \frac{3}{2}x$

c $y = \frac{-x-5}{3}$

d $y = -1 - \frac{1}{x}, x \neq 0$

e $y = \frac{1}{x+5}, x \neq -5$

f $y = x^2 + 1$

g $y = x^3$

h $y = \sqrt[3]{x+2}, x \geq -2$

i $y = \frac{x+3}{x-2}, x \neq 2$

C Reversible and irreversible changes

- How do you know when two functions are inverses of each other?
- Is the inverse of a function always a function?

Exploration 2

1 Draw the graph of the quadratic function $y = x^2$. State the largest possible domain and the corresponding range of the function $y = x^2$.

2 Find the inverse of the function algebraically.

▶ Continued on next page

3. Draw the inverse on the same graph. Draw the line $y = x$ and confirm that your graph shows the inverse of the function.
4. Determine if the inverse of the function is also a function.
5. Determine how you can restrict the domain of $y = x^2$ so that its inverse is also a function.
6. State the domain and range of your new function and the domain and range of its inverse function.

Reflect and discuss 4

- Is there more than one way of restricting the domain of the function $y = x^2$ so that its inverse is also a function?
- Is the function $y = x^2$ one-to-one and onto?
- When you restrict the domain of $y = x^2$ so that its inverse is also a function, does it become one-to-one and onto?

If the inverse of a function is not a function, you may be able to restrict the domain of the function so that its inverse *is* a function.

Theorem

The inverse of a function f is also a function, if and only if f is both an onto and a one-to-one function.

The proof of this theorem is beyond the level of this course.

Example 4

a Find the inverse of $y = x^2 - 1$ algebraically, and confirm graphically.

b If the inverse is not a function, restrict the domain of the function so that its inverse is also a function, and confirm graphically.

a $y = x^2 - 1$ ──────────────────────── Find the inverse.

$x^2 = y + 1$

$x = \pm\sqrt{y+1}$

$y = \pm\sqrt{x+1}$

$y = \pm\sqrt{x+1}$ is the reflection of $y = x^2 - 1$ in the line $y = x$.

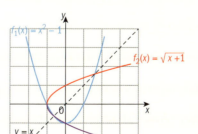

Graph both functions and confirm that one is a reflection of the other in the line $y = x$.

▶ Continued on next page

b

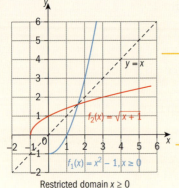

Restricted domain $x \geq 0$

Try dividing the original function into pieces that are both one-to-one and onto.

Define the restricted domain for each section.

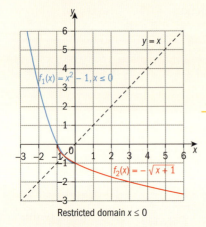

Restricted domain $x \leq 0$

Write the inverse using function notation.

Therefore, for the domain $x \geq 0$ of f, $f^{-1}(x) = \sqrt{x+1}$.

For the domain $x \leq 0$ of f, $f^{-1}(x) = -\sqrt{x+1}$.

Example 5

a Find the inverse of $f(x) = x^2 - 6x + 2$ algebraically, and confirm graphically.

b Restrict the domain of f so that its inverse is also a function, and state its domain and range.

c State the domain and range of f^{-1}.

d Confirm graphically that the inverse function is indeed a function.

a
$y = x^2 - 6x + 2$

$x^2 - 6x = y - 2$

$x^2 - 6x + 9 = y - 2 + 9$

$(x - 3)^2 = y + 7$

$x - 3 = \pm\sqrt{y+7}$

$x = 3 \pm \sqrt{y+7}$

$y = 3 \pm \sqrt{x+7}$

You cannot immediately isolate x.

Complete the square to solve for x. (You could also use the quadratic formula to solve for x.)

Interchange x and y.

▶ Continued on next page

Since one graph is the reflection of the other in the line $y = x$, they are inverses of each other.

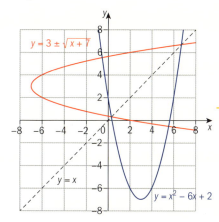

Draw the graphs and the line $y = x$.

b The vertex is $(3, -7)$. Therefore for the inverse to be a function, the domain of f is either $x \geq 3$ or $x \leq 3$.

Identify the domain that would make the function onto and one-to-one by finding the vertex of the parabola either graphically, or using the formula $x = -\frac{b}{a}$, $y = f\left(-\frac{b}{a}\right)$.

c The domain of f is $x \geq 3$.
Then the domain of f^{-1} is $x \geq -7$.
and the range of f^{-1} is $y \geq 3$.

Select a domain such that the inverse is a function, and graph it.

d

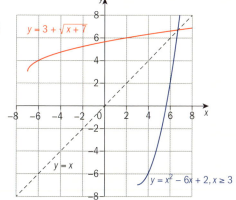

Check that the graphs are symmetrical about the line $y = x$.

The graphs are reflections of each other in $y = x$.

Practice 3

For each function in questions **1** to **8**:

a Find its inverse algebraically, and confirm graphically.

b If the inverse is not a function, restrict the domain of the original function so that its inverse is also a function and write the inverse function with its restricted domain.

1 $f(x) = 5x - 2$ **2** $f(x) = 7 - \frac{1}{2}x$ **3** $f(x) = x^2 - 3$

4 $f(x) = 1 - x^2$ **5** $f(x) = \frac{x+1}{x}$, $x \neq 0$ **6** $f(x) = (x - 2)^2$

7 $f(x) = x^2 + 2x - 4$ **8** $f(x) = 2x^2 - 4x + 5$

Problem solving

9 Below left is the graph of the absolute value function $f(x) = |x|$. The domain is the set of real numbers. The range is the set of positive real numbers and zero. Reflecting the graph of $f(x) = |x|$ in the line $y = x$ gives this graph of the inverse, below right.

$|a| = a$ for $a > 0$.
$|a| = -a$ for $a < 0$.

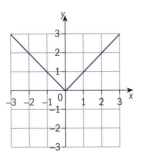 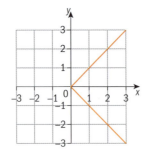

Explain how you could restrict the domain of f so that its inverse is a function. Write the inverse function for f with your restricted domain.

Objective D: Applying mathematics in real-life contexts
iii. apply the selected mathematical strategies successfully to reach a solution

*Apply what you have learned about inverse functions in solving question **10**.*

10 In sports like bowling or golf, players often have a 'handicap' (h). A handicap is a number added to a player's score to allow comparisons with players of very different levels. For example, if you bowl a 150 and your handicap is 30, while Davinia bowls a 170 and her handicap is 5, then you would win, 180 to 175. Handicaps are usually calculated based on a bowler's average score (s).

In one bowling league, handicaps are calculated using:

$$h(s) = 0.9(220 - s)$$

a Write down the domain of $h(s)$. Explain why it needs to be restricted.

b Find the inverse of $h(s)$ and confirm your result graphically. Use your graph to determine if the inverse also a function.

c Find the average score for an opponent whose handicap is 15.

In bowling, a handicap is never less than zero.

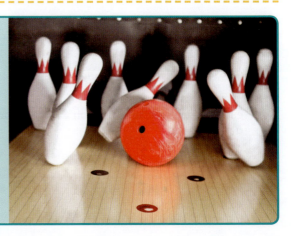

A 'perfect' game in ten-pin bowling is achieved by a score of 300. This can be done only by getting 12 strikes in a row, where a strike is knocking down all ten pins with one ball. However, when you bowl an optimal strike, the ball itself hits just four pins. The rest of the pins are knocked down by other pins, rather than the ball, and any pin will fall if it tilts just 9 degrees from upright. A right-handed bowler's ball will connect with the number 1, 3, 5 and 9 pins, and a left-handed bowler's ball will connect with just the 1, 2, 5 and 8 pins. So, a perfect game can be the result of the ball connecting with just 48 pins.

This mapping diagram shows a function and its inverse.

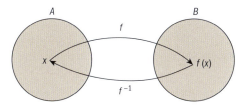

f maps x onto $f(x)$ and then f^{-1} maps $f(x)$ to x.

So the composite function $f^{-1} \circ f(x) = x$. Similarly $f \circ f^{-1}(x) = x$.

> If two functions f and f^{-1} are inverses of each other, then $f \circ f^{-1}(x) = x$ and $f^{-1} \circ f(x) = x$.

Example 6

Confirm algebraically that the given functions are inverses:

a $f(x) = 2x + 3$ and $f^{-1}(x) = \dfrac{x-3}{2}$

b $f(x) = \dfrac{1}{x-2} + 3$, $x \neq 2$ and $f^{-1}(x) = \dfrac{2x-5}{x-3}$, $x \neq 3$

a $f \circ f^{-1}(x) = 2\left(\dfrac{x-3}{2}\right) + 3$ — Find $f \circ f^{-1}(x)$ and simplify.

$\qquad = x - 3 + 3 = x$

$f^{-1} \circ f(x) = \dfrac{2x + 3 - 3}{2} = \dfrac{2x}{2} = x$ — Find $f^{-1} \circ f(x)$ and simplify.

$f \circ f^{-1}(x) = x$ and $f^{-1} \circ f(x) = x$, so f and f^{-1} are inverse functions. — Check that both composite functions are equal to x.

b $f^{-1} \circ f(x) = \dfrac{2\left(\dfrac{1}{x-2} + 3\right) - 5}{\left(\dfrac{1}{x-2} + 3\right) - 3} = \dfrac{\dfrac{2}{x-2} + 1}{\dfrac{1}{x-2}} = \dfrac{\dfrac{x}{x-2}}{\dfrac{1}{x-2}} = x$ — Find $f^{-1} \circ f(x)$ and simplify.

$f \circ f^{-1}(x) = \dfrac{1}{\dfrac{2x-5}{x-3} - 2} + 3 = \dfrac{1}{\dfrac{1}{x-3}} + 3$ — Find $f \circ f^{-1}(x)$ and simplify.

$\qquad = x - 3 + 3 = x$

Since $f^{-1} \circ f(x) = x = f \circ f^{-1}(x)$, the functions are mutual inverses. — Check that both composite functions are equal to x.

Practice 4

In questions **1** to **8**, determine algebraically whether or not the pairs of functions are inverses of each other.

1. $f(x) = x - 3$; $g(x) = x + 3$
2. $f(x) = 4x + 8$; $g(x) = -\frac{1}{4}x - 2$
3. $f(x) = -\frac{2}{5}x - 2$; $g(x) = -5 - \frac{5}{2}x$
4. $f(x) = \sqrt{x+2} - 2, x \geq -2$; $g(x) = (x - 2)^2 - 2$
5. $f(x) = \frac{-4 + \sqrt{4x}}{2}, x \geq 0$; $g(x) = (x + 2)^2, x \geq -2$
6. $f(x) = \frac{1}{x-2} - 2, x \neq 2$; $g(x) = \frac{5 + 2x}{2 + x}, x \neq -2$
7. $f(x) = \frac{1-x}{1+x}, x \neq -1$; $g(x) = \frac{1-x}{1+x}, x \neq -1$

Problem solving

8. $f(x) = -2x^2 - 2$; $g(x) = \sqrt{\frac{-x-2}{2}}$

Exploration 3

In Practice 4, question **8**, many students think that *f* and *g* are mutual inverses when, in fact, they are not.

1. Graph the functions and state why this is the case.
2. State, therefore, the difference between $f \circ g$ and $g \circ f$ in this example.
3. State what additional information must be given when determining whether or not two functions are mutual inverses.

Exploration 4

1. For the functions $f(x) = 3x - 1$ and $g(x) = 2x + 3$, find:
   a. $f^{-1}(x)$ and $g^{-1}(x)$
   b. $f \circ g(x)$
   c. $[f \circ g(x)]^{-1}$
   d. $f^{-1} \circ g^{-1}(x)$
   e. $g^{-1} \circ f^{-1}(x)$

2. Repeat step **1** for these functions:
   a. $f(x) = 2x + 5$; $g(x) = \frac{3x-1}{2}$
   b. $f(x) = x^2 + x - 1$; $g(x) = x + 1$

3. Make a conjecture about the relationship between $(f \circ g)^{-1}$ and $g^{-1} \circ f^{-1}$.

Reflect and discuss 6

Which is easier, finding $(f \circ g)^{-1}$ or $g^{-1} \circ f^{-1}$?

For any two functions *f* and *g*, $f \circ g^{-1}(x) = g^{-1} \circ f^{-1}(x)$.

You should have discovered this in Exploration 4. The proof of this result is beyond the level of this course.

ALGEBRA

D When doing and undoing produces the same result

- Are there functions that are self-inverses?
- Can everything be undone?

The inverse of the function $f(x) = 2 - x$ is $f^{-1}(x) = 2 - x$, so f is its own inverse. You can say that f is self-inverse.

> A function f is self-inverse if $f^{-1}(x) = f(x)$.

Exploration 5

1 Determine algebraically and graphically whether or not any linear function of the form $f(x) = k - x$, $k \in \mathbb{R}$, is self-inverse.

2 Determine if $f(x) = \dfrac{k}{x}$, $k \in \mathbb{R}$, is self-inverse. Are there values of k for which the function is not self-inverse?

3 Determine if $f(x) = \dfrac{k}{x-1}$, $x \neq 1$, is self-inverse. Are there values for k other than 1 for which $f(x)$ is not self-inverse?

4 Determine if $f(x) = \dfrac{x-k}{x-1}$, $x \neq 1$ is self-inverse. Are there values of k for which $f(x)$ is not self-inverse?

Summary

For a function $f: A \to B$:

- each element in the domain A must have an image in the range B.
- each element in the domain A can have only one image in the range B.

A function $f: A \to B$ is **onto** if every element in B is an image of an element in A.

A function $f: A \to B$ is a **one-to-one** function if every element in the range is the image of exactly one element in the domain.

The horizontal line tests:

A function f is onto, if and only if its graph intersects any horizontal line drawn on it at least once.

A function f is one-to-one, if and only if its graph intersects any horizontal line drawn on it at exactly one point.

If the inverse of a function is not a function, you may be able to restrict the domain of the function so that its inverse is also a function.

1 The inverse of a function $f: A \to B$ is the relation formed when the domain and range of the function are interchanged.

2 The graph of the inverse of a function is the reflection of the graph of the original function in the line $y = x$.

3 If the inverse relation is also a function, then it is the inverse function of the original function, $f^{-1}: B \to A$. We say the function is *invertible*.

Theorem

The inverse of a function f is also a function if and only if f is both an onto and a one-to-one function.

If two functions f and f^{-1} are inverses of each other, then $f \circ f^{-1}(x)$ and $f^{-1} \circ f(x) = x$.

For any two functions f and g,
$(f \circ g)^{-1}(x) = g^{-1} \circ f^{-1}(x)$.

A function f is self-inverse if $f^{-1}(x) = f(x)$.

E2.2 Doing and undoing

Mixed practice

1 i **Determine** if these relations are functions.
  ii **Find** the inverse of the relation.
  iii **Determine** if the inverse is a function.
   a $R = \{(1, 2), (2, 3), (3, 4), (4, 5)\}$
   b $S = \{(-2, 0), (3, 1), (-5, 1), (6, 7), (0, 0)\}$
   c $T = \{(0, -1), (3, 1), (-4, -1), (-3, 6), (3, -1)\}$

2 Determine whether each function is one-to-one, onto, neither, or both.

a $f : \mathbb{R} \to \mathbb{R}$

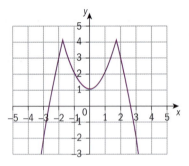

b $f : \mathbb{R} \to \mathbb{R}$

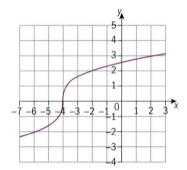

c $f : \mathbb{R} \to \mathbb{R}$

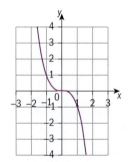

3 Justify which of these functions have inverses that are also functions.

a $f : \mathbb{R} \to \mathbb{R}$

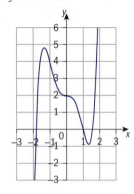

b $f : \{x : x \geq 1\} \to \{y : y \geq 1\}$

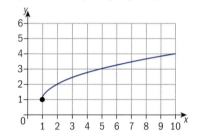

c $f : \mathbb{R} \to \mathbb{R}$

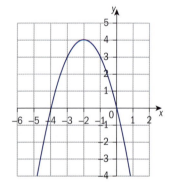

d $f : \mathbb{R} \to \{y : y \geq -1\}$

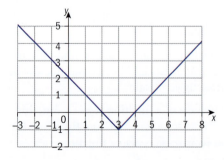

4 a For each function below, **find** its inverse algebraically, and confirm graphically.

b If its inverse is not a function, restrict the domain of the original function so that its inverse is also a function, and **write down** the inverse function with its restricted domain.

i $f(x) = -3x + 8$ **ii** $f(x) = \frac{3}{4}x - 1$

iii $f(x) = 4 - x^2$ **iv** $f(x) = 3x^2 - 2$

v $f(x) = \frac{x}{2x-1}$, $x \neq \frac{1}{2}$

5 Find the inverse of each function algebraically. **Determine** whether or not the inverse of the function is also a function. If the inverse relation is not a function, restrict the domain of the function so that its inverse is also a function.

a $f(x) = 5x$ **b** $f(x) = 6x + 2$

c $f(x) = \frac{2}{3}x - \frac{1}{2}$ **d** $g(x) = \frac{x-2}{5} + 1$

e $h(x) = x^2 - 3$ **f** $p(x) = \frac{1}{x} + 2$, $x \neq 0$

g $q(x) = \frac{x+2}{x-3}$, $x \neq 3$ **h** $s(x) = \sqrt{x+3}$, $x \geq -3$

6 Justify algebraically that these pairs of functions are inverses of each other.

a $f(x) = -7x + 4$; $g(x) = -\frac{1}{7}x + \frac{4}{7}$

b $f(x) = \frac{-2}{x-1} + 3$, $x \neq 1$; $g(x) = \frac{x-5}{x-3}$, $x \neq 3$

c $f(x) = \frac{-3 + \sqrt[3]{2x}}{2}$, $x \geq 0$; $g(x) = \frac{(2x+3)^3}{2}$

d $f(x) = 2 + \sqrt{3x - 4}$, $x \geq \frac{4}{3}$; $g(x) = \frac{(x-2)^2 + 4}{3}$

e $f(x) = x^2 + 3x - 1$, $x \leq -\frac{3}{2}$; $g(x) = -\frac{\sqrt{4x+13}}{2} - 3$

7 The weight w in kg of a certain type of fish is related to its length l in cm, and can be modelled by the function $w = (9.4 \times 10^{-6}) \times l^3$.
By obtaining the inverse of the model, **determine** the length of a fish that weighs 0.58 kg.

> **Objective D:** Applying mathematics in real-life contexts
> **iii.** apply the selected mathematical strategies successfully to reach a solution
>
> *To correctly answer the questions, you will have to decide which function is best to use, the original or the inverse.*

8 A computer company uses the formula $d(t) = 700 - 12t$ to find the depreciated value in d dollars of computers t months after they are put on sale.

a Calculate $d(24)$ and state what this means.

b Find t such that $d(t) = 50$, and explain what this value means.

c Write a formula for $d^{-1}(t)$, and explain what the formula means in this context.

d Find after how many months a computer will be worth $500.

e Find t such that $d^{-1}(t) = 100$, and explain what this means.

Review in context

1 Aryabhata, a 6th century Hindu mathematician, came up with this puzzle:

- 5 is added to a certain number.
- The result is divided by 2.
- That result is multiplied by 6.
- Then 8 is subtracted from that result.
- The answer is 34.
- Find the number.

Starting with 34, **write down** the steps, in order, necessary to arrive at the original number.

2 Choose a number, add 6, multiply the result by 2, subtract 12, and finally divide by 2.

a Justify the result you obtain.

b Invert the process and **justify** the result.

3 Ask a classmate to think of a positive integer. Then perform the following operations in order on the given number:

- Square the number.
- Multiply the result by 2.
- Add 3 to the result.
- Ask for the final result.

Write the process above in the form of a function, then write its inverse so that you can identify the number that your classmate started out with. **Find** the original number. **State** for which numbers this puzzle would have no solution.

4 Make up your own puzzle whose result is the same as the number you start with.

E3.1 How can we get there?

Global context: Scientific and technical innovation

Objectives
- Describing translations using vectors
- Adding and subtracting vectors
- Multiplying a vector by a scalar
- Finding the magnitude of a vector
- Finding the dot product of two vectors
- Using the dot product to find the angle between two vectors
- Solving geometric problems using vectors

Inquiry questions

F
- What is a vector?
- How do you find the magnitude of a vector?

C
- How do you perform mathematical operations on vectors?

D
- How can vectors be used to solve geometric problems?
- Can logic solve everyday problems?

ATL Critical-thinking

Consider ideas from multiple perspectives

GEOMETRY AND TRIGONOMETRY

You should already know how to:

• find the distance between two points	**1** Find the distance between each pair of points: **a** (4, 7) and (11, 5) **b** (−3, −2) and (1, 1)
• solve linear simultaneous equations	**2** Solve the simultaneous equations: $3x + 5y = 5$ $2x + y = 1$
• use the cosine rule	**3** Find the size of angle A.

F Introduction to vectors

- What is a vector?
- How do you find the magnitude of a vector?

The map shows part of Salt Lake City, USA, which has a grid layout.

You can give directions in this area of Salt Lake City by giving the number of blocks to travel in an east-west direction, and in the north-south direction.

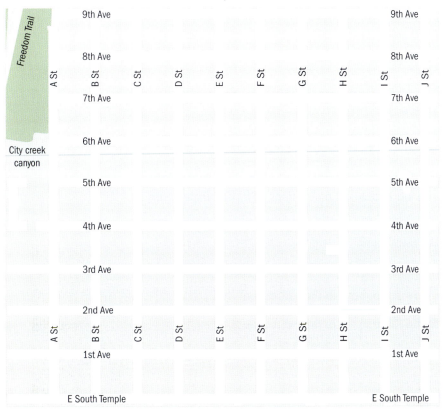

A block is the distance between two adjacent intersections.

E3.1 How can we get there?

Exploration 1

1 Using the map of Salt Lake City, write down instructions to get from:

  a the intersection of A Street and 1st Avenue to the intersection of H Street and 3rd Avenue

  b the intersection of D Street and 2nd Avenue to the intersection of G Street and 7th Avenue

  c the intersection of F Street and 3rd Avenue to the intersection of A Street and 6th Avenue.

2 You can use a column vector to represent a translation from one point to another.

  The column vector $\begin{pmatrix} 3 \\ 4 \end{pmatrix}$ means 3 units right and 4 units up.

  Describe these translations in words:

  a $\begin{pmatrix} 1 \\ 3 \end{pmatrix}$ **b** $\begin{pmatrix} 8 \\ 7 \end{pmatrix}$ **c** $\begin{pmatrix} 2 \\ 4 \end{pmatrix}$ **d** $\begin{pmatrix} 4 \\ 2 \end{pmatrix}$

3 The column vector $\begin{pmatrix} -3 \\ -4 \end{pmatrix}$ means 3 units left and 4 units down.

  Using the map, write down the column vector for the journey from:

  a the intersection of A Street and 3rd Avenue to the intersection of C Street and 8th Avenue

  b the intersection of J Street and 4th Avenue to the intersection of B Street and 8th Avenue

  c the intersection of C Street and 8th Avenue to the intersection of G Street and 3rd Avenue.

4 For each vector below, start at the intersection of F Street and 5th Avenue. Write down your final position after a translation of:

  a $\begin{pmatrix} 2 \\ 2 \end{pmatrix}$ **b** $\begin{pmatrix} 3 \\ -1 \end{pmatrix}$ **c** $\begin{pmatrix} 4 \\ 0 \end{pmatrix}$ **d** $\begin{pmatrix} -2 \\ 4 \end{pmatrix}$ **e** $\begin{pmatrix} -3 \\ -4 \end{pmatrix}$ **f** $\begin{pmatrix} 0 \\ -3 \end{pmatrix}$

You will usually use column vectors to describe translations on a coordinate grid.

A **column vector** describes a translation.

A column vector has an x component and a y component: $\begin{pmatrix} x \\ y \end{pmatrix}$

The x component tells you how far to move in the x-direction.

The y component tells you how far to move in the y-direction.

▶ Continued on next page

GEOMETRY AND TRIGONOMETRY

$\overrightarrow{AB}$ is the vector representing the translation from *A* to *B*.

On this grid, $\overrightarrow{AB} = \begin{pmatrix} 3 \\ 1 \end{pmatrix}$

The **position vector** of a point is the vector from the origin *O* to the point.

On this grid, $\overrightarrow{OA} = \begin{pmatrix} 1 \\ 2 \end{pmatrix}$

$\begin{pmatrix} 0 \\ 0 \end{pmatrix}$ is the **zero vector**. It is the position vector of the origin.

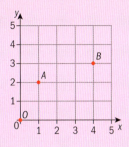

Example 1

A is the point (4, 9) and *B* is the point (6, 4).
Find $\overrightarrow{AB}$.

$\overrightarrow{AB} = \begin{pmatrix} 6-4 \\ 4-9 \end{pmatrix} = \begin{pmatrix} 2 \\ -5 \end{pmatrix}$ —— From *A* to *B* you move 2 units right and 5 units down.

Reflect and discuss 1

- How does the vector $\overrightarrow{AB} = \begin{pmatrix} 2 \\ -5 \end{pmatrix}$ relate to the coordinates of points *A* and *B*?
- What operation would you perform to find the vector $\overrightarrow{AB}$ if you knew the coordinates of *A* and *B*?
- How could you find the vector $\overrightarrow{BA}$ from the coordinates of *A* and *B*?

Practice 1

1 **a** Write down the column vectors for these translations on the grid.

    i $\overrightarrow{AB}$ **ii** $\overrightarrow{AC}$ **iii** $\overrightarrow{AF}$
    iv $\overrightarrow{GF}$ **v** $\overrightarrow{OD}$ **vi** $\overrightarrow{OC}$

  b Write down the column vectors for:

    i $\overrightarrow{AD}$ and $\overrightarrow{DA}$ **ii** $\overrightarrow{EG}$ and $\overrightarrow{GE}$

  c Conjecture the relationship between vectors of the form $\overrightarrow{PQ}$ and $\overrightarrow{QP}$.

  d Write down the position vector of:

    i *A*   **ii** *B*   **iii** *C*   **iv** *O*

  e Write down the point with position vector:

    i $\begin{pmatrix} -1 \\ -1 \end{pmatrix}$ **ii** $\begin{pmatrix} -3 \\ 3 \end{pmatrix}$ **iii** $\begin{pmatrix} 2 \\ -4 \end{pmatrix}$

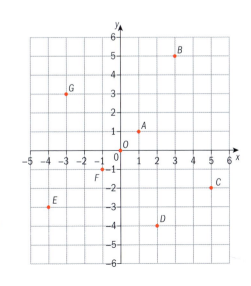

E3.1 How can we get there?

2 On this grid, find two points so that the translation from one to the other is:

   a $\begin{pmatrix} 0 \\ 4 \end{pmatrix}$ b $\begin{pmatrix} 3 \\ -1 \end{pmatrix}$ c $\begin{pmatrix} -5 \\ -1 \end{pmatrix}$ d $\begin{pmatrix} 4 \\ -3 \end{pmatrix}$

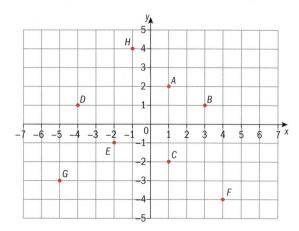

3 On a clean grid, start at the point with coordinates (3, 4). Write down the coordinates of the point after each translation:

   a $\begin{pmatrix} 2 \\ 5 \end{pmatrix}$ b $\begin{pmatrix} -2 \\ 4 \end{pmatrix}$ c $\begin{pmatrix} -3 \\ -1 \end{pmatrix}$ d $\begin{pmatrix} 11 \\ -9 \end{pmatrix}$

4 Write down the vector that translates:

   a $M(-2, 4)$ to $N(1, 6)$ b $P(4, 6)$ to $Q(11, 3)$
   c $R(11, -2)$ to $S(7, 5)$ d $T(-5, 1)$ to $U(-6, -3)$

Problem solving

5 $\overrightarrow{CD} = \begin{pmatrix} -2 \\ 4 \end{pmatrix}$. Write down the column vector $\overrightarrow{DC}$.

6 Point A has position vector $\begin{pmatrix} 4 \\ 2 \end{pmatrix}$. $\overrightarrow{AB} = \begin{pmatrix} 2 \\ -6 \end{pmatrix}$ and $\overrightarrow{BC} = \begin{pmatrix} -3 \\ 5 \end{pmatrix}$.

   a Show points A, B and C on suitable axes.
   b Write down the position vector of point C.

7 Write down three pairs of points, where one transforms to the other by the translation $\begin{pmatrix} -5 \\ 3 \end{pmatrix}$.

A journey of $\begin{pmatrix} 2 \\ 2 \end{pmatrix}$ is shown on this section of the Salt Lake city map.
The length of the journey is 4 blocks.

The larger map, on the next page, shows different journeys of length 4 blocks. Each is labelled with a vector: **a**, **b**, **c**, etc.

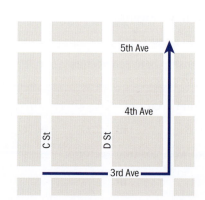

58   E3 Logic

GEOMETRY AND TRIGONOMETRY

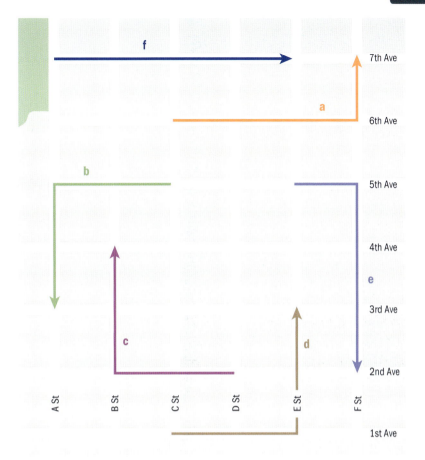

Tip

Single letter vectors are printed in **bold**. When writing vectors by hand, in some countries the convention is to underline them, e.g. a, b, and in others they are written with an arrow above, e.g. $\vec{a}, \vec{b}$.

ATL **Reflect and discuss 2**

- Are any of the journeys shown on the larger map the same as the one on the smaller map?
- What does it mean for two journeys to be the same as each other?

Vectors are equal if their components are equal. This means that they are equal if they represent journeys that have exactly the same instructions. Two vectors are not equal if they point in different directions, even if the length of the journey is the same.

The starting position of the vector doesn't matter. Walking two blocks east and two blocks south has vector $\begin{pmatrix} 2 \\ -2 \end{pmatrix}$, no matter where you start from.

> Two vectors are equal if their components are equal.
>
> If $\mathbf{v}_1 = \begin{pmatrix} x_1 \\ y_1 \end{pmatrix}$ and $\mathbf{v}_2 = \begin{pmatrix} x_2 \\ y_2 \end{pmatrix}$ then $\mathbf{v}_1 = \mathbf{v}_2$ if and only if $x_1 = x_2$ and $y_1 = y_2$.

E3.1 How can we get there?

Exploration 2

1 A hiker walks 3 miles due east and then 4 miles due north.

  a Sketch her walk.

  b Write down the total distance she walks.

  c Find the shortest distance between her starting position and ending position.

2 $\vec{AB} = \begin{pmatrix} 5 \\ -12 \end{pmatrix}$

The diagram represents the vector $\vec{AB}$.

Find the shortest distance from A to B.

3 $\vec{CD} = \begin{pmatrix} x \\ y \end{pmatrix}$

  a Sketch a diagram to represent this vector.

  b Find the shortest distance from C to D in terms of x and y.

The length of a vector, or the (shortest) distance from one end of the vector to the other, is called the **magnitude** of the vector. The magnitude of vector **v** is written |**v**|.

The components of a vector represent perpendicular distances, so you can use Pythagoras' theorem to find the vector's magnitude.

|**v**|, the magnitude of $\mathbf{v} = \begin{pmatrix} x \\ y \end{pmatrix}$, is given by $|\mathbf{v}| = \sqrt{x^2 + y^2}$

Example 2

$A = (4, 11)$ and $B = (-3, 7)$. Find $|\vec{AB}|$. Give your answer correct to 3 s.f.

$\vec{AB} = \begin{pmatrix} -7 \\ -4 \end{pmatrix}$

$|\vec{AB}| = \sqrt{(-7)^2 + (-4)^2}$ 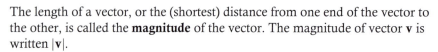 $|\mathbf{v}| = \sqrt{x^2 + y^2}$

$= \sqrt{65}$

$= 8.06$ (3 s.f.)

GEOMETRY AND TRIGONOMETRY

Vectors **a** and **b** both have the same magnitude:

$|\mathbf{a}| = \sqrt{5^2 + 5^2} = \sqrt{50}$ and $|\mathbf{b}| = \sqrt{7^2 + (-1)^2} = \sqrt{50}$

But **a** and **b** are not equal because they have different components:

$\mathbf{a} = \begin{pmatrix} 5 \\ 5 \end{pmatrix}$ and $\mathbf{b} = \begin{pmatrix} 7 \\ -1 \end{pmatrix}$.

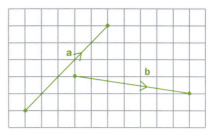

Even though their lengths are the same, they point in different directions.

When two vectors **a** and **b** are parallel but point in opposite directions, $\mathbf{a} = -\mathbf{b}$

> A vector has both magnitude and direction. A scalar quantity does not have a direction. For two vectors to be equal, they must have the same magnitude and direction.

Numbers are scalar quantities. Speed is a scalar quantity, but velocity (defined as speed in a given direction) is a vector quantity. Force is also a vector quantity. The force you experience due to gravity has a size or magnitude (your weight) and a direction (down).

Practice 2

1 Find the magnitude of each vector:

   a $\begin{pmatrix} 7 \\ 24 \end{pmatrix}$ **b** $\begin{pmatrix} 3 \\ 4 \end{pmatrix}$ **c** $\begin{pmatrix} 8 \\ -6 \end{pmatrix}$

2 Find the magnitude of each vector. Give your answers in exact form.

   a $\begin{pmatrix} -3 \\ 4 \end{pmatrix}$ **b** $\begin{pmatrix} 1 \\ 3 \end{pmatrix}$ **c** $\begin{pmatrix} 2 \\ 4 \end{pmatrix}$

   d $\begin{pmatrix} -5 \\ 5 \end{pmatrix}$ **e** $\begin{pmatrix} -5 \\ 12 \end{pmatrix}$ **f** $\begin{pmatrix} 16 \\ 5 \end{pmatrix}$

3 Find the magnitude of each vector. Give your answers correct to 3 s.f.

   a $\begin{pmatrix} 4 \\ 5 \end{pmatrix}$ **b** $\begin{pmatrix} 2 \\ -8 \end{pmatrix}$ **c** $\begin{pmatrix} -4 \\ -2 \end{pmatrix}$

   d $\begin{pmatrix} 13 \\ -2 \end{pmatrix}$ **e** $\begin{pmatrix} 8 \\ 4 \end{pmatrix}$ **f** $\begin{pmatrix} -6 \\ -11 \end{pmatrix}$

4 For each vector $\overrightarrow{AB}$ find $|\overrightarrow{AB}|$.

   a $\overrightarrow{AB} = \begin{pmatrix} 0 \\ -6 \end{pmatrix}$ **b** $\overrightarrow{AB} = \begin{pmatrix} -10 \\ 24 \end{pmatrix}$ **c** $\overrightarrow{AB} = \begin{pmatrix} 24 \\ -7 \end{pmatrix}$

   d $\overrightarrow{AB} = \begin{pmatrix} 0.28 \\ -0.96 \end{pmatrix}$ **e** $\overrightarrow{AB} = \begin{pmatrix} 8 \\ -6 \end{pmatrix}$ **f** $\overrightarrow{AB} = \begin{pmatrix} -4 \\ 3 \end{pmatrix}$

5 For each pair of points A and B, find $|\overrightarrow{AB}|$ correct to 3 s.f.

   a $A(1, 5)$ and $B(2, -4)$ **b** $A(4, 6)$ and $B(3, 3)$

   c $A(-1, -6)$ and $B(2, 7)$ **d** $A(3, 5)$ and $B(5.8, 11.2)$

   e $A(11, 4)$ and $B(4.5, 23.8)$ **f** $A(8.7, 5.3)$ and $B(2.1, -14.6)$

E3.1 How can we get there?

Problem solving

6 Four points, $A(1, 7)$, $B(6, 2)$, $C(3, -3)$ and $D(2, 4)$ lie in a plane.
Show that $|\vec{AB}| = |\vec{CD}|$.

7 Points A and B lie in a plane, and A has coordinates $(6, -2)$.
Find four possible positions for B such that $|\vec{AB}| = \sqrt{13}$.

8 The vectors $\mathbf{u} = \begin{pmatrix} 2x \\ y \end{pmatrix}$ and $\mathbf{v} = \begin{pmatrix} 7x-10 \\ 3-x \end{pmatrix}$ are equal.

Determine the values of x and y.

9 The vectors $\mathbf{u} = \begin{pmatrix} x^2 \\ 3x \end{pmatrix}$ and $\mathbf{v} = \begin{pmatrix} 7x-10 \\ x^2+2 \end{pmatrix}$ are equal.

Determine the value of x.

C Operations with vectors

- How do you perform mathematical operations on vectors?

The diagram shows two journeys in Salt Lake City:

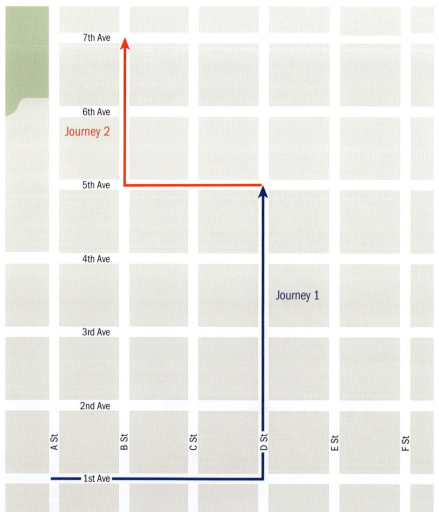

Journey 1 then Journey 2 = Whole journey

$\begin{pmatrix} 3 \\ 4 \end{pmatrix}$ then $\begin{pmatrix} -2 \\ 2 \end{pmatrix} = \begin{pmatrix} 1 \\ 6 \end{pmatrix}$

Reflect and discuss 3

What do you notice about:
- the relationship between the *x* components of the LHS and RHS?
- the relationship between the *y* components of the LHS and RHS?

A translation from *A* to *B* followed by a translation from *B* to *C* is the same as the translation from *A* to *C*: $\overrightarrow{AB} + \overrightarrow{BC} = \overrightarrow{AC}$

$\overrightarrow{AC}$ is sometimes called the **resultant** of $\overrightarrow{AB}$ and $\overrightarrow{BC}$.

To add vectors you simply add their components.

The vector addition law:

If $\overrightarrow{AB} = \begin{pmatrix} x_1 \\ y_1 \end{pmatrix}$ and $\overrightarrow{BC} = \begin{pmatrix} x_2 \\ y_2 \end{pmatrix}$ then

$\overrightarrow{AC} = \overrightarrow{AB} + \overrightarrow{BC} = \begin{pmatrix} x_1 \\ y_1 \end{pmatrix} + \begin{pmatrix} x_2 \\ y_2 \end{pmatrix} = \begin{pmatrix} x_1 + x_2 \\ y_1 + y_2 \end{pmatrix}$

You can also draw diagrams to add vectors using a graphical method. Draw the two vectors so that the second vector begins where the first vector ends. The resultant is the vector that joins the starting point to the tip of the second vector:

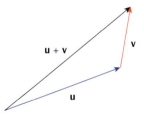

This shows the result of performing both translations.

Using the vector addition law: $\overrightarrow{OA} + \overrightarrow{AB} = \overrightarrow{OB}$.

You can rearrange this to $\overrightarrow{AB} = \overrightarrow{OB} - \overrightarrow{OA}$.

You can also see this on the diagram. Another route from *A* to *B* is:

    A to *O*, then *O* to *B*

or $\overrightarrow{AO} + \overrightarrow{OB} = -\overrightarrow{OA} + \overrightarrow{OB} = \overrightarrow{OB} - \overrightarrow{OA}$

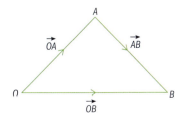

E3.1 How can we get there?

For any two points A and B, $\overrightarrow{AB} = \overrightarrow{OB} - \overrightarrow{OA}$.

Reflect and discuss 4
Explain why $\overrightarrow{AB} + \overrightarrow{BA}$ must be equal to the zero vector.
Hence explain why $\overrightarrow{AB} = -\overrightarrow{BA}$.

Example 3

A is the point $(2, 11)$ and B is the point $(-3, 6)$. Find $\overrightarrow{AB}$.

$\overrightarrow{OA} = \begin{pmatrix} 2 \\ 11 \end{pmatrix} \quad \overrightarrow{OB} = \begin{pmatrix} -3 \\ 6 \end{pmatrix}$ — Write down $\overrightarrow{OA}$ and $\overrightarrow{OB}$.

$\overrightarrow{AB} = \overrightarrow{OB} - \overrightarrow{OA} = \begin{pmatrix} -3 \\ 6 \end{pmatrix} - \begin{pmatrix} 2 \\ 11 \end{pmatrix} = \begin{pmatrix} -5 \\ -5 \end{pmatrix}$

Example 4

A has coordinates $(a, 3a)$, B has coordinates $(b, b-6)$, and $\overrightarrow{AB} = \begin{pmatrix} 4 \\ -8 \end{pmatrix}$.
Find the values of a and b.

$\overrightarrow{OA} = \begin{pmatrix} a \\ 3a \end{pmatrix} \quad \overrightarrow{OB} = \begin{pmatrix} b \\ b-6 \end{pmatrix}$

$\overrightarrow{AB} = \overrightarrow{OB} - \overrightarrow{OA} = \begin{pmatrix} b-a \\ b-6-3a \end{pmatrix} = \begin{pmatrix} 4 \\ -8 \end{pmatrix}$

$b - a = 4$
$b - 6 - 3a = -8$ — The components of $\overrightarrow{AB}$ are equal to the components of $\overrightarrow{AB}$ given in the question.
$b - 3a = -2$

$2a = 6$ — Solve the simultaneous equations.
$a = 3, b = 7$

Practice 3

1 Find $\overrightarrow{AC}$ for each pair of vectors:

a $\overrightarrow{AB} = \begin{pmatrix} 3 \\ -1 \end{pmatrix}$ and $\overrightarrow{BC} = \begin{pmatrix} -2 \\ 6 \end{pmatrix}$
b $\overrightarrow{AB} = \begin{pmatrix} -2 \\ 4 \end{pmatrix}$ and $\overrightarrow{BC} = \begin{pmatrix} 10 \\ -5 \end{pmatrix}$

c $\overrightarrow{AB} = \begin{pmatrix} 0 \\ 6 \end{pmatrix}$ and $\overrightarrow{BC} = \begin{pmatrix} -5 \\ 11 \end{pmatrix}$
d $\overrightarrow{AB} = \begin{pmatrix} -9 \\ 12 \end{pmatrix}$ and $\overrightarrow{BC} = \begin{pmatrix} 9 \\ -12 \end{pmatrix}$

e $\overrightarrow{AB} = \begin{pmatrix} -8 \\ -3 \end{pmatrix}$ and $\overrightarrow{BC} = \begin{pmatrix} 2 \\ -1 \end{pmatrix}$

2 Find the vectors $\overrightarrow{AB}$ and $\overrightarrow{BA}$ for each pair of coordinates:

  a $A(0, 7)$ and $B(-1, 0)$ **b** $A(-1, 3)$ and $B(4, 2)$

  c $A(5, -2)$ and $B(-6, 8)$ **d** $A(3, -9)$ and $B(12, -9)$

  e $A(-2, 0)$ and $B(-11, -7)$

3 $\overrightarrow{AB} = \begin{pmatrix} -3 \\ -7 \end{pmatrix}$ and point A is $(-12, 3)$. Find the coordinates of point B.

Problem solving

4 $\overrightarrow{AB} = \begin{pmatrix} -3 \\ -7 \end{pmatrix}$. Find the values of a and b for these coordinate pairs:

  a $A(2a, a - 1)$ and $B(3b, b - 7)$

  b $A(-3a + 1, a + 4)$ and $B(b - 3, b)$

  c $A(2a + 4, 3a - 5)$ and $B(2b + 1, -2b + 3)$

Exploration 3

1 On squared paper, draw the vectors $\mathbf{a} = \begin{pmatrix} 3 \\ 6 \end{pmatrix}$, $\mathbf{b} = \begin{pmatrix} 5 \\ 10 \end{pmatrix}$ and $\mathbf{c} = \begin{pmatrix} 8 \\ 16 \end{pmatrix}$.

State what the vectors have in common.

Draw a fourth vector, $\mathbf{d}$, that has the same property.

2 Explain why for any non-zero vector $\mathbf{v}$, $\mathbf{v} + \mathbf{v}$ must be parallel to $\mathbf{v}$.

3 a Determine which of these vectors are parallel to each other:

$\begin{pmatrix} 4 \\ 5 \end{pmatrix}$ $\begin{pmatrix} 2 \\ -8 \end{pmatrix}$ $\begin{pmatrix} 26 \\ -4 \end{pmatrix}$ $\begin{pmatrix} 3 \\ -12 \end{pmatrix}$ $\begin{pmatrix} -4 \\ -2 \end{pmatrix}$ $\begin{pmatrix} -16 \\ -20 \end{pmatrix}$

$\begin{pmatrix} 13 \\ -2 \end{pmatrix}$ $\begin{pmatrix} 8 \\ 4 \end{pmatrix}$ $\begin{pmatrix} -6 \\ -11 \end{pmatrix}$ $\begin{pmatrix} -5 \\ 20 \end{pmatrix}$ $\begin{pmatrix} 30 \\ 15 \end{pmatrix}$

b Hence find which vector is not parallel to any of the others.

4 For any vector $\mathbf{v} = \begin{pmatrix} x \\ y \end{pmatrix}$, $k\mathbf{v} = \begin{pmatrix} kx \\ ky \end{pmatrix}$.

Explain why $k\mathbf{v}$ and $\mathbf{v}$ are parallel to each other.

Scalar multiplication of a vector:

For any vector $\mathbf{v} = \begin{pmatrix} x \\ y \end{pmatrix}$ and any scalar value k, $k\mathbf{v} = \begin{pmatrix} kx \\ ky \end{pmatrix}$.

For any vectors $\mathbf{u}$ and $\mathbf{v}$, if $\mathbf{u} = k\mathbf{v}$ then $\mathbf{u}$ and $\mathbf{v}$ are parallel.

Example 5

Two points have coordinates $A(-1, 4)$ and $B(2b, b+2)$. $\overrightarrow{AB}$ is parallel to $\mathbf{v} = \begin{pmatrix} 3 \\ 4 \end{pmatrix}$. Find the value of b.

$\overrightarrow{OA} = \begin{pmatrix} -1 \\ 4 \end{pmatrix}$ $\overrightarrow{OB} = \begin{pmatrix} 2b \\ b+2 \end{pmatrix}$ — Write the position vectors.

$\overrightarrow{AB} = \overrightarrow{OB} - \overrightarrow{OA}$

$\overrightarrow{AB} = \begin{pmatrix} 2b+1 \\ b-2 \end{pmatrix}$ — Find the vector $\overrightarrow{AB}$.

$\begin{pmatrix} 2b+1 \\ b-2 \end{pmatrix} = k \begin{pmatrix} 3 \\ 4 \end{pmatrix}$ — Parallel vectors are multiples of each other.

$2b + 1 = 3k$
$b - 2 = 4k$ — Write the system of simultaneous equations and solve for b.
$b = -2$

Example 6

Four points, $A(4, 6)$, $B(-2, 16)$, $C(-5, 12)$ and $D(-2, 7)$ lie in a plane. Show that $ABCD$ forms a trapezoid.

$\overrightarrow{AB} = \begin{pmatrix} -6 \\ 10 \end{pmatrix}$ $\overrightarrow{BC} = \begin{pmatrix} -3 \\ -4 \end{pmatrix}$

$\overrightarrow{CD} = \begin{pmatrix} 3 \\ -5 \end{pmatrix}$ $\overrightarrow{DA} = \begin{pmatrix} 6 \\ -1 \end{pmatrix}$ — Find the vectors for each side of the trapezoid: $\overrightarrow{AB}, \overrightarrow{BC}, \overrightarrow{CD}$ and $\overrightarrow{DA}$.

$\overrightarrow{AB} = \begin{pmatrix} -6 \\ 10 \end{pmatrix} = -2 \begin{pmatrix} 3 \\ -5 \end{pmatrix} = (-2) \times \overrightarrow{CD}$ — Look for parallel vectors.

$\overrightarrow{AB}$ and $\overrightarrow{CD}$ are parallel. None of the other sides are parallel.

$ABCD$ has one pair of parallel sides so it is a trapezoid.

Practice 4

Problem solving

1 The vectors $\mathbf{u} = \begin{pmatrix} a-2 \\ a+1 \end{pmatrix}$ and $\mathbf{v} = \begin{pmatrix} a \\ 3a+3 \end{pmatrix}$ are parallel.

Determine the value of a.

2 Two points have coordinates $A(2, 5)$ and $B(b-2, 1-b)$. $\overrightarrow{AB}$ is parallel to $\mathbf{v} = \begin{pmatrix} 1 \\ -3 \end{pmatrix}$. Find the value of b.

> In some parts of the world (and in this book) a trapezoid is defined as a quadrilateral with exactly *one* pair of opposite sides parallel. In other parts of the world this is called a trapezium. The confusion dates back to 1795, to a reference in a mathematical dictionary published in the USA that directly reversed the accepted meanings of the day.

GEOMETRY AND TRIGONOMETRY

3 Two points have coordinates $P(2a, a-4)$ and $Q(3b+1, 2a-3b)$. $\overrightarrow{PQ} = \begin{pmatrix} -7 \\ 5 \end{pmatrix}$. Find the position vector of the midpoint of the line segment PQ.

> The **position vector** of a point is the vector from the origin O to the point.

4 Four points have coordinates $A(3, 7)$, $B(2, 11)$, $C(15, 3)$ and $D(17, -5)$.

  a Use vectors to show that the lines AB and CD are parallel.

  b Determine if the line segments AB and CD are equal in length.

5 Four points have coordinates $A(4, 6)$, $B(-1, 9)$, $C(1, 1)$ and $D(11, -5)$. Show that $\overrightarrow{AB}$ and $\overrightarrow{DC}$ are parallel.

6 Four points have position vectors:
$\overrightarrow{OA} = 9\mathbf{u} + 9\mathbf{v}$, $\overrightarrow{OB} = 19\mathbf{u} + 14\mathbf{v}$, $\overrightarrow{OC} = 25\mathbf{u} + 5\mathbf{v}$ and $\overrightarrow{OD} = 15\mathbf{u}$.

  a Find $\overrightarrow{AB}$ and $\overrightarrow{DC}$ in terms of $\mathbf{u}$ and $\mathbf{v}$.

  b Hence show that $ABCD$ is a parallelogram.

  c Given that $\mathbf{u} = \begin{pmatrix} 2 \\ 1 \end{pmatrix}$ and $\mathbf{v} = \begin{pmatrix} -1 \\ 1 \end{pmatrix}$, show that $ABCD$ is a rhombus.

> Three points P, Q and R are on a straight line if the vectors $\overrightarrow{PQ}$ and $\overrightarrow{QR}$ are parallel.

7 From the five points $A(-3, 1)$, $B(1, 3)$, $C(4, 3)$, $D(5, 1)$ and $E(-1, -2)$, find two sets of four points, each of which forms a trapezoid.

8 Use vectors to show that the three points $P(5, -4)$, $Q(10, 0)$ and $R(25, 12)$ lie on a straight line.

Reflect and discuss 5

Why does showing that the vectors formed by three points are parallel also show that the three points lie on a straight line?

Exploration 4

1 On suitable axes, draw the points $A(2, 4)$ and $B(6, 1)$.

Label the vectors $\overrightarrow{OA}$, $\overrightarrow{OB}$ and $\overrightarrow{AB}$ on your diagram.

2 The angle between $\overrightarrow{OA}$ and $\overrightarrow{OB}$ is $\angle AOB$.

Describe how you could use the cosine rule to find the size of $\angle AOB$.

3 Find $|\overrightarrow{OA}|$ and $|\overrightarrow{OB}|$. Hence find the size of $\angle AOB$, correct to the nearest degree.

4 Use the method in steps **1** to **3** to find the angle between the vectors $\begin{pmatrix} 3 \\ 8 \end{pmatrix}$ and $\begin{pmatrix} 6 \\ 3 \end{pmatrix}$.

▶ Continued on next page

E3.1 How can we get there?

5 The diagram shows vectors **u** and **v**.

Write down an expression for **v** − **u** in terms of u_x, u_y, v_x and v_y.

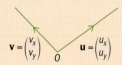

6 Write down expressions for |**u**| and |**v**|.
Show that $|\mathbf{v} - \mathbf{u}| = \sqrt{u_x^2 + u_y^2 + v_x^2 + v_y^2 - 2u_x v_x - 2u_y v_y}$.

7 Use the cosine rule to show that $-2u_x v_x - 2u_y v_y = -2\sqrt{u_x^2 + u_y^2}\sqrt{v_x^2 + v_y^2}\cos\theta$

and hence that $\cos\theta = \dfrac{u_x v_x + u_y v_y}{|\mathbf{u}||\mathbf{v}|}$

For any two vectors $\mathbf{u} = \begin{pmatrix} u_x \\ u_y \end{pmatrix}$ and $\mathbf{v} = \begin{pmatrix} v_x \\ v_y \end{pmatrix}$, the quantity $u_x v_x + u_y v_y$ is called the **dot product**. It is written $\mathbf{u} \cdot \mathbf{v}$.

From the equation you have formed above, you can see that $u_x v_x + u_y v_y = |\mathbf{u}||\mathbf{v}|\cos\theta$.

Therefore $\mathbf{u} \cdot \mathbf{v} = u_x v_x + u_y v_y = |\mathbf{u}||\mathbf{v}|\cos\theta$.

Another name for the dot product is the **scalar product**, because the result is a scalar quantity.

Example 7

Find the angle between the vectors $\begin{pmatrix} 1 \\ 4 \end{pmatrix}$ and $\begin{pmatrix} -2 \\ 5 \end{pmatrix}$.

$|\mathbf{u}| = \sqrt{1^2 + 4^2} = \sqrt{17}$ — Find |**u**| and |**v**|.

$|\mathbf{v}| = \sqrt{(-2)^2 + 5^2} = \sqrt{29}$

$\mathbf{u} \cdot \mathbf{v} = |\mathbf{u}||\mathbf{v}|\cos\theta$

$\begin{pmatrix} 1 \\ 4 \end{pmatrix} \cdot \begin{pmatrix} -2 \\ 5 \end{pmatrix} = \sqrt{17}\sqrt{29}\cos\theta$ — Substitute in the vectors and their magnitudes.

$1 \times -2 + 4 \times 5 = \sqrt{17}\sqrt{29}\cos\theta$ — Use the definition of the dot product.

$\cos\theta = \dfrac{18}{\sqrt{17}\sqrt{29}}$

$\theta = 35.9°$ — Give your answer to a suitable degree of accuracy.

Example 8

Vectors $\mathbf{a} = \begin{pmatrix} 4 \\ 7 \end{pmatrix}$ and $\mathbf{b} = \begin{pmatrix} 2x-2 \\ 2-x \end{pmatrix}$ are perpendicular. Find the value of x.

$\mathbf{a} \cdot \mathbf{b} = |\mathbf{a}||\mathbf{b}|\cos 90° = 0$ — The angle between **a** and **b** is 90°, and $\cos 90° = 0$.

$\begin{pmatrix} 4 \\ 7 \end{pmatrix} \cdot \begin{pmatrix} 2x-2 \\ 2-x \end{pmatrix} = 0$ — Substitute in **a** and **b**.

$(8x - 8) + (14 - 7x) = 0$

$\Rightarrow x = -6$ — Evaluate the scalar product.

GEOMETRY AND TRIGONOMETRY

> If two vectors **a** and **b** are perpendicular then **a · b** = 0.
>
> Equally, if **a · b** = 0 then either $\mathbf{a} = \begin{pmatrix} 0 \\ 0 \end{pmatrix}$, $\mathbf{b} = \begin{pmatrix} 0 \\ 0 \end{pmatrix}$ or **a** and **b** are perpendicular.

Practice 5

1 Find the angle between each pair of vectors, to the nearest degree.

   a $\begin{pmatrix} 4 \\ 5 \end{pmatrix}$ and $\begin{pmatrix} 2 \\ -8 \end{pmatrix}$

   b $\begin{pmatrix} 26 \\ -4 \end{pmatrix}$ and $\begin{pmatrix} 3 \\ -12 \end{pmatrix}$

   c $\begin{pmatrix} -4 \\ -2 \end{pmatrix}$ and $\begin{pmatrix} -16 \\ -20 \end{pmatrix}$

   d $\begin{pmatrix} 13 \\ -2 \end{pmatrix}$ and $\begin{pmatrix} 8 \\ 4 \end{pmatrix}$

2 Show that $\begin{pmatrix} 4 \\ -8 \end{pmatrix}$ and $\begin{pmatrix} 2 \\ 1 \end{pmatrix}$ are perpendicular.

Problem solving

3 Two speedboats are travelling with velocities $\begin{pmatrix} 20 \\ -15 \end{pmatrix}$ ms^{-1} and $\begin{pmatrix} 7 \\ 24 \end{pmatrix}$ ms^{-1} respectively, on an east-north coordinate grid.

   a Show that both boats have the same speed.

   b Find the angle between their directions of travel.

4 Triangle *ABC* has vertices *A*(6, 3), *B*(2, 1) and *C*(4, 5). Use the dot product to find ∠*ABC*.

5 Vectors **u** and **v** are perpendicular:

   $\mathbf{u} = \begin{pmatrix} -2 \\ 5 \end{pmatrix}$, $\mathbf{v} = \begin{pmatrix} 3m+3 \\ 2m-2 \end{pmatrix}$

   Find the value of *m*.

6 Vectors $\begin{pmatrix} a \\ a+2 \end{pmatrix}$ and $\begin{pmatrix} b \\ 2-b \end{pmatrix}$ are perpendicular.

   a Write an equation involving *a* and *b*.

   b Explain why you need more information to determine the values of *a* and *b*.

   c Given that *b* = 1 − *a*, determine the values of *a* and *b*.

Problem solving

7 Find three different vectors perpendicular to the vector $\begin{pmatrix} 8 \\ -2 \end{pmatrix}$.

8 Parallelogram *OABC* has vertices *O*(0, 0), *A*(4, 9) and *B*(6, 6).

   a Find the position vector of *C*.

   b Find the size of ∠*OAC*.

9 A rhombus $OABC$ has $\overrightarrow{OA} = \mathbf{a}$ and $\overrightarrow{OC} = \mathbf{c}$. $\angle AOC = 120°$ and $|\mathbf{a}| = 4$.

Find the value of $\mathbf{a} \cdot \mathbf{c}$.

10 For the vectors $\mathbf{a} = \begin{pmatrix} 3 \\ 5 \end{pmatrix}$, $\mathbf{b} = \begin{pmatrix} 2 \\ -2 \end{pmatrix}$ and $\mathbf{c} = \begin{pmatrix} 5 \\ 1 \end{pmatrix}$, verify that

$\mathbf{a} \cdot (\mathbf{b} + \mathbf{c}) = \mathbf{a} \cdot \mathbf{b} + \mathbf{a} \cdot \mathbf{c}$

11 Consider the vectors $\mathbf{a} = \begin{pmatrix} a_x \\ a_y \end{pmatrix}$, $\mathbf{b} = \begin{pmatrix} b_x \\ b_y \end{pmatrix}$ and $\mathbf{c} = \begin{pmatrix} c_x \\ c_y \end{pmatrix}$.

Prove that $\mathbf{a} \cdot (\mathbf{b} + \mathbf{c}) = \mathbf{a} \cdot \mathbf{b} + \mathbf{a} \cdot \mathbf{c}$ for any 2D vectors $\mathbf{a}$, $\mathbf{b}$ and $\mathbf{c}$.

D Vector geometry

- How can vectors be used to solve geometric problems?
- Can logic solve everyday problems?

You can use vectors in geometric problems even when you do not know their components. Remember that you can add vectors pictorially.

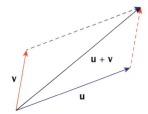
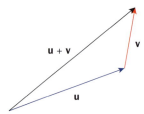

ATL Exploration 5

1 The diagram shows a pair of vectors $\mathbf{a}$ and $\mathbf{b}$.

Copy the diagram and draw the vectors:

a $3\mathbf{a}$ b $2\mathbf{b}$

c $\mathbf{a} + \mathbf{b}$ (the resultant vector of $\mathbf{a}$ and then $\mathbf{b}$)

d $\mathbf{a} - \mathbf{b}$ e $-\mathbf{a} + \mathbf{b}$ f $-\mathbf{a} - \mathbf{b}$

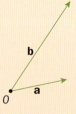

Start with a new copy for each diagram.

2 In this diagram, M is the midpoint of AB and N is the midpoint of AC.

a Given that $\overrightarrow{AM} = \mathbf{a}$ and $\overrightarrow{AN} = \mathbf{b}$, explain why $\overrightarrow{AB} = 2\mathbf{b}$ and write down a similar expression for $\overrightarrow{AC}$.

b Explain why $\overrightarrow{MN} = \mathbf{b} - \mathbf{a}$.
Form a similar expression for $\overrightarrow{BC}$.

c Hence show that $MNCB$ is a trapezoid.

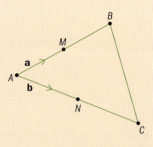

Tip

Remember that $\overrightarrow{MN} = \overrightarrow{MA} + \overrightarrow{AN}$.

▶ Continued on next page

GEOMETRY AND TRIGONOMETRY

3 Four points A, B, C and D have position vectors $\mathbf{a} + \mathbf{b}$, $\mathbf{a} - \mathbf{b}$, $-\mathbf{a} - \mathbf{b}$, and $-\mathbf{a} + \mathbf{b}$ respectively, where $\mathbf{a}$ and $\mathbf{b}$ are non-parallel non-zero vectors.

   a This sketch shows one possible pair of vectors $\mathbf{a}$ and $\mathbf{b}$.
   Copy the sketch and draw points A, B, C and D on the diagram.

   b This sketch shows another possible pair of vectors $\mathbf{a}$ and $\mathbf{b}$.
   Copy the sketch and draw points A, B, C and D on the diagram.

   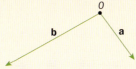

   c Find expressions for $\overrightarrow{AB}$ and $\overrightarrow{DC}$.

   d Hence show that $ABCD$ is a parallelogram.

4 This diagram shows points A, B, C and D.
$\overrightarrow{AD} = \mathbf{a} + \mathbf{b}$
$\overrightarrow{BD} = 2\mathbf{a} - \mathbf{a}$
$\overrightarrow{BC} = 3\mathbf{a}$

   a Find $\overrightarrow{AB}$.

   b Find $\overrightarrow{DC}$.

   c Hence show that $\overrightarrow{AB}$ and $\overrightarrow{DC}$ are parallel.

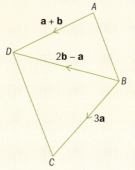

The word 'vector' was originally a Latin word, meaning 'to carry'. You can think of a vector in mathematics as something that carries you from one place to another. We also use the word in other contexts: mosquitos and some other insects are referred to as vectors because they transmit disease from one organism to another. In sociology, a vector is a person that passes folklore on from one generation to the next. And in aeronautics, a vector is a set of appropriate headings to guide an aircraft in flight.

Reflect and discuss 6

- How does your answer to Exploration 5, step **3c** show you that $ABCD$ is a parallelogram regardless of the direction of vectors $\mathbf{a}$ and $\mathbf{b}$?
- Why was it necessary in step **3** to state that $\mathbf{a}$ and $\mathbf{b}$ were non-zero?
- Why was it necessary to state that $\mathbf{a}$ and $\mathbf{b}$ were not parallel?

E3.1 How can we get there?

Practice 6

1 Using the diagram, express the vectors below it in terms of **u**, **v** and **w**:

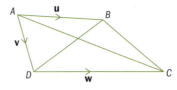

   a $\vec{BD}$ **b** $\vec{CA}$ **c** $\vec{BC}$

Problem solving

2 The diagram shows a regular hexagon *ABCDEF*.

Find in terms of **a** and **b**:

   a $\vec{AB}$ **b** $\vec{OC}$ **c** $\vec{FO}$ **d** $\vec{AC}$ **e** $\vec{DB}$

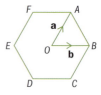

3 Points *P*, *Q* and *R* form a straight line and have position vectors **p**, **q**, and **r** respectively. $|\vec{QR}| = 4|\vec{PQ}|$.

   a Express **r** in terms of **p** and **q**.

   b Hence show that $\mathbf{q} = \frac{1}{5}(4\mathbf{p} + \mathbf{r})$.

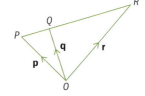

4 Points *S*, *T* and *U* form a straight line and have position vectors **s**, **t** and **u** respectively, where $|\vec{SU}| = 3|\vec{TU}|$.

Express **s** in terms of **t** and **u**.

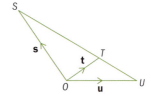

5 In this diagram, *M* is the midpoint of *AB*. $\vec{OA} = 3\mathbf{a}$ and $\vec{OB} = 2\mathbf{b}$.

   a Find $\vec{OM}$.

   b The point *N* has position vector $\vec{ON}$ where $\vec{ON} = 2\vec{OM}$.

      Show that $\vec{AN} = 2\mathbf{b}$.

   c Given that $|\mathbf{a}| = |\mathbf{b}|$ and $\mathbf{a} \cdot \mathbf{b} = 0$, fully describe the quadrilateral *OANB*.

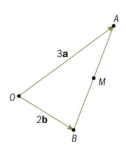

GEOMETRY AND TRIGONOMETRY

Summary

A **column vector** describes a translation.

A column vector has an x component and a y component: $\begin{pmatrix} x \\ y \end{pmatrix}$

The x component tells you how far to move in the x-direction. The y component tells you how far to move in the y-direction.

$\overrightarrow{AB}$ is the vector representing the translation from A to B.

The **position vector** of a point is the vector from the origin O to the point.

$\begin{pmatrix} 0 \\ 0 \end{pmatrix}$ is the **zero vector**.

Two vectors are equal if their components are equal.

If $\mathbf{v}_1 = \begin{pmatrix} x_1 \\ y_1 \end{pmatrix}$ and $\mathbf{v}_2 = \begin{pmatrix} x_2 \\ y_2 \end{pmatrix}$ then $\mathbf{v}_1 = \mathbf{v}_2$ if and only if $x_1 = x_2$ and $y_1 = y_2$.

The **magnitude** of a vector is its length.

$|\mathbf{v}|$, the magnitude of $\mathbf{v} = \begin{pmatrix} x \\ y \end{pmatrix}$, is given by

$|\mathbf{v}| = \sqrt{x^2 + y^2}$.

A vector has both magnitude and direction.

A scalar quantity does not have a direction.

For two vectors to be equal, they must have the same magnitude and direction.

A translation from A to B followed by a translation from B to C is the same as the translation from A to C: $\overrightarrow{AB} + \overrightarrow{BC} = \overrightarrow{AC}$

$\overrightarrow{AC}$ is sometimes called the **resultant** of $\overrightarrow{AB}$ and $\overrightarrow{BC}$.

The vector addition law:

If $\overrightarrow{AB} = \begin{pmatrix} x_1 \\ y_1 \end{pmatrix}$ and $\overrightarrow{BC} = \begin{pmatrix} x_2 \\ y_2 \end{pmatrix}$ then

$\overrightarrow{AC} = \overrightarrow{AB} + \overrightarrow{BC} = \begin{pmatrix} x_1 \\ y_1 \end{pmatrix} + \begin{pmatrix} x_2 \\ y_2 \end{pmatrix} = \begin{pmatrix} x_1 + x_2 \\ y_1 + y_2 \end{pmatrix}$

For any two points A and B, $\overrightarrow{AB} = \overrightarrow{OB} - \overrightarrow{OA}$ and $\overrightarrow{AB} = -\overrightarrow{BA}$.

Scalar multiplication of a vector:

For any vector, $\mathbf{v} = \begin{pmatrix} x \\ y \end{pmatrix}$ and any scalar value k:

$k\mathbf{v} = \begin{pmatrix} kx \\ ky \end{pmatrix}$

If $\mathbf{u} = k\mathbf{v}$ then $\mathbf{u}$ and $\mathbf{v}$ are parallel.

For any two vectors $\mathbf{u} = \begin{pmatrix} u_x \\ u_y \end{pmatrix}$ and $\mathbf{v} = \begin{pmatrix} v_x \\ v_y \end{pmatrix}$, the quantity $u_x v_x + u_y v_y$ is called the **dot product**.

$\mathbf{u} \cdot \mathbf{v} = u_x v_x + u_y v_y = |\mathbf{u}| |\mathbf{v}| \cos \theta$

If two vectors $\mathbf{a}$ and $\mathbf{b}$ are perpendicular then $\mathbf{a} \cdot \mathbf{b} = 0$.

Equally, if $\mathbf{a} \cdot \mathbf{b} = 0$ then either $\mathbf{a} = \begin{pmatrix} 0 \\ 0 \end{pmatrix}$, $\mathbf{b} = \begin{pmatrix} 0 \\ 0 \end{pmatrix}$ or $\mathbf{a}$ and $\mathbf{b}$ are perpendicular.

Mixed practice

1 **Write down** column vectors that represent:

   a a translation of 5 right and 4 up

   b a translation of 2 left and 3 up

   c a translation from (4, 5) to (7, 9)

   d a translation from (9, 14) to (−2, −5).

2 **Find** the vectors $\overrightarrow{AB}$ and $\overrightarrow{BA}$ given the points:

   a $A(1, 3)$ and $B(-2, 0)$

   b $A(4, 2)$ and $B(-3, 5)$

   c $A(-7, 1)$ and $B(-4, -4)$

   d $A(-2, 7)$ and $B(11, -7)$

3 Find $\overrightarrow{AC}$ given:

   a $\overrightarrow{AB} = \begin{pmatrix} -2 \\ 4 \end{pmatrix}$ and $\overrightarrow{BC} = \begin{pmatrix} -1 \\ -7 \end{pmatrix}$

   b $\overrightarrow{AB} = \begin{pmatrix} 0 \\ -8 \end{pmatrix}$ and $\overrightarrow{BC} = \begin{pmatrix} 9 \\ 2 \end{pmatrix}$

   c $\overrightarrow{AB} = \begin{pmatrix} -1 \\ -3 \end{pmatrix}$ and $\overrightarrow{BC} = \begin{pmatrix} 4 \\ -2 \end{pmatrix}$

   d $\overrightarrow{AB} = \begin{pmatrix} -6 \\ 15 \end{pmatrix}$ and $\overrightarrow{BC} = \begin{pmatrix} 4 \\ 0 \end{pmatrix}$

4 Given that $\vec{AB} = \begin{pmatrix} 2 \\ -1 \end{pmatrix}$, **find** the values of a and b given the coordinates:
   a $A(a-3, 3a+2)$ and $B(-b, 2b-1)$
   b $A(2a, -4+a)$ and $B(-2b, b)$
   c $A(a-6, -2a+10)$ and $B(2b+9, b+3)$
   d $A(-3a, a-2)$ and $B(11+b, 2b+1)$

5 **Find** the magnitude of each of these vectors, giving your answer in exact form.
   a $\begin{pmatrix} 5 \\ 2 \end{pmatrix}$ b $\begin{pmatrix} -4 \\ 3 \end{pmatrix}$ c $\begin{pmatrix} 12 \\ 15 \end{pmatrix}$ d $\begin{pmatrix} 13 \\ -16 \end{pmatrix}$

6 $M(-3, 2)$ and $N(2, 5)$.
   Find a $\vec{ON}$ b $\vec{MN}$ c $\vec{NM}$

7 **Find**, correct to the nearest degree, the angle between:
   a $\begin{pmatrix} 3 \\ 7 \end{pmatrix}$ and $\begin{pmatrix} -2 \\ 1 \end{pmatrix}$ b $\begin{pmatrix} 1 \\ 1 \end{pmatrix}$ and $\begin{pmatrix} -4 \\ 5 \end{pmatrix}$
   c $\begin{pmatrix} 3 \\ -2 \end{pmatrix}$ and $\begin{pmatrix} 2 \\ 4 \end{pmatrix}$

Problem solving

8 From the list of nine vectors below, **select** pairs of vectors which are parallel to each other. **Determine** which vector is not parallel to any of the others.

$\begin{pmatrix} -6 \\ 20 \end{pmatrix}$ $\begin{pmatrix} 16 \\ 20 \end{pmatrix}$ $\begin{pmatrix} 12 \\ -18 \end{pmatrix}$ $\begin{pmatrix} 36 \\ 45 \end{pmatrix}$ $\begin{pmatrix} 9 \\ -30 \end{pmatrix}$

$\begin{pmatrix} 15 \\ 25 \end{pmatrix}$ $\begin{pmatrix} 5 \\ 8 \end{pmatrix}$ $\begin{pmatrix} -9 \\ -15 \end{pmatrix}$ $\begin{pmatrix} -4 \\ 6 \end{pmatrix}$

9 **Show that** the vectors $\begin{pmatrix} 1 \\ 4 \end{pmatrix}$ and $\begin{pmatrix} 8 \\ -2 \end{pmatrix}$ are perpendicular to each other.

Problem solving

10 The vectors $\mathbf{u} = \begin{pmatrix} 5a+2 \\ a-4 \end{pmatrix}$ and $\mathbf{v} = \begin{pmatrix} a+7 \\ 8-2a \end{pmatrix}$ are parallel.

   Determine the possible values of a.

11 For the points with coordinates $A(-3, 2)$ and $B(2b+1, b-4)$:
   $\vec{AB}$ is parallel to $\mathbf{v} = \begin{pmatrix} -2 \\ -9 \end{pmatrix}$. **Find** the value of b.

12 The diagram shows five points O, A, B, C and X.
   $\vec{OX} = \vec{XB} = \mathbf{a}$ and
   $\vec{CX} = \vec{XA} = \mathbf{b}$
   Show that $OABC$ is a parallelogram.

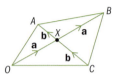

Problem solving

13 The diagram shows a quadrilateral $OABC$.
   $\vec{AB} = \mathbf{a}$, $\vec{OC} = 2\mathbf{a}$ and $\vec{OA} = \mathbf{b}$. Lines $\overleftrightarrow{OB}$ and $\overleftrightarrow{AC}$ intersect at X.

   a **Explain** why $OABC$ is a trapezoid.
   b **Show that** $\vec{OB} = \mathbf{a} + \mathbf{b}$.
   c **Find** $\vec{AC}$.
   d Use the fact that X lies on $\overleftrightarrow{OB}$ to **explain** why $\vec{OX} = k(\mathbf{a} + \mathbf{b})$ for some constant k.
   e By considering $\vec{OA} + \vec{AX}$, **show that** $\vec{OX} = \mathbf{b} + m(2\mathbf{a} - \mathbf{b})$ for some constant m.
   f Hence **find** $\vec{OX}$ and **determine** the ratio in which X divides $\vec{AC}$.

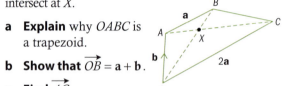

Review in context

Objective D: Applying mathematics in real-life contexts
ii. select appropriate mathematical strategies when solving authentic real-life situations

Think about how magnitude, direction, the dot product, etc. relate to the questions being asked.

In these questions, a floor cleaning robot has been placed on the floor of a large room. It measures its progress across the floor in 10 cm units, so the vector $\begin{pmatrix} 1 \\ 0 \end{pmatrix}$ represents a movement of 10 cm in the x-direction.

1 **Show that** if the robot travels on a vector of $\begin{pmatrix} 3 \\ 4 \end{pmatrix}$, it covers a distance of 50 cm.

2 **Find**, correct to the nearest centimeter, the distance travelled by the robot as it moves on each vector.

a $\begin{pmatrix} 3 \\ 7 \end{pmatrix}$ b $\begin{pmatrix} 2 \\ -4 \end{pmatrix}$ c $\begin{pmatrix} 6 \\ 3 \end{pmatrix}$

d $\begin{pmatrix} 9 \\ -3 \end{pmatrix}$ e $\begin{pmatrix} -15 \\ 25 \end{pmatrix}$ f $\begin{pmatrix} 40 \\ -20 \end{pmatrix}$

3 In a rectangular room, the robot is positioned so that the *x*- and *y*-directions are the edges of the room.
The robot starts in one corner at (0, 0), travels on $\begin{pmatrix} 25 \\ 0 \end{pmatrix}$ until it reaches another corner and then on $\begin{pmatrix} 0 \\ 60 \end{pmatrix}$ until it reaches a third corner.

  a **Find** the area of the room in m².

  b **Write down** the vector which would return the robot to its starting point.

  c The robot travels along two sides of the room and then straight back to the start. **Find** the total distance travelled.

4 The robot is put in the L-shaped room below, *ABCDEF*. It starts at *O*, which is somewhere inside the room. Its *x*-direction is parallel to *AF*, *BC* and *DE* and its *y*-direction is parallel to *AB*, *CD* and *FE*.

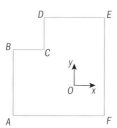

In its mapping process, the robot determines that $\overrightarrow{OA} = \begin{pmatrix} -14 \\ -30 \end{pmatrix}$, $\overrightarrow{OC} = \begin{pmatrix} -2 \\ 15 \end{pmatrix}$ and $\overrightarrow{OE} = \begin{pmatrix} 22 \\ 25 \end{pmatrix}$.

  a **Find** vector $\overrightarrow{AE}$ and the distance from *A* to *E*.

  b **Find** the position vector $\overrightarrow{OF}$.
  Hence find the distance from *C* to *F*.

  c **Determine** which is longer, *AC* or *CE*.

  d **Calculate** the perimeter and area of the room.

5 The robot is placed in another room, *PQRS*, but its axes are not lined up with the sides.

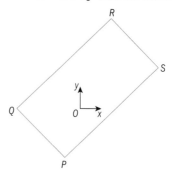

The robot finds the location of the room's corners to be as follows:

$\overrightarrow{OP} = \begin{pmatrix} -10 \\ -30 \end{pmatrix}$ $\overrightarrow{OQ} = \begin{pmatrix} -34 \\ 2 \end{pmatrix}$

$\overrightarrow{OR} = \begin{pmatrix} 25 \\ 45 \end{pmatrix}$ $\overrightarrow{OS} = \begin{pmatrix} 50 \\ 15 \end{pmatrix}$

  a **Find** $\overrightarrow{PQ}$ and $\overrightarrow{PS}$.

  b **Show that** walls *PQ* and *PS* are perpendicular.

  c **Find** $\overrightarrow{SR}$.

  d **Determine** whether *SR* and *PQ* are parallel. **Hence determine** whether the room is rectangular. **Justify** your answer.

6 The robot is placed at *O*, somewhere in a four-sided room. It finds the four corners of the room, *U*, *V*, *W* and *X*.

You are given the following vector information:

$\overrightarrow{OW} = \begin{pmatrix} 15 \\ -25 \end{pmatrix}$ $\overrightarrow{OV} = \begin{pmatrix} 40 \\ 0 \end{pmatrix}$

$\overrightarrow{OX} = \begin{pmatrix} -32 \\ 22 \end{pmatrix}$ $\overrightarrow{UX} = \begin{pmatrix} -50 \\ 0 \end{pmatrix}$

  a **Find** $\overrightarrow{UV}$ and $\overrightarrow{XW}$.

  b **Hence show that** the room is a trapezoid.

  c **Show that** $\overleftrightarrow{VW}$ is perpendicular to both $\overleftrightarrow{UV}$ and $\overleftrightarrow{XW}$.

  d **Find** the area of the room.

  e **Find** the size of ∠*VUX*.

E3.1 How can we get there?

E4.1 How to stand out from the crowd

Global context: Identities and relationships

Objectives

- Making inferences about data, given the mean and standard deviation
- Using different forms of the standard deviation formula
- Understanding the normal distribution
- Making inferences about normal distributions
- Using the standard deviation and the mean
- Using unbiased estimators of the population mean and standard deviation

Inquiry questions

F
- How do you calculate the standard deviation of a data set?
- What is the normal distribution?

C
- How is the meaning of 'standard deviation' represented in its different formulas?

D
- Can samples give reliable results?
- Do we want to be like everybody else?

RELATIONSHIPS

| ATL | Communication |

Make inferences and draw conclusions

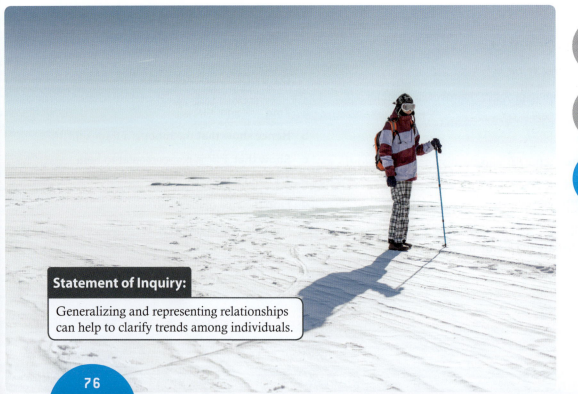

4.3

12.3

E4.1

Statement of Inquiry:

Generalizing and representing relationships can help to clarify trends among individuals.

STATISTICS AND PROBABILITY

You should already know how to:

• find the mean from frequency tables of discrete and grouped data	**1 a** This table gives the exam grades for 200 students. Calculate the mean grade. | Grade	1	2	3	4	5 | |---|---|---|---|---|---| | Frequency | 8 | 29 | 33 | 35 | 95 | **b** This table gives the weight in grams of 80 packets of chia seeds. Calculate an estimate for the mean weight. | Weight, w (g) | Frequency | |---|---| | $70 \leq w < 75$ | 10 | | $75 \leq w < 80$ | 22 | | $80 \leq w < 85$ | 38 | | $85 \leq w < 90$ | 10 |
• understand the difference between a population and a sample	**2** Define these terms: **a** population **b** sample

F The ultimate measure of dispersion

* How do you calculate the standard deviation of a data set?
* What is the normal distribution?

Exploration 1

The temperature on the first day of the first five months of the year was recorded in La Crieux, France and Betzdorf, Luxembourg.

	Temperature (°C)	
	La Crieux, France	Betzdorf, Luxembourg
January	10	18
February	16	10
March	14	13.5
April	12	14
May	18	14.5

▶ Continued on next page

E4.1 How to stand out from the crowd

1 For this data, calculate the range and the three measures of central tendency (mean, mode and median). What do you notice?

When two sets of data in an experiment have the same mean and range, there is another statistical measure that you can use to help discover differences between the data sets. You will now explore this measure, called the standard deviation.

2 Copy and complete this table for the La Crieux data. Calculate the mean of the differences, and the mean of the squares of the differences.

Temperature (°C)	Difference between recorded temperature and mean temperature	Square of the difference
10	−4	16
16		
14		
12		
18		
	Mean of differences =	
		Mean of squares of differences =
		Square root of the mean of squares of differences =

> A tardigrade is a tiny animal that lives in a variety of environments, from the tropics to the polar regions. Also known as Water Bears, these tiny creatures can withstand temperatures ranging from −273 °C to 150 °C.

3 Determine the units for the mean of squares and for the square root of the mean of squares. Hence, explain why you need to take the square root of the mean of the squares of differences.

4 Decide, with reasons, whether the mean of the differences or the mean of the squares of the differences best represents the difference (deviation) from the mean of the original temperature values.

5 Generalize this algorithm using mathematical notation and vocabulary.

Reflect and discuss 1

- Why do we square the differences between the data and the mean?
- Squaring large numbers makes them even larger. How does that impact the mean? Why is that important?

STATISTICS AND PROBABILITY

In Exploration 1, squaring the differences from the mean gives positive values to represent the deviation from the mean. It also gives more weighting to the larger deviations, which is important because the data further away from the mean may be more significant. The square root of the mean of these squared deviations is a measure of dispersion called the standard deviation.

> The **standard deviation** is a measure of dispersion that gives an idea of how close the original data values are to the mean.
>
> A small standard deviation shows that the data values are close to the mean. The units of standard deviation are the same as the units of the original data.

The standard deviation shows how representative the mean is of the data.

The notation used for the standard deviation depends on whether it is calculated for a whole population or for a sample. The convention is:

Use Greek letters for populations

- μ = mean of the population
- σ = standard deviation of the population

Use Roman letters for samples

- $\bar{x}$ = mean of the sample
- s_x = standard deviation of the sample

Standard deviation formulae also use the Greek symbol Σ (upper case sigma) to mean 'the sum of'.

> Σx means the sum of all the x values.
>
> Using this notation (called sigma notation), the mean of a set of values is $\mu = \dfrac{\Sigma x}{n}$, the sum of all the values divided by the number of values.

In Exploration 1 you found the standard deviation by calculating:

$(x - \mu)$	the differences between the data values, x, and the mean, μ
$(x - \mu)^2$	the squares of the differences
$\dfrac{\Sigma(x - \mu)^2}{n}$	the mean of the squares of the differences
$\sqrt{\dfrac{\Sigma(x - \mu)^2}{n}}$	the square root of the mean of the squares

Tip

You can calculate standard deviation with your GDC. Most GDCs use σ for the sample standard deviation, instead of $\bar{x}$.

> A formula to calculate the standard deviation of a population is: $\sigma = \sqrt{\dfrac{\Sigma(x - \mu)^2}{n}}$

E4.1 How to stand out from the crowd

Example 1

At a summer carnival, a small crowd of people are trying to guess the weight of a small goat. The first nine people to guess give the following weights, in kilograms.

18.3, 21.6, 22.2, 24.0, 20.4, 17.4, 14.9, 26.7, 23.1

a Find the mean weight.
b Find the standard deviation of the weights.
c Write down what the standard deviation means in this context.

a Mean $\mu = \frac{\Sigma x}{n} = \frac{188.6}{9} = 20.96$ (2 d.p.) —— The nine weights are the population, so use Greek letters.

b $\sigma = \sqrt{\frac{\Sigma(x-\mu)^2}{n}}$ —— Make a table to organize your calculations.

x	μ	$(x-\mu)$	$(x-\mu)^2$
18.3	20.96	−2.66	7.07
21.6	20.96	0.64	0.4096
22.2	20.96	1.24	1.5376
24.0	20.96	3.04	9.2419
20.4	20.96	−0.56	0.3136
17.4	20.96	−3.56	12.674
14.9	20.96	−6.06	36.724
26.7	20.96	5.74	32.948
23.1	20.96	2.14	4.5696
			$\Sigma(x-\mu)^2 = 105.5024$

$n = 9$

$\sigma = \sqrt{\frac{\Sigma(x-\mu)^2}{n}} = \sqrt{\frac{105.5024}{9}} = 3.4238 = 3.42$ kg (3 s.f.)

c This implies that the mean weight estimate is 20.96 kilograms, but with a spread of up to 3.42 kilograms heavier or lighter than the mean weight.

No, you're not looking at a photo that's been retouched and you're not seeing things either. This is an Argania tree, almost exclusive to Morocco, and the goats who climb them do so because of their love for the nuts that grow on its thorny, twisted branches. When the nuts are ripe, the Argania trees are often seen with a dozen or more goats in them at a time.

Practice 1

1 For each set of data, first find the mean and the standard deviation by hand, then with a calculator. Explain what the standard deviation represents in each context.

a
2	2	4	4	4
5	6	6	8	9

b
13.1	20.4	17.4	16.5
21.0	14.8	12.6	

c
15 cm	17 cm	14 cm
12 cm	14 cm	18 cm
14 cm	13 cm	18 cm

d
44.3 kg	41.5 kg	36.5 kg
41.0 kg	33.6 kg	41.8 kg
51.2 kg	39.2 kg	37.9 kg

2 The duration, in minutes, for a train journey is recorded on 15 consecutive days:

21	24	30	21	25
27	24	27	25	29
23	22	26	28	29

    a Find the mean and standard deviation of the times for this train journey.

    b State the inferences that you can draw from the data.

3 Beatriz has a choice of routes to school. She timed her journeys along each route on five occasions. The times, recorded to the nearest minute, were:

Motorway	15	16	20	28	21
Country roads	19	21	20	22	18

    a Calculate the mean and standard deviation for each route.

    b State the route you would recommend and explain with reasons.

4 Find the mean and standard deviation of the set of integers 1, 2, 3, 4, …, 15.

5 The heights (in meters), of a squad of 12 basketball players are:

1.65	1.62	1.75	1.80	1.75	1.72
1.81	1.72	1.79	1.63	1.69	1.75

    a Find the mean and standard deviation of the heights of the players.

    b A new player whose height is 1.98 m joins the squad. Find the new mean and standard deviation of the heights of the squad.

E4.1 How to stand out from the crowd

Problem solving

6 a Calculate the mean and the standard deviation of this set of data:

> 3   6   7   9   10

Each number in the data set is then increased by 4.

b Predict what will happen to the mean and standard deviation.

c Calculate the new mean and standard deviation.

d Comment on your answers.

One of the numbers in the original data set has changed, so the data set is now: 3, 6, 7, 9, x. The new mean, y, is unknown but the standard deviation is known to be $\sqrt{34}$.

e Calculate the two possible values of the unknown values x and y.

Using Σ notation, the **mean** of a set of discrete data values in a frequency table is $\frac{\Sigma xf}{\Sigma f}$. This is the sum of all the xf values divided by the total frequency.

A formula to calculate the standard deviation for discrete data presented in a frequency table is:

$$\sigma = \sqrt{\frac{\Sigma f(x-\mu)^2}{\Sigma f}}$$

Reflect and discuss 2

How is the formula for the standard deviation of discrete data in a frequency table similar to the formula for the standard deviation of a population? How is it different?

Example 2

The scores for 25 players in a golf tournament are given in this frequency table. Find the mean and the standard deviation.

x	f
66	2
67	3
68	4
69	5
70	4
71	4
72	3

▶ Continued on next page

STATISTICS AND PROBABILITY

x	f	xf
66	2	132
67	3	201
68	4	272
69	5	345
70	4	280
71	4	284
72	3	216
	$\Sigma f = 25$	$\Sigma xf = 1730$

$\mu = \dfrac{1730}{25} = 69.2$ ——— Use $\mu = \dfrac{\Sigma xf}{\Sigma f}$ to find the mean.

x	f	$(x-\mu)$	$(x-\mu)^2$	$f(x-\mu)^2$
66	2	−3.2	10.24	20.48
67	3	−2.2	4.84	14.53
68	4	−1.2	1.44	5.76
69	5	−0.2	0.04	0.2
70	4	0.8	0.64	2.56
71	4	1.8	3.24	123.96
72	3	2.8	7.84	23.52
	$\Sigma f = 25$			$\Sigma f(x-\mu)^2 = 80$

$\sigma = \sqrt{\dfrac{\Sigma f(x-\mu)^2}{\Sigma f}} = \sqrt{\dfrac{80}{25}} = 1.789$ ——— To find the standard deviation, use the formula.

The mean score is 69.2 and the standard deviation is 1.79 (3 s.f.)

Example 3

The table gives the frequency distribution of total penalty minutes in a season for all 20 players on a hockey team. Find the mean and the standard deviation.

Total penalty minutes (t)	Number of players
$0 \le t < 10$	3
$10 \le t < 20$	9
$20 \le t < 30$	6
$30 \le t < 40$	2

▶ Continued on next page

E4.1 How to stand out from the crowd

Total penalty minutes (t)	Mid-value (t)	f	tf
$0 \leq t < 10$	5	3	15
$10 \leq t < 20$	15	9	135
$20 \leq t < 30$	25	6	150
$30 \leq t < 40$	35	2	70
		$\Sigma f = 20$	$\Sigma tf = 370$

$\mu = \dfrac{370}{20} = 18.5$ —— Use $\mu = \dfrac{\Sigma xf}{\Sigma f}$ to calculate an estimate for the mean.

Total penalty minutes (t)	Mid-value (t)	f	$t - \mu$	$(t-\mu)^2$	$f(t-\mu)^2$
$0 \leq t < 10$	5	3	−13.5	182.25	546.75
$10 \leq t < 20$	15	9	−3.5	12.25	110.25
$20 \leq t < 30$	25	6	6.5	42.25	253.5
$30 \leq t < 40$	35	2	16.5	272.25	544.5
		$\Sigma f = 20$			$\Sigma f(t-\mu)^2 = 1455$

$\sigma = \sqrt{\dfrac{\Sigma f(t-\mu)^2}{\Sigma f}} = \sqrt{\dfrac{1455}{20}} = 8.529$ —— Use the standard deviation formula.

The mean number of penalties is 18.5 and the standard deviation is 8.53 (3 s.f.)

Practice 2

1 Find the mean and the standard deviation for each set of data.

a

x	1	2	3	4	5
f	3	4	4	5	2

b

Interval	1–4	5–8	9–12	13–16	17–20	21–24	25–28
f	2	0	3	3	4	2	1

c

Interval	$0 \leq x < 5$	$5 \leq x < 10$	$10 \leq x < 15$	$15 \leq x < 20$
Frequency	10	12	15	12

STATISTICS AND PROBABILITY

2 The weights of 90 cyclists in a bicycle race were recorded:

Weight (kg)	Number of cyclists
$40 \leq w < 50$	8
$50 \leq w < 60$	27
$60 \leq w < 70$	30
$70 \leq w < 80$	25

a Calculate an estimate for

   i the mean weight

   ii the standard deviation.

b A cyclist must wear a weight belt if his or her weight is 1 standard deviation or more below the mean. Find the minimum weight to compete without a weight belt.

Problem solving

3 Copy and complete this table:

Σf	Σfx	$\Sigma f(x-\mu)^2$	μ	σ
10	50			$\sqrt{2}$
20			6.15	$\sqrt{2.06}$
	197	80.34	5.05	
43	223		5.186	
		39.56	5	$\sqrt{1.72}$

4 Copy the table below and fill in the missing values, then calculate the mean and standard deviation for the data.

Length (cm)	Mid-value (x)	f	xf	$x - \mu$	$f(x-\mu)^2$
$0 < x \leq 5$	2.5	3	7.5	−11.4	389.88
$5 < x \leq 10$	7.5	4	30	−6.4	163.84
$10 < x \leq 15$	12.5	6	75	−1.4	
$15 < x \leq 20$	17.5	7	122.5		90.72
$20 < x \leq 25$	22.5			8.6	369.8
		$\Sigma f =$	$\Sigma fx = 347.5$		$\Sigma f(x-\mu)^2 =$

Exploration 2

32 students' test scores were recorded:

Test score	1	2	3	4	5	6	7	8	9
Frequency	1	2	3	6	9	6	3	2	1

1 The data is shown in this histogram. The curve joins the midpoints of the bars. Describe the shape of the curve.

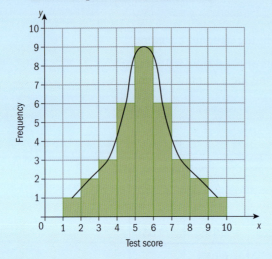

2 Calculate the mean and standard deviation for the data, to 1 d.p.

3 Determine where the mean lies on the curve.

4 From the original data:

   a State how many data items are more than 1 standard deviation from the mean.

   b State how many data items are within 1 standard deviation of the mean.

   c Find the percentage of the data items that are within 1 standard deviation of the mean.

The data in Exploration 2 has a normal distribution.

> The **normal distribution** is a symmetric distribution, with most values close to the mean and tailing off evenly in either direction. Its frequency graph is a bell-shaped curve.

Data on natural attributes, such as height, weight, and age generally follow a normal distribution. The normal frequency distribution curve looks like this:

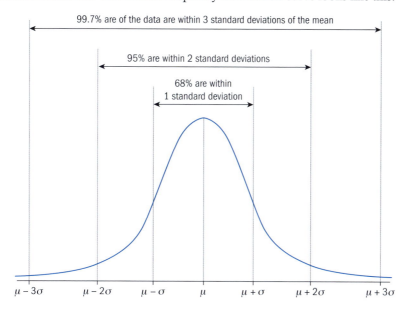

If you know that the frequency distribution follows a normal distribution then you can infer that:

- 68% of the data values lie within ±1 standard deviation of the mean.
- 95% lie within ±2 standard deviations of the mean.
- 99.7% lie within ±3 standard deviations of the mean.

This is extremely useful for making inferences about sets of data.

Reflect and discuss 3

Employees at La Salsa del Erizo have a mean starting salary of €45 000 with a standard deviation of €5000. Employees at rival company La Salsa del Zorro have a mean starting salary of €41 000 with a standard deviation of €8000. Which company would you rather work for?

C Different representations of a formula for different purposes

- How is the meaning of 'standard deviation' represented in its different formulas?

Sometimes you may not be given the whole set of data, but just some summary statistics taken from the raw data. Usually this includes Σx (the sum of all the data values) and Σx^2 (the sum of the squares of all the data values).

Tip

Most GDCs have a summary statistics page, which gives Σx and Σx^2 when you input a set of data.

Exploration 3

Here are the data values for two populations:

Population 1	Population 2
2, 4, 4, 5, 7	4, 6, 9, 10, 11, 14

1. Find μ_1, the mean of Population 1.
2. Find the mean of the squared values of Population 1.
3. Find μ_2, the mean of Population 2.
4. Find the mean of the squared values of Population 2.
5. Generalize your findings using sigma notation:
   If $\mu = \frac{\Sigma x}{n}$, then $\mu^2 =$ _____
6. Find μ_{1+2}, the mean of the combined populations (Population 1 + Population 2).
7. Generalize your findings using sigma notation:

   $\mu_{1+2} =$ _____

When you can manipulate this mathematical notation (called sigma notation), you can analyze the standard deviation formula and represent it in a variety of ways.

Exploration 4

1. Start with the formula for the population standard deviation: $\sigma = \sqrt{\frac{\Sigma(x-\mu)^2}{n}}$

2. Expand the numerator: $\sigma = \sqrt{\frac{\Sigma(x^2 - \Box + \Box)}{n}}$

3. Distribute the sigma notation: $\sigma = \sqrt{\frac{\Sigma\Box + \Sigma\Box + \Sigma\Box}{n}}$

4. Distribute the n: $\sigma = \sqrt{\frac{\Sigma\Box}{n} + \frac{\Sigma\Box}{n} + \frac{\Sigma\Box}{n}}$

5. Substitute $\mu = \frac{\Sigma x}{n}$ and $\frac{\Sigma \mu^2}{n} = \mu^2$.

6. Collect like terms.

7. Substitute $\mu = \frac{\Sigma x}{n}$ again.

8. Write your new formula: $\sigma =$ _____

Reflect and discuss 4

What does your new formula for standard deviation mean?
What information do you need to use to calculate standard deviation?

STATISTICS AND PROBABILITY

> A formula to calculate standard deviation from the summary statistics
> Σx and Σx^2 is: $\sigma = \sqrt{\dfrac{\Sigma x^2}{n} - \left(\dfrac{\Sigma x}{n}\right)^2}$

ATL Example 4

The length of time t, in minutes, for a skier to travel down a mountain follows a normal distribution and is recorded on 15 runs. The summary statistics are: $\Sigma x = 129$ and $\Sigma x^2 = 1181$.

a Find the mean and standard deviation of the times taken for this ski run.

b Make a generalization about the times for the ski run.

a Mean $= \dfrac{\Sigma x}{n} = \dfrac{129}{15} = 8.6$ min

Standard deviation $\sigma = \sqrt{\dfrac{\Sigma x^2}{n} - \left(\dfrac{\Sigma x}{n}\right)^2} = \sqrt{\dfrac{1181^2}{15} - (8.6)^2}$ ——— $\left(\dfrac{\Sigma x}{n}\right)^2 = \mu^2$

$= 2.18$ min

b $8.6 - 2.18 = 6.42$

$8.6 + 2.18 = 10.78$

68% of the runs will take between 6.42 min and 10.78 min.

> Calculate the values 1 standard deviation on either side of the mean. 68% of the data values lie between these.

ATL Example 5

The summary statistics for a class of 15 students' test marks are $\Sigma x = 930$ and $\Sigma x^2 = 59146$.

a Calculate the mean and standard deviation of the marks for this class.

b A second class of 25 students had a mean of 63.5 and standard deviation of 12.5 marks for the same test. Calculate Σx^2 for the second class.

c Calculate the mean mark of all the students in the two classes.

d Compare the performance of the two classes on this test.

(Assume test marks follow the normal distribution.)

a $\mu = \dfrac{930}{15} = 62$ marks ——— The mean is $\mu = \dfrac{\Sigma x}{n}$.

$\sigma = \sqrt{\dfrac{\Sigma x^2}{n} - \left(\dfrac{\Sigma x}{n}\right)^2} = \sqrt{\dfrac{59146}{15} - (62)^2} = 9.95$ marks

▶ Continued on next page

E4.1 How to stand out from the crowd

b $\sigma = \sqrt{\dfrac{\Sigma x^2}{n} - \left(\dfrac{\Sigma x}{n}\right)^2}$ ———— Substitute known values into the formula for σ and solve for Σx^2.

$12.5 = \sqrt{\dfrac{\Sigma x^2}{25} - (63.5)^2}$

$156.25 = \dfrac{\Sigma x^2}{25} - 4032.25$

$4188.5 = \dfrac{\Sigma x^2}{25}$

$104\,712.5 = \Sigma x^2$

c $\dfrac{\Sigma x_1 + \Sigma x_2}{n_1 + n_2} = \mu_{1+2}$

$\dfrac{930 + (63.5 \times 25)}{15 + 25} = \mu_{1+2}$ ———— $63.5 = \mu_2 = \dfrac{\Sigma x_2}{n_2} = \dfrac{\Sigma x_2}{25}$

so $\Sigma x_2 = 63.5 \times 25$

Therefore, $\mu_{1+2} = 62.9$ (3 s.f.)

d The mean and standard deviation were both slightly lower for the class of 15 than for the class of 25.

Assuming the test marks are normally distributed, 68% of the students' marks in the first class would be between 52.05 and 71.95. For the second class, 68% of the students' marks would be between 51 and 76.

ATL ### Practice 3

In these questions, assume that all the data follow normal distributions.

1 The mute swan is the national bird of Denmark and has been a protected species since 1926. In a breeding farm in Copenhagen, data on the weights of 50 fully grown mute swans was collected:

$\Sigma x = 550$ kg $\qquad \Sigma x^2 = 6162.5$ kg

Find the mean and standard deviation for the data collected and make a generalization about the weights of mute swans.

2 The mean and standard deviation of the heights of a squad of 15 football players are 1.80 m and 0.0432 m respectively.

a Find Σx and Σx^2.

b A new player whose height is 1.98 m joins the squad. Find the new mean and standard deviation of the heights of the squad.

c Provided this data is normally distributed, make generalizations from this data.

3 In Houston, Texas, the mean annual rainfall is 1264 mm with standard deviation of 155 mm.

> Exceptional results are more than 1 standard deviation from the mean.

a One year the rainfall was 1100 mm. Assuming that the rainfall in Houston follows a normal distribution, determine whether or not this was an exceptional year.

b The next year saw a total of 1400 mm of rain. Determine whether or not this was an exceptional year.

STATISTICS AND PROBABILITY

4 In a year group, 45 students each recorded the number of minutes, x, they spent on their homework one night. The total was $\Sigma x = 2230$.

  a Find the mean number of minutes spent on homework.

  b Three new students joined the year group and reported that they spent 42 minutes, 25 minutes and 30 minutes respectively on their homework. Calculate the new mean including these three students.

  c State what extra information you would need to find the standard deviation.

Problem solving

5 Twenty five students were asked about the amount of monthly pocket money they received. The summary values were calculated to be $\Sigma x = \$2237$ and $\Sigma x^2 = \$228\,361$.

Two additional students added their monthly pocket money: $85 and $125.

Find the mean and standard deviation for all this data.

 ## Sample versus population

- Can samples give reliable results?
- Do we want to be like everybody else?

A key purpose of statistics is to make inferences about a population by using data from a sample of the population. For a sample, you can calculate the sample mean $\bar{x}$ and the sample standard deviation s_x.

If, for example, you want to make inferences about a whole population of Elstar apples you can use the mean as an estimator for the average weight of all Elstar apples. But to make predictions about the standard deviation of the whole population you need to take into account that the population is much larger than the sample and, therefore, there is more likely to be a larger spread.

The weights (in grams) of a sample of 25 Elstar apples selected at random are:

132	122	132	125	134
129	130	131	133	129
126	132	133	133	131
133	138	135	135	134
142	140	136	132	135

The sample mean is $\bar{x} = \dfrac{\Sigma x}{n} = 132.48$ g

The sample standard deviation is $s_n = \sqrt{\dfrac{\Sigma(x-\bar{x})^2}{n}} = 4.28$ g

In statistics, you want the sample values (mean, standard deviation, etc.) to be good estimates of the same values for the entire population. When they are, these values are called 'unbiased estimators'. For example, the sample mean is an unbiased estimator of the population mean. The notation $\hat{\mu}$ shows that the value is an unbiased estimator for the mean. For the apples, $\hat{\mu} = 132.48$ g.

E4.1 How to stand out from the crowd

When you calculate a population standard deviation, you divide the sum of squared deviations from the mean by the number of items in the population n.

When you calculate an estimate of the population standard deviation from a sample, you divide the sum of squared deviations from the mean by the number of items in the sample less one, $n - 1$. This gives an unbiased estimator of the population's standard deviation.

> For a sample of n items:
> - an unbiased estimate of the mean for the whole population is $\hat{\mu}$, the sample mean
> - an unbiased estimate of the standard deviation for the whole population is:
> $$s_{n-1} = \sqrt{\frac{\Sigma(x - \bar{x})^2}{n-1}}$$

As a result, the estimate of the standard deviation for the population is slightly higher than if you used the population standard deviation formula. This gives a better, and unbiased, estimate of the population's standard deviation by taking into account that the sample may not exactly represent the whole range of the population.

> Dividing by $n - 1$ compensates for the fact that you are working only with a sample rather than with the whole population.

> The alternative formula for calculating an unbiased estimate of the population standard deviation from summary statistics is:
> $$s_{n-1} = \sqrt{\frac{\Sigma x^2}{n-1} - \frac{n}{n-1}\left(\frac{\Sigma x}{n}\right)^2}$$

Reflect and discuss 5
- Would you trust inferences made from a sample? Explain.
- Are there times when sampling would be absolutely necessary? Give specific examples.

Example 6

A machine produces 1000 ball bearings each day with a mean diameter of 1 cm. A sample of 8 ball bearings is taken from the production line and the diameters measured. The results in centimeters are:

1.0   1.1   1.0   0.8   1.4   1.3   0.9   1.1

Determine the standard deviation of the diameters of the ball bearings produced by the machine that day.

▶ Continued on next page

STATISTICS AND PROBABILITY

$$\bar{x} = \frac{1.0+1.1+1.0+0.8+1.4+1.3+0.9+1.1}{8}$$

$$= 1.075$$

x	$(x-\bar{x})$	$(x-\bar{x})^2$
1.0	(1.0 − 1.075)	0.005625
1.1	(1.1 − 1.075)	0.000625
1.0	(1.0 − 1.075)	0.005625
0.8	(0.8 − 1.075)	0.075625
1.4	(1.4 − 1.075)	0.105625
1.3	(1.3 − 1.075)	0.050625
0.9	(0.9 − 1.075)	0.030625
1.1	(1.1 − 1.075)	0.000625
		$\Sigma(x-\bar{x})^2 = 0.275$

Calculate $\bar{x}$ and $\Sigma(x-\bar{x})^2$.

$$s_{n-1} = \sqrt{\frac{\Sigma(x-\bar{x})^2}{n-1}} = \sqrt{\frac{0.275}{7}} = 0.1982 \text{ cm (4 d.p.)}$$

To estimate the standard deviation of the population (day's production) from a sample use:

$$s_{n-1} = \sqrt{\frac{\Sigma(x-\bar{x})^2}{n-1}}$$

Objective D: Applying mathematics in real-life contexts
iii. apply the selected mathematical strategies successfully to reach a solution

You must decide whether you need the standard deviation of the data set or an estimate for the population standard deviation, and then correctly apply the appropriate formula.

ATL Practice 4

1 Each day for 10 days, Auberon recorded how many minutes late his bus had arrived. The times in minutes are shown here:

10   12   5   0   14   2   5   8   9   6

Find an estimate for the standard deviation of how many minutes late buses arrive in the whole town.

2 On a farm, 25 baby rabbits are born one week. Their weights in grams are:

450	453	452	480	501
462	475	460	470	430
485	435	425	465	456
475	435	466	482	455
462	435	462	478	455

From this information, make a prediction about the mean and standard deviations of the weights for the whole year.

E4.1 How to stand out from the crowd

3 The summary statistics for a sample of the life length of batteries are given in the table. Calculate the mean and standard deviation of the batteries in the factory.

n	25
Σx	38 750
Σx^2	60 100 000

Problem solving

4 Katie recorded the number of passengers in a carriage on her train each day for 50 days. The results are shown here:

1	7	6	7	7	6	7	6	7	8
8	2	10	6	10	10	5	12	5	8
6	8	10	8	9	3	7	12	9	5
8	6	7	5	9	11	12	4	9	6
7	8	9	7	9	11	7	13	14	15

a Construct a tally chart to represent the data.

b Comment on the shape of the distribution of the representation.

c Estimate the mean and standard deviation of the whole train.

d Make some inferences for Katie to report on her findings.

Summary

The **standard deviation** is a measure of dispersion that gives an idea of how close the original data values are to the mean, and thus how representative the mean is of the data. A small standard deviation shows that the data values are close to the mean. The units of standard deviation are the same as the units of the original data.

Σx means the sum of all the x values.

Using this notation, the mean of a set of values is $\mu = \dfrac{\Sigma x}{n}$, the sum of all the values divided by the number of values.

A formula to calculate the standard deviation of a population is:

$$\sigma = \sqrt{\dfrac{\Sigma(x-\mu)^2}{n}}$$

Using Σ notation, the mean of a set of discrete data values in a frequency table is $\dfrac{\Sigma xf}{\Sigma f}$, the sum of all the xf values divided by the total frequency.

A formula to calculate the standard deviation for discrete data presented in a frequency table is:

$$\sigma = \sqrt{\dfrac{\Sigma f(x-\mu)^2}{\Sigma f}}$$

The **normal distribution** is a symmetric distribution, with most values close to the mean and tailing off evenly in either direction. Its frequency graph is a bell-shaped curve.

A formula to calculate standard deviation from the summary statistics Σx and Σx^2 is:

$$\sigma = \sqrt{\dfrac{\Sigma x^2}{n} - \left(\dfrac{\Sigma x}{n}\right)^2}$$

STATISTICS AND PROBABILITY

For a sample of n items:
- an unbiased estimate of the mean for the whole population is $\hat{\mu}$, the sample mean, $\bar{x}$.
- an unbiased estimate of the standard deviation for the whole population is:

$$s_{n-1} = \sqrt{\frac{\Sigma(x-\bar{x})^2}{n-1}}$$

The alternative formula for calculating an unbiased estimate of the population standard deviation from summary statistics is:

$$s_{n-1} = \sqrt{\frac{\Sigma x^2}{n-1} - \frac{n}{n-1}\left(\frac{\Sigma x}{n}\right)^2}$$

Mixed practice

In questions **1** and **2** the data provided is for the population.

1. **Find** the mean and the standard deviation of each data set:

   a 2, 3, 3, 4, 4, 5, 5, 6, 6, 6

   b 21 kg, 21 kg, 24 kg, 25 kg, 27 kg, 29 kg

   c

x	3	4	5
f	2	3	2

   d

Interval	Frequency
1–5	2
6–10	4
11–15	4
16–20	5
21–25	2

2. Charlotte buys 20 strawberry plants for her garden. She feeds half each day with water and the other half with a special strawberry plant food. After one month she records the amount of strawberries on each plant.

Number of strawberries on each plant fed with water								
15	13	12	15	17	16	12	10	11	15

Number of strawberries on each plant fed with special strawberry plant food								
14	15	17	17	19	19	12	14	15	15

   Analyze the data to find whether there is an effect of using the special strawberry plant food.

3. Bernie recorded the weight of food in grams his hamster ate each day. Here are his results:

Day	1	2	3	4	5	6	7	8	9
Food (g)	55	64	45	54	60	50	59	61	49

   a **Find** the mean and standard deviation of the weight of food the hamster ate over the 9 days.

   b Bernie assumes that his hamster's food consumption is normally distributed. **Write down** the inferences he can make.

   c One day the hamster eats 25g of food. Should Bernie be worried?

4. The summary statistics for the weights of 100 lambs on a farm were:

   $\Sigma x = 425$ kg $\Sigma x^2 = 2031.25$ kg

   a **Show that** $\mu = 4.25$ and $\sigma = 1.5$.

   b The farmer wanted to produce guidelines for his staff to see if any lambs needed special attention. He decided that any lamb born weighing between 2 and 6.5 kg was fine. **Calculate** how many standard deviations away from the mean these values are.

E4.1 How to stand out from the crowd

5 For a particular data set, the summary statistics are: $\Sigma f = 20$, $\Sigma fx = 563$, and $\Sigma fx^2 = 16143$. **Find** the values of the mean and the standard deviation.

6 In a biscuit factory a sample of 10 packets of biscuits were weighed.

| Mass (g) | 196 | 197 | 199 | 200 | 200 | 200 | 202 | 203 | 203 | 205 |

  a **Calculate** the mean and standard deviation.

  b To make predictions about the mean and standard deviations of all the packets of biscuits in the factory, which values of the mean and standard deviation would you use?

Problem solving

7 Samples of water were taken from a river near a chemical plant to see if there were raised levels of lead. Ten samples were collected, shown in the table below. The units are μg per liter.

| 6.3 | 9.6 | 12.2 | 12.3 | 10.3 |
| 12.1 | 10.3 | 8.4 | 9.2 | 4.3 |

  a **Use** this data to **predict** the mean and standard deviation for the amount of lead in the river.

  b Over 10 μg per liter of lead in water is dangerous. **Comment** on whether the lead levels should be investigated further.

Review in context

1 Dr Roussianos asked his cardiology patients to measure their systolic blood pressure each Saturday morning for ten weeks.

> Systolic pressure is the pressure in the arteries between the beats of the heart, while the heart muscle is resting and refilling with blood.

Here is the data for three patients.

Patient 1

| 95 | 95 | 87 | 92 | 100 |
| 84 | 87 | 90 | 87 | 96 |

Patient 2

| 78 | 76 | 75 | 80 | 76 |
| 75 | 81 | 80 | 82 | 80 |

Patient 3

| 65 | 100 | 76 | 98 | 58 |
| 75 | 76 | 98 | 66 | 75 |

This table shows the treatments recommended for different values of the yearly mean and standard deviation.

Mean	Standard deviation	Treatment
<75	<10	low blood pressure treatment
75–85	<10	no treatment
>85	<10	high blood pressure treatment
any value	>10	more tests needed

Calculate the mean and use the sample data to **estimate** the yearly statistics and therefore help to **determine** the treatment for each patient.

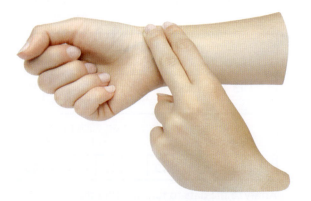

STATISTICS AND PROBABILITY

2 80 people were asked to measure their pulse rate as soon as they woke up in the morning.

They were asked to repeat the measurements after a 5 km run, and just before going to bed.

The summary statistics are: $\Sigma x = 6000$ and $\Sigma x^2 = 452\,450$.

From this information, **find** the mean and the standard deviation of each data set. **State** the inferences that can be drawn from the data.

3 Birth weight follows a normal distribution. The summary statistics for the weights of 50 babies in a hospital were $\Sigma x = 160$ kg and $\Sigma x^2 = 524$ kg.

   a From this data, **state** the inferences that can be drawn from the data.

   b If a baby's weight is less than 2.5 times the standard deviation below the mean, the baby is at risk. **Find** a value x, so that babies who weigh less than x kg are considered to be at risk.

4 In one town, 300 mothers were asked their age at the birth of their first child. The results are given in the table.

Age at birth of first child	Number of mothers
18–20	15
21–23	20
24–26	30
27–29	42
30–32	53
33–35	52
36–38	43
39–41	25
42–44	12
45–47	8

   a **Predict** a mean and standard deviation for a mother's age at the birth of her first child, for this town.

   b The summary statistics for 30 years ago, for a different sample of 300 women were $\Sigma x = 7500$ and $\Sigma x^2 = 190\,000$. **State** the inferences that can be drawn from the data.

Reflect and discuss

How have you explored the statement of inquiry? Give specific examples.

Statement of Inquiry:

Generalizing and representing relationships can help to clarify trends among individuals.

E4.1 How to stand out from the crowd

E5.1 Super powers

Objectives
- Evaluating numerical expressions with a positive or negative fractional exponent
- Writing numerical expressions with fractional exponents as radicals
- Using the rules of indices to simplify expressions that contain radicals and/or fractional exponents

Inquiry questions

F
- How do you evaluate a number with a fractional exponent?
- How do you use the rules of indices to simplify expressions with fractional exponents?

C
- How are the rules of indices and the rules of radicals related?
- How do the rules of indices help simplify radical expressions?

D
- Can expressions with decimal exponents be simplified?
- Is simpler always better?

| ATL | Critical-thinking |

Analyse complex concepts and projects into their constituent parts and synthesize them to create new understanding

5.1

10.3

E5.1

Conceptual understanding:
Forms can be changed through simplification.

NUMBER

You should already know how to:

• use the multiplicative rule for exponents	**1** Simplify: **a** $(x^3)^2$ **b** $(2^4)^2$ **c** $(2x^2)^3$
• use the rules of radicals to simplify expressions	**2** Simplify completely: **a** $\sqrt{45}$ **b** $\dfrac{\sqrt{128}}{\sqrt{8}}$ **c** $\sqrt{60} \times \sqrt{3}$
• evaluate expressions with unit fraction exponents	**3** Evaluate: **a** $4^{\frac{1}{2}}$ **b** $8^{\frac{1}{3}}$ **c** $32^{\frac{1}{5}}$ **d** $64^{\frac{1}{2}}$ **e** $64^{\frac{1}{3}}$ **f** $(0.001)^{\frac{1}{3}}$
• evaluate expressions with negative exponents	**4** Evaluate, leaving your answer as a fraction where necessary: **a** 4^{-1} **b** 6^{-2} **c** 2^{-3} **d** $\left(\dfrac{1}{2}\right)^{-1}$ **e** $\left(\dfrac{2}{3}\right)^{-1}$ **f** $\left(\dfrac{2}{5}\right)^{-2}$

F Fractional exponents

- How do you evaluate a number with a fractional exponent?
- How do you use the rules of indices to simplify expressions with fractional exponents?

> For any positive real number x, and integer n, where $n \neq 0$, $x^{\frac{1}{n}} = \sqrt[n]{x}$.

ATL Example 1

Evaluate:

a $\left(\dfrac{9}{25}\right)^{\frac{1}{2}}$ **b** $\left(\dfrac{1}{8}\right)^{-\frac{1}{3}}$ **c** $\left(\dfrac{7}{81}\right)^{-\frac{1}{2}}$

a $\left(\dfrac{9}{25}\right)^{\frac{1}{2}} = \sqrt{\dfrac{9}{25}} = \dfrac{3}{5}$

b $\left(\dfrac{1}{8}\right)^{-\frac{1}{3}} = 8^{\frac{1}{3}} = \sqrt[3]{8} = 2$

> Using the multiplicative rule for exponents,
> $(x^a)^b = x^{ab}$: $\left(\dfrac{1}{8}\right)^{-\frac{1}{3}} = \left(\left(\dfrac{1}{8}\right)^{-1}\right)^{\frac{1}{3}} = 8^{\frac{1}{3}}$

c $\left(\dfrac{7}{81}\right)^{-\frac{1}{2}} = \left(\dfrac{81}{7}\right)^{\frac{1}{2}} = \dfrac{9}{\sqrt{7}} = \dfrac{9\sqrt{7}}{7}$

> Here, the denominator has been rationalized.

Practice 1

1 Work out the value of:

a $\left(\dfrac{1}{4}\right)^{\frac{1}{2}}$ **b** $\left(\dfrac{1}{16}\right)^{\frac{1}{4}}$ **c** $\left(\dfrac{1}{8}\right)^{\frac{1}{3}}$ **d** $\left(\dfrac{1}{25}\right)^{-\frac{1}{2}}$

e $\left(\dfrac{4}{9}\right)^{\frac{1}{2}}$ **f** $27^{-\frac{1}{3}}$ **g** $\left(\dfrac{4}{25}\right)^{-\frac{1}{2}}$ **h** $\left(\dfrac{27}{1000}\right)^{-\frac{1}{3}}$

E5.1 Super powers

2 Simplify and leave your final answer as a power of x.

a $(4x)^{\frac{1}{2}}$ b $(27x^3)^{\frac{1}{3}}$ c $(8x^6)^{-\frac{1}{3}}$ d $\left(\dfrac{x^4}{100}\right)^{-\frac{1}{2}}$

e $x^{-\frac{1}{2}} \times x^{\frac{1}{2}}$ f $x^{\frac{1}{3}} \div x^{\frac{1}{2}}$ g $x^{\frac{1}{4}} \times x^{-\frac{1}{2}}$ h $(9x^2)^{-\frac{1}{2}} \times 9x^2$

> Assume that $x \neq 0$ for all expressions.

Problem solving

3 Write each of these using exponents instead of the radicals.

a $\dfrac{1}{\sqrt{x}}$ b $\sqrt[3]{x}$ c $\dfrac{1}{\sqrt[4]{x}}$

4 Solve:

a $\left(\dfrac{1}{x}\right)^{-\frac{1}{3}} = 2$ b $\left(\dfrac{y}{9}\right)^{-\frac{1}{2}} = 0.3$

Using the multiplicative rule for exponents $(x^a)^b = x^{ab}$:

$(x^{\frac{1}{2}})^3 = x^{\frac{3}{2}}$

$(x^{\frac{1}{4}})^3 = x^{\frac{4}{3}}$

What do expressions with fractional exponents like $x^{\frac{2}{3}}$ or $x^{\frac{4}{3}}$ mean?

ATL

Exploration 1

1 Copy and complete this table:

Using integer and unit fraction exponents	Using a radical	Using fractional exponents
$(x^{\frac{1}{2}})^3$	$(\sqrt{x})^3$	$x^{\frac{3}{2}}$
$(x^{\frac{1}{2}})^5$		
$(x^3)^{\frac{1}{2}}$	$\sqrt{x^3}$	
	$\sqrt[3]{x^4}$	
$(x^{\frac{1}{5}})^2$		$x^{\frac{2}{5}}$
		$x^{\frac{2}{3}}$

> A unit fraction has numerator 1.

2 Simplify $(x^{\frac{1}{3}})^2$ and $(x^2)^{\frac{1}{3}}$. Are $(x^{\frac{1}{3}})^2$ and $(x^2)^{\frac{1}{3}}$ equivalent?

Generalize your result for $(x^{\frac{1}{n}})^m$ where m and n are integers and $n \neq 0$.

3 Copy and complete, using your result from step **2**:

$(\sqrt[n]{x})^{\square} = (x^{\frac{1}{n}})^{\square} = x^{\frac{m}{n}} = (x^m)^{\square} = \sqrt[\square]{x^{\square}}$

NUMBER

Reflect and discuss 1

- Did everyone in the class get the same results for the last two rows in the table in Exploration 1?
- Is there more than one way of writing $x^{\frac{2}{3}}$ and $x^{\frac{3}{4}}$ in the other two columns?

For any positive real number x, and integers m and n, where $n \neq 0$,
$$x^{\frac{m}{n}} = (\sqrt[n]{x})^m = \sqrt[n]{x^m}$$

ATL

Example 2

Simplify $16^{\frac{3}{4}}$ without using a calculator.

$16^{\frac{3}{4}} = (16^{\frac{1}{4}})^3$ — Split the exponent into a unit fraction and an integer.

$= (2)^3$ — $16^{\frac{1}{4}} = \sqrt[4]{16} = 2$

$= 8$

Reflect and discuss 2

You could also simplify $16^{\frac{3}{4}}$ like this: $16^{\frac{3}{4}} = (16^3)^{\frac{1}{4}}$. Complete the calculation and verify that this gives the same answer as Example 2.

Which was easier – this method, or the one in Example 2? Explain why.

Write a hint to help you simplify expressions with fractional exponents.

Practice 2

1 Simplify these expressions without using a calculator.

   a $8^{\frac{2}{3}}$ **b** $27^{\frac{4}{3}}$ **c** $25^{\frac{3}{2}}$ **d** $64^{\frac{2}{3}}$ **e** $125^{\frac{2}{3}}$ **f** $343^{\frac{2}{3}}$

Problem solving

2 Write each of these using exponents instead of the radicals.

   a $\sqrt[3]{x^2}$ **b** $(\sqrt[4]{x})^3$ **c** $(\sqrt{x})^3$ **d** $\sqrt{x^5}$

3 Simplify. Where applicable, leave your final answer as a power of x.

   a $(x^{\frac{2}{3}})^2$ **b** $x^{\frac{1}{3}} \times x^{\frac{4}{}}$ **c** $x^{\frac{1}{2}} \times x^{\frac{2}{3}}$ **d** $(3^4)^{\frac{1}{2}}$ **e** $(x^{\frac{2}{5}})^{\frac{2}{3}}$ **f** $(32^2)^{\frac{2}{5}}$

E5.1 Super powers

Example 3

ATL

Simplify $\left(\dfrac{27}{8}\right)^{-\frac{2}{3}}$

$\left(\dfrac{27}{8}\right)^{-\frac{2}{3}} = \left(\dfrac{8}{27}\right)^{\frac{2}{3}}$

$= \left(\left(\dfrac{8}{27}\right)^{\frac{1}{3}}\right)^{2}$

$= \left(\dfrac{2}{3}\right)^{2}$

$= \dfrac{4}{9}$

Using the multiplicative rule for exponents, $(x^a)^b = x^{ab}$, gives:

$\left(\dfrac{27}{8}\right)^{-\frac{2}{3}} = \left(\left(\dfrac{27}{8}\right)^{-1}\right)^{\frac{2}{3}}$

$x^{\frac{m}{n}} = (x^{\frac{1}{n}})^m$

Reflect and discuss 3

ATL

The method in Example 3 dealt with the negative sign in the exponent first, by taking the reciprocal. Confirm that taking the cube root first, or squaring first, all give the same final answer. Which method is easier?

If necessary, add to the hint you wrote in **Reflect and discuss 2**, to help you simplify expressions with fractional exponents.

Practice 3

1 Find the value of each expression. Leave your answer as a simplified fraction.

  a $25^{-\frac{3}{2}}$
  b $16^{-\frac{3}{4}}$
  c $\left(\dfrac{9}{16}\right)^{-\frac{3}{2}}$

  d $\left(\dfrac{1}{64}\right)^{-\frac{4}{3}}$
  e $\left(\dfrac{36}{25}\right)^{-\frac{3}{2}}$
  f $\left(\dfrac{81}{16}\right)^{-\frac{3}{4}}$

2 Simplify these expressions completely.

  a $125^{\frac{1}{4}} \times 125^{\frac{1}{12}}$
  b $16^{\frac{1}{6}} \times 16^{\frac{1}{12}}$
  c $49^{\frac{3}{2}} \times 343^{\frac{1}{3}}$

  d $8^{\frac{3}{4}} \div 8^{\frac{1}{12}}$
  e $27^2 \div 27^{\frac{2}{3}}$
  f $45^{\frac{1}{2}} \div 15^{\frac{1}{2}}$

Problem solving

3 Write down an equivalent expression that contains fractional exponents.

  a 2
  b 5
  c $\dfrac{1}{2}$
  d $\dfrac{2}{3}$

4 Simplify:

  a $3^2 \times 9^{-2}$
  b $4^{\frac{1}{4}} \times 2^{\frac{3}{2}}$
  c $3^{\frac{1}{2}} \times 27^{\frac{1}{2}}$
  d $16^{\frac{1}{3}} \div 2^{\frac{1}{3}}$

In question **4**, first write both numbers as powers of the same base. For example: $3^2 \times 9^{-2} = 3^2 \times (3^2)^{-2}$

C Rules of indices and radicals

- How are the rules of indices and the rules of radicals related?
- How do the rules of indices help simplify radical expressions?

Writing a radical as an exponent can help you simplify some expressions.

Example 4

Write $32\sqrt{2}$ as a power of 2.

$32\sqrt{2} = 2^5 \times 2^{\frac{1}{2}}$ ⎯⎯⎯⎯⎯⎯⎯⎯⎯⎯⎯⎯⎯⎯⎯⎯⎯⎯⎯ Write each number as a power of 2, then simplify.

$= 2^{\frac{11}{2}}$

Exploration 2

1 Use the rules of radicals to simplify:

   a $\sqrt{8} \times \sqrt{2}$ **b** $\dfrac{\sqrt{8}}{\sqrt{2}}$

2 Here are the expressions from step **1** written using exponents. Simplify them by writing 8 as a power of 2, and using the rules of exponents:

   a $8^{\frac{1}{2}} \times 2^{\frac{1}{2}}$ **b** $8^{\frac{1}{2}} \div 2^{\frac{1}{2}}$

3 Justify why the two methods for simplifying expressions with square roots are equivalent.

4 Use the rules of exponents to show that for $a, b \geq 0$, $\sqrt[3]{ab} = \sqrt[3]{a} \times \sqrt[3]{b}$. Generalize your findings.

> Write down a rule for the nth root of ab.

5 Use the rules of exponents to show that for $a \geq 0$ and $b > 0$, $\sqrt[4]{\dfrac{a}{b}} = \dfrac{\sqrt[4]{a}}{\sqrt[4]{b}}$. Generalize your findings.

> Why can b not equal 0?

6 Use the rules of exponents to show that $\sqrt[n]{a^n} = a$, where $a, n \geq 0$.

> For $a, b \geq 0$: $\sqrt[n]{ab} = \sqrt[n]{a} \times \sqrt[n]{b}$; for $a \geq 0, b > 0$: $\sqrt[n]{\dfrac{a}{b}} = \dfrac{\sqrt[n]{a}}{\sqrt[n]{b}}$.

Practice 4

1 Write as a single power of 2:

   a $8\sqrt{2}$ **b** $4\sqrt[3]{2}$ **c** $\dfrac{16}{\sqrt{2}}$ **d** $\dfrac{\sqrt[3]{2}}{4}$

Problem solving

2 Write as powers of prime numbers:

   a $\sqrt[3]{7} \times \sqrt[5]{7}$ **b** $\sqrt{9} \times \sqrt[4]{9}$ **c** $\dfrac{\sqrt[3]{8}}{\sqrt[4]{8}}$ **d** $\dfrac{2\sqrt[4]{8}}{\sqrt{2^3}}$

> In part **b**, first write 9 as a power of 3.

3 Write as products of powers of prime numbers:

   a $\sqrt{6}$ **b** $\sqrt[3]{14}$ **c** $\sqrt[4]{162}$ **d** $\sqrt[4]{100}$
   e $\sqrt[4]{50}$ **f** $\sqrt[6]{72}$

Objective C: Communicating
iii. move between different forms of mathematical representation

In Q4, write radical expressions as exponents to simplify them.

4 Write these expressions as powers of prime numbers. Use the rules of exponents to simplify. Give your answers as powers of prime numbers.

- **a** $\sqrt{2} \times \sqrt[3]{2}$
- **b** $\sqrt[3]{9} \times \sqrt[4]{9}$
- **c** $\sqrt[5]{7} \times \sqrt[10]{7}$
- **d** $\dfrac{\sqrt{6}}{\sqrt[3]{6}}$
- **e** $\dfrac{\sqrt[3]{4}}{\sqrt[5]{4}}$
- **f** $\dfrac{\sqrt[n]{10}}{\sqrt[m]{10}}$
- **g** $\sqrt[n]{a} \times \sqrt[m]{a}$
- **h** $\dfrac{\sqrt[n]{a}}{\sqrt[m]{a}}$

In parts **g** and **h**, assume *a* is prime.

Reflect and discuss 4

Use the rules of exponents to justify why these expressions cannot be simplified:

$\sqrt[3]{4} \times \sqrt[5]{3}$ $\sqrt[4]{144} \times \sqrt{125}$ $\dfrac{\sqrt{5}}{\sqrt[3]{2}}$ $\dfrac{\sqrt[4]{64}}{\sqrt{27}}$

Write a general rule for when you can simplify products and quotients of radicals.

When radicals have the same radicand but different indices, write them with fractional exponents to simplify them. For example:

- $\sqrt[n]{a} \times \sqrt[m]{a} = a^{\frac{1}{n}} \times a^{\frac{1}{m}} = a^{\frac{n+m}{nm}}$
- $\dfrac{\sqrt[n]{a}}{\sqrt[m]{a}} = \dfrac{a^{\frac{1}{n}}}{a^{\frac{1}{m}}} = a^{\frac{1}{n} - \frac{1}{m}} = a^{\frac{n-m}{nm}}$

The radicand is the number under the square root sign.

ATL Example 5

Simplify $\dfrac{\sqrt[4]{8}}{\sqrt{2}}$

$\dfrac{\sqrt[4]{8}}{\sqrt{2}} = \dfrac{\sqrt[4]{2^3}}{\sqrt{2}}$ — Write the radicands as exponents with the same base.

$= \dfrac{2^{\frac{3}{4}}}{2^{\frac{1}{2}}}$

$= 2^{\frac{3}{4} - \frac{1}{2}}$

$= 2^{\frac{1}{4}}$ — Give the answer as a radical, with prime number radicand.

$= \sqrt[4]{2}$

When radicals have different radicands, but one is a power of the other, write them as exponents with the same base to simplify them.

E5 Simplification

ATL **Example 6**

Write $10\sqrt[3]{5} \times 3\sqrt[3]{75}$ as a simplified radical, by expressing the radicands as powers of prime numbers.

$10\sqrt[3]{5} \times 3\sqrt[3]{75} = 10 \times \sqrt[3]{5} \times 3 \times \sqrt[3]{25 \times 3}$ — Write the radicands as multiples of primes.

$= 10 \times 3 \times \sqrt[3]{5} \times \sqrt[3]{5^2} \times \sqrt[3]{3}$

$= 30 \times 5^{\frac{1}{3}} \times 5^{\frac{2}{3}} \times 3^{\frac{1}{3}}$ — Write radicands using exponents.

$= 30 \times 5^1 \times 3^{\frac{1}{3}}$

$= 150\sqrt[3]{3}$

Practice 5

1 Simplify these completely, leaving your answer as a simplified radical.

a $\sqrt{2} \times \sqrt[3]{2} \times \sqrt[6]{2}$ **b** $\sqrt[3]{3^2} \times \sqrt[4]{3} \times \sqrt[12]{3}$ **c** $\dfrac{\sqrt{5}}{\sqrt[3]{5}}$ **d** $\dfrac{\sqrt[3]{11}}{\sqrt{11}}$ **e** $\dfrac{\sqrt{7^3}}{\sqrt{7}}$

f $\dfrac{\sqrt[3]{3^4}}{\sqrt{3}}$ **g** $3\sqrt{2} \times 5\sqrt[3]{2}$ **h** $5\sqrt{3^3} \times 2\sqrt[4]{3}$ **i** $\dfrac{2\sqrt{5}}{\sqrt[3]{5}}$ **j** $\dfrac{4\sqrt{3^5}}{8\sqrt{3}}$

2 Write as simplified radicals, with prime number radicands.

a $\dfrac{4}{\sqrt[3]{2}}$ **b** $\sqrt[4]{12} \times \sqrt[3]{3}$ **c** $\sqrt{12} \times \sqrt[3]{18}$ **d** $\dfrac{\sqrt[3]{16}}{\sqrt{2}}$ **e** $\dfrac{\sqrt[3]{7}}{\sqrt{343}}$

f $\dfrac{\sqrt[4]{125}}{\sqrt[3]{5}}$ **g** $\dfrac{4\sqrt{6}}{\sqrt{3}}$ **h** $\dfrac{5\sqrt[4]{125}}{\sqrt[4]{5}}$ **i** $\dfrac{\sqrt[3]{81}}{\sqrt[3]{24}}$ **j** $\sqrt[5]{49} \times \sqrt[5]{343}$

Problem solving

3 Solve these equations, where the unknown is always a power of 3.

a $\dfrac{\sqrt{27}}{x} = \sqrt{3}$ **b** $\dfrac{y}{3} = \dfrac{1}{\sqrt[3]{9}}$ **c** $\dfrac{\sqrt[3]{9}}{z} = \sqrt[6]{3}$

D Other exponents

- Can expressions with decimal exponents be simplified?
- Is simpler always better?

A rational number is a number that can be written as a fraction.

> Think: 'rational' has the word 'ratio' in it, and a ratio can be expressed as a fraction.

Exploration 3

1 Convert these decimals into fractions:

a 0.1 **b** 0.375 **c** 1.6 **d** $0.\dot{3}$

2 Write the decimal exponents as fractions, and hence write these expressions as radicals:

a $5^{0.75}$ **b** $5^{0.1}$ **c** $5^{0.375}$ **d** $5^{1.6}$ **e** $5^{0.\dot{3}}$ **f** $5^{0.8}$

3 Use your answers to step **2** to justify why any number with a finite decimal exponent can be written as a radical.

You cannot *evaluate* an expression like 3^π or $5^{\sqrt{2}}$ without a calculator, because the exponents are irrational numbers. However, you can *simplify* expressions like $3^\pi \times 3^4$ or $(7^\pi)^2$, using the rules for exponents. Examples:

$3^\pi \times 3^4 = 3^{\pi+4}$

$(7^\pi)^2 = 7^{2\pi}$

> An irrational number cannot be written as a fraction. How could you get an approximation for $5^{\sqrt{2}}$?

Reflect and discuss 5

The definition of a radical is this: $\sqrt[n]{a} = x$ means that $a = x^n$.

Suggest why it is difficult to evaluate $\sqrt[\pi]{a}$.

Practice 6

1. Write these expressions as radicals of powers of prime numbers:

   a $9^{0.2}$ b $16^{1.25}$ c $100^{1.5}$

   d $(3^2)^{0.3}$ e $4^{1.5}$ f $10000^{0.75}$

Problem solving

2. Evaluate $32^{0.2}$. Explain your method.

3. Copy and complete:

   a $4^{0.2} \times 4 = 2^\square$ b $27^{0.2} \times 81^{0.1} = 3^\square$ c $32^{0.4} \times 16^{0.5} \times 64^{0.1} = 2^\square$

4. Solve $x^{0.1} = 2$.

5. Simplify:

   a $\dfrac{27^{0.4}}{3^{0.2}}$ b $\dfrac{6^2 \times 2^{0.6}}{9 \times 4^{0.8}}$

Summary

For any positive real number x, and integer n, where $n \neq 0$, $x^{\frac{1}{n}} = \sqrt[n]{x}$.

For any positive real number x, and integers m and n, where $n \neq 0$, $x^{\frac{m}{n}} = (\sqrt[n]{x})^m = \sqrt[n]{x^m}$.

For $a, b \geq 0$: $\sqrt[n]{ab} = \sqrt[n]{a} \times \sqrt[n]{b}$

For $a \geq 0, b > 0$: $\sqrt[n]{\dfrac{a}{b}} = \dfrac{\sqrt[n]{a}}{\sqrt[n]{b}}$

> The radicand is the number under the square root sign.

When radicals have the same radicand but different indices, write them with fractional exponents to simplify them.

For example: $\sqrt[n]{a} \times \sqrt[m]{a} = a^{\frac{1}{n}} \times a^{\frac{1}{m}} = a^{\frac{n+m}{nm}}$

$\dfrac{\sqrt[n]{a}}{\sqrt[m]{a}} = \dfrac{a^{\frac{1}{n}}}{a^{\frac{1}{m}}} = a^{\frac{1}{n} - \frac{1}{m}} = a^{\frac{n-m}{nm}}$

When radicals have different radicands, but one is a power of the other, write them as exponents with the same base to simplify them.

When radicals have different radicands, but one is *not* a power of the other, then they can not be simplified into one radical.

NUMBER

Mixed practice

1 Work out the value of:

a $\left(\frac{1}{9}\right)^{\frac{1}{2}}$ b $\left(\frac{1}{9}\right)^{-\frac{1}{2}}$ c $25^{-\frac{1}{2}}$ d $\left(\frac{1}{27}\right)^{-\frac{1}{3}}$

e $\left(\frac{9}{16}\right)^{\frac{1}{2}}$ f $\left(\frac{25}{100}\right)^{-\frac{1}{2}}$ g $\left(\frac{8}{125}\right)^{-\frac{1}{3}}$

2 Simplify:

a $(49x^6)^{\frac{1}{2}}$ b $(64x^3)^{-\frac{1}{3}}$ c $(4x^4)^{-\frac{1}{2}} \times 3x$

3 Work out the value of:

a $10000^{\frac{3}{4}}$ b $1000^{\frac{2}{3}}$ c $27^{-\frac{2}{3}}$ d $9^{-\frac{3}{2}}$

4 Find the value of each of these expressions. Leave your answers as rational numbers.

a $\left(\frac{1}{8}\right)^{-\frac{2}{3}}$ b $\left(\frac{4}{9}\right)^{-\frac{3}{2}}$ c $\left(\frac{125}{64}\right)^{\frac{2}{3}}$ d $\left(\frac{5}{9}\right)^{\frac{3}{2}} \times \sqrt{5}$

5 Simplify these expressions completely. Leave your answers as rational powers of prime numbers.

a $64^{\frac{1}{3}} \times 64^{\frac{1}{12}}$ b $81^{\frac{1}{4}} \times 9^{\frac{1}{8}}$ c $\dfrac{16^{\frac{5}{12}}}{16^{\frac{1}{6}}}$ d $\dfrac{36^{\frac{4}{3}}}{6^{\frac{1}{3}}}$

6 Simplify:

a $25^2 \times 5^{-2}$ b $16^{\frac{1}{2}} \times 4^{-\frac{3}{2}}$ c $7^{-\frac{1}{2}} \times 343^{\frac{1}{2}}$

7 Simplify these expressions completely. Leave your answers as simplified radicals.

a $\sqrt[3]{16} \times \sqrt[3]{8}$ b $\sqrt[4]{12} \times \sqrt[4]{48}$ c $\dfrac{\sqrt[4]{64}}{\sqrt[4]{4}}$

d $\dfrac{3\sqrt[3]{9}}{\sqrt[3]{81}}$ e $\sqrt[4]{4} \times \sqrt[8]{4} \times \sqrt[4]{2}$ f $\sqrt{8} \times \sqrt[3]{8} \times \sqrt[6]{8}$

g $\dfrac{\sqrt[4]{12}}{\sqrt[3]{12}}$ h $\dfrac{\sqrt[3]{9}}{\sqrt[4]{27}}$

Problem solving

8 Write down an equivalent expression that contains fractional exponents.

a 4 b 3 c $\dfrac{1}{4}$ d $\dfrac{3}{2}$

9 Write as simplified radicals with prime number radicands.

a $2\sqrt[3]{16}$ b $\sqrt{24} \times \sqrt[4]{36}$ c $\dfrac{4^2}{(\sqrt{8})^3}$ d $\dfrac{\sqrt[3]{5}}{2\sqrt[4]{10}}$

Problem solving

10 Solve these equations:

a $8^{-x} = \dfrac{1}{2}$ b $32^{\frac{x}{5}} = 4$ c $\left(\dfrac{x}{64}\right)^{\frac{2}{3}} = \dfrac{9}{16}$

11 Simplify, leaving your answers as simplified radicals of prime numbers:

a $25^{0.4} \times 5$ b $125^{0.3} \times 5^{0.1}$ c $9^{0.4} \times 81^{0.2} \times 3^{0.4}$

d $\dfrac{50^{0.9}}{125^{0.3}}$ e $\dfrac{12^3 \cdot 9^{0.2}}{30^{0.4}}$

12 The quotation below has been encoded using simplified surds and fractional exponents.

Without using a calculator, simplify the following expressions completely. **Use** your results and the key to decode a quote by Stefan Banach.

> Key: $2\sqrt{3} = L$, $\dfrac{1}{2} = I$, $2 = A$, $5\sqrt{2} = O$,
> $3 = M$, $8 = D$, $\sqrt{3} = S$, $1 = N$

'*Mathematics* $\underline{}$ $\underline{}$ $\underline{}$ $\underline{}$ $\underline{}$

$\left(\dfrac{1}{32}\right)^{\frac{1}{5}}$ $\left(\dfrac{1}{27^{\frac{1}{3}}}\right)^{-\frac{1}{2}}$ $\dfrac{15^{\frac{1}{2}} \times 20^{\frac{1}{2}}}{5\sqrt{3}}$ $\dfrac{\sqrt{18}\sqrt{10}}{\sqrt{4}\sqrt{15}}$ $\left(\dfrac{10^2}{2}\right)^{\frac{1}{2}}$

$\underline{}$ $\underline{}$ $\underline{}$ $\underline{}$ $\underline{}$ $\underline{}$ $\underline{}$,

$12^{\frac{1}{2}}$ $\dfrac{64^{\frac{2}{3}}}{\sqrt[3]{8}}$ $\dfrac{3 \times \sqrt[3]{64}}{\sqrt{2}\sqrt{18}}$ $\dfrac{3^{\frac{1}{4}} \times 6^{\frac{1}{4}}}{2^{\frac{1}{4}}}$ $81^{\frac{1}{4}}$ $\dfrac{64^{\frac{1}{2}}}{\sqrt[4]{256}}$ $\left(\dfrac{144^{\frac{1}{3}} \times 6^{\frac{1}{3}}}{6 \times 2^{\frac{2}{3}}}\right)^{-\frac{1}{2}}$

Reflect and discuss

How have you explored the statement of conceptual understanding? Give specific examples.

Conceptual understanding:

Forms can be changed through simplification.

E5.1 Super powers   107

E6.1 Ideal work for lumberjacks

Global context: Orientation in space and time

Objectives

- Evaluating logarithms with and without a calculator
- Writing an exponential statement as a logarithmic statement
- Solving exponential equations
- Writing and solving exponential equations from real-life situations
- Using natural logarithms

Inquiry questions

F
- What does $\log_a b$ mean?
- How can you find logarithms using a calculator?

C
- How do logarithms make problems easier to solve?

D
- How can anything in mathematics be considered natural?
- Is change measurable and predictable?

RELATIONSHIPS

| ATL | Critical-thinking |

Consider ideas from multiple perspectives

Statement of Inquiry:

Generalizing changes in quantity helps establish relationships that can model duration, frequency and variability.

E6.1

E10.1

E12.1

You should already know how to:

• use index laws	1 Simplify $x^2 \times x^5$. 2 Evaluate: a 5^3 b 2^{-1} c $16^{\frac{1}{2}}$
• rewrite surds and roots as quantities with exponents	3 Write with an exponent: a $\sqrt{6}$ b $\sqrt[3]{9}$ c $\sqrt{5^3}$
• model a real-life problem with an exponential equation	4 In 2007, the Zika virus was growing at a rate of 18% per week on Yap Island. Two people were infected initially. Use an exponential model to determine the number of people infected after 52 weeks.

F Introduction to logarithms

- What does $\log_a b$ mean?
- How can you find logarithms using a calculator?

An exponential equation is in the form $a^x = b$.

- a is the base
- x is the exponent or power

Exploration 1

Write each statement as an exponential equation, and solve to find x. The first one is done for you.

1 Find the exponent of 10 that gives the answer:

100	$10^x = 100$	$x = 2$
1000		
10 000		
0.1		
0.0001		

2 Find the exponent of 2 that gives the answer:

4		
8		
16		
128		
$\frac{1}{2}$		

▶ Continued on next page

3 Find the exponent of 6 that gives the answer:

36		
216		
$\frac{1}{216}$		

4 What patterns do you notice?

Determine what would happen to the answer if you add 1 to the exponent.

Determine what would happen to the answer if you subtract 1 from the exponent.

In steps **1** to **3** the exponents are all integers. When you can't spot the answer, you can use trial and improvement.

5 To solve $10^x = 150$, estimate the value of x and try it. Improve your estimate to find x correct to 2 decimal places.

6 Solve $2^x = 45$. Give your answer to 2 decimal places.

7 Solve $5^x = 77$. Give your answer to 2 significant figures.

8 You can also use graphing software to solve exponential equations graphically.

For example, to solve $10^x = 60$, draw the graph of the exponential function $y = 10^x$ and the graph of $y = 60$.

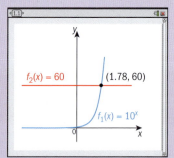

Read the value of x at the intersection to 2 decimal places.

a Solve $2^x = 6$ graphically. Give your answer to 2 decimal places.

b Solve $7^x = 14$ graphically. Give your answer to 2 significant figures.

The three methods for solving exponential equations in Exploration 1 are:
- the inspection method
- trial and improvement
- graphical

Reflect and discuss 1

- Would you use the trial and improvement method or the graphical method to solve the equation $9^x = 85$? Explain your choice.
- What are the advantages and disadvantages of the three methods?

The trial and improvement method and the graphing method give two different perspectives on solving exponential equations.

Here is another method. In the exponential equation $10^x = 500$, the quantity x is defined as 'the logarithm to base 10 of 500'.

You can write $\log_{10} 500 = x$.

Using a calculator: $\log_{10} 500 = 2.69897...$

$x = 2.70$ (2 d.p.)

You can use the equivalent statements $10^x = 500$ and $\log_{10} 500 = x$ to find the unknown exponent.

Rewriting exponential statements as logarithmic statements changes the unknown variable from an exponent to a regular variable. Whenever you see an exponential or logarithmic statement you can use the equivalence concept to consider it from another perspective.

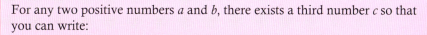

> **Tip**
>
> The log button on your calculator is for '$\log_{10}$'. You do not need to enter the number 10.

Before calculators, logarithms of different values were published in log tables. Modern calculators calculate logarithms at the touch of a button.

> For any two positive numbers a and b, there exists a third number c so that you can write:
>
> the exponential statement $a^c = b$
>
> or the equivalent logarithmic statement $\log_a b = c$; $a, b > 0$.
>
> We say that c is 'the logarithm to base a of b', or 'the exponent of a that gives the answer b'. In the logarithmic statement $\log_a b = c$, a is called the **base** of the logarithm and b is called the **argument** of the logarithm.

When solving equations involving exponents or logarithms, it can be helpful to rewrite the equation using the other form. This allows you to see the equation from a new perspective.

Example 1

Solve $\log_7 x = 2$.

$7^2 = x$ Rewrite as an exponential statement.
$x = 49$ $\log_a b = c$ means $a^c = b$.

Logarithms with base 10 are often called 'common' logs and written $\log x$, without a base.

When you see 'log' written without a base, you can assume the base is 10.

On most calculators you can choose the base of the logarithm. If your calculator does not allow this, you will need to use the **change of base formula**.

Change of base formula: $\log_a b = \dfrac{\log_c b}{\log_c a}$, where c is any new base.

> The change of base formula is derived in E12.1.

Example 2

Solve $2^x = 50$ using a calculator and the change of base formula.

$2^x = 50$ — Rewrite as a logarithmic statement.

$\log_2 50 = x$

$\log_2 50 = \dfrac{\log_{10} 50}{\log_{10} 2}$ — Use the change of base formula with $c = 10$.

$= 5.64385619$ — Work out $\dfrac{\log 50}{\log 2}$ on your calculator.

$= 5.64$ (3 s.f.)

Example 3

$\log_7 7\sqrt{7} = x$. Find the value of x.

$\log_7 (7 \times 7^{\frac{1}{2}}) = x$ — Write surds as exponents.

$\log_7 (7^{\frac{3}{2}}) = x$

$7^x = 7^{\frac{3}{2}}$ — Rewrite as an exponential statement.

$x = \dfrac{3}{2}$ — The bases are equal, so the exponents are equal.

Practice 1

1. Write down the equivalent logarithmic statement for:

   a $10^x = 500$ b $10^x = 150$ c $10^x = 60$ d $2^x = 45$ e $2^x = 6$

   Hence, using your calculator, find the value of x in each case. Give your answers to 3 s.f.

2. Write down an equivalent exponential statement for:

   a $\log_8 64 = 2$ b $\log_8 4 = \dfrac{2}{3}$ c $\log 0.1 = -1$

   d $\log_5 x = 5$ e $\log_b b = 4$ f $\log_x y = 0$

3. Write down an equivalent logarithmic statement for:

   a $3^2 = 9$ b $4^{-\frac{2}{3}} = 0.125$ c $1000^{\frac{1}{3}} = 10$

Use this table to help you answer questions **4 – 6**.

Powers of 2		Powers of 3		Powers of 4		Powers of 5	
2^{-3}	0.125	3^{-3}	$0.0\dot{3}\dot{7}$	4^{-3}	0.15625	5^{-3}	0.008
2^{-2}	0.25	3^{-2}	$0.\dot{1}$	4^{-2}	0.0625	5^{-2}	0.04
2^{-1}	0.5	3^{-1}	$0.\dot{3}$	4^{-1}	0.25	5^{-1}	0.2
2^{0}	1	3^{0}	1	4^{0}	1	5^{0}	1
2^{1}	2	3^{1}	3	4^{1}	4	5^{1}	5
2^{2}	4	3^{2}	9	4^{2}	16	5^{2}	25
2^{3}	8	3^{3}	27	4^{3}	64	5^{3}	125
2^{4}	16	3^{4}	81	4^{4}	256	5^{4}	625
2^{5}	32	3^{5}	243	4^{5}	1024	5^{5}	3125

Tip

Learn these results to help you solve logarithmic and exponential equations without a calculator.

4 Without using a calculator, find the value of x.
  a $\log_5 625 = x$ **b** $\log_3 243 = x$ **c** $\log_4 0.625 = x$ **d** $\log_4 1024 = x$
  e $\log 0.00001 = x$ **f** $\log_2 0.125 = x$ **g** $\log_3 1 = x$ **h** $\log_5 0.04 = x$

5 Without using a calculator, find the value of x.
  a $\log_5 \sqrt{5} = x$ **b** $\log_3 3\sqrt{3} = x$ **c** $\log_4 \sqrt{256} = x$ **d** $\log_2 \sqrt{256} = x$

6 Find the value of x to the nearest hundredth.
  a $\log_2 17 = x$ **b** $\log_6 121 = x$ **c** $\log_3 31 = x$ **d** $\log_8 5 = x$

7 Solve, using a calculator:
  a $3^x = 40$ **b** $5^x = 100$ **c** $4^x = 85$ **d** $2^x = 90$

Logarithms are used in astronomy to measure the apparent magnitude of stars; in geology to measure the intensity of earthquakes; in music, to measure semitones; and perhaps most beautifully, in nature, as seen in the inside of nautilus shells whose chambers form a logarithmic spiral.

Problem solving

8 Solve these exponential equations.
  a $2^{x+1} = 4^{2x}$
  b $3^{x+2} = 9^{2x-2}$
  c $2^{x+1} = \left(\dfrac{1}{2}\right)^{2x}$
  d $\left(\dfrac{1}{3}\right)^{x+2} = 3^{2x-2}$
  e $\left(\dfrac{1}{3}\right)^{x+2} = 9^{2x-2}$

In **8**, write them with the same base and use the laws of exponents.

C Using logarithms to solve equations

- How do logarithms make problems easier to solve?

You can use logarithms to solve complex exponential equations. Rewriting an exponential statement as an equivalent logarithmic statement changes the equation into one you can solve.

Example 4

Solve $6^{2x+1} = 20$, accurate to 2 decimal places.

$\log_6 20 = 2x + 1$ ——— Rewrite as a logarithmic statement.

$1.67195... = 2x + 1$ ——— Use a calculator to find $\log_6 20 = \dfrac{\log_{10} 20}{\log_{10} 6}$.

$x = 0.33567 = 0.34$ (2 d.p.) ——— Solve for x.

Practice 2

Solve each equation, giving your answer to the degree of accuracy stated.

1 $2^x = 5$ (2 d.p.)
2 $2^{2x+1} = 5$ (3 d.p.)
3 $5^{x-1} = 3$ (2 s.f.)
4 $3^{2x+1} = 5$ (3 s.f.)
5 $4 \times 2^x = 6$ (2 d.p.)
6 $2 \times 2^{x-1} = 6$ (2 d.p.)
7 $\dfrac{2^x}{3} = 6$ (2 s.f.)
8 $\dfrac{2^{2x-3}}{3} = 16$ (2 s.f.)

You can also use logarithms to solve equations that result from real-life growth and decay problems.

> **Objective D:** Applying mathematics in real-life contexts
> **ii.** select appropriate mathematical strategies when solving authentic real-life situations

> *In Exploration 2, you will write an equation to describe a real-life situation and then use an appropriate strategy to solve it.*

Exploration 2

An amoeba is a single-celled organism that reproduces by cell division. A certain amoeba splits into two separate amoebas every hour. There is one amoeba on a microscope slide at the beginning of an experiment. Determine how long it will take for there to be 1000 amoebas on the slide.

1 The number of amoebas depends on time t. Copy and complete the table:

Time (hours)	$t=0$	$t=1$	$t=2$	$t=3$	$t=4$
Number of amoebas	$A(0)=$	$A(1)=$	$A(2)=$	$A(3)=$	$A(4)=$

2 Write the number of amoebas as powers of 2.
3 Write $A(t)$ as an exponential function.
4 Write and solve an exponential equation to find t when $A(t) = 100$.

Reflect and discuss 2

- How do you know when to use logarithms to solve an equation?
- How could you verify that your answer makes sense?

Example 5

The number of drosophila (fruit flies) in a laboratory at the beginning of an experiment is 200. The population increases by 8.5% each day.

a Find the population at the end of the first and second days.

b Write an exponential function for the number of drosophila after x days.

c Find the number of days until the population reached 100 000 drosophila.

a

Number of days (x)	Number of drosophila (D)
0	200
1	200×1.085
2	200×1.085^2

Growth factor $b = 1 + r = 1 + 0.085 = 1.085$

b $a = 200$, $b = 1.085$

Exponential growth is modelled by $y = a(1+r)^x$ where x is the number of days.

$D = 200 \times 1.085^x$

c $100\,000 = 200 \times 1.085^x$ — Rearrange to isolate the term in x.

$$\frac{100\,000}{200} = 1.085^x$$

$500 = 1.085^x$ — Rewrite as a logarithmic statement.

$\log_{1.085} 500 = x$

$x = 76.18$

After 77 days there would be more than 100 000 drosophila.

E6.1 Ideal work for lumberjacks

Practice 3

1 At the end of the year 2000 the population of a city was 300 000. The population then increased by 1.3% per year.

  a Find the population at the end of 2001 and 2002.

  b Write an exponential function for $P(t)$, the population after t years.

  c Predict the year in which the population should exceed 350 000.

2 A group of 15 snow foxes are introduced into a nature reserve. The number of snow foxes N can be modelled by the exponential equation $N = 15 \times 3^{0.4t}$ where t is the number of years since their introduction.

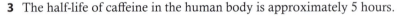

Find how many years after their introduction when there should be at least 100 snow foxes.

3 The half-life of caffeine in the human body is approximately 5 hours.

Copy and complete the table for a student who consumes 120 mg of caffeine at 08:00 one morning.

Time	Time period (t hours)	Amount of caffeine (C mg)
08:00	$t = 0$	120
13:00	$t = 1$	60
18:00	$t = 2$	
23:00		

  a Find the common ratio for the geometric sequence.

  b Find the exponential equation for the half-life of caffeine in terms of t.

  c Calculate the amount of caffeine in the student's body after 35 hours.

  d The effect of caffeine is negligible below 0.02 mg. Determine the number of hours after which there is only a negligible effect from the caffeine.

> Each time period represents 5 hours.

Problem solving

4 Coco is training for a race and records her 10 km times each week. Before she starts training, she can run 10 km in 72 minutes.

Week (w)	Time (t, min) to run 10 km
1	70.56
2	69.149
3	67.766
4	66.411
5	65.082

> The times form a geometric sequence. Use $w = 0$ and $w = 1$ to find the rate of decrease.

Assuming her time continues to decrease at the same rate, calculate how many weeks it will be before Coco can run 10 km in less than one hour.

5 A large lake has a population of 100 000 perch. The number of perch is decreasing at a rate of 10% per year due to pollution.

  a Make a table to show the number of perch P after 1, 2 and 3 years.

  b Write an exponential model for the number of perch P after t years.

  c When the population of perch falls below 25 000 it is at a 'critical level'. Calculate how long it will take for this lake to be at a critical level.

 Natural logarithms

- How can anything in mathematics be considered natural?
- Is change measurable and predictable?

Exploration 6 of MYP Standard 14.1 explored how $1 invested in a bank for exactly one year would grow with the interest compounded at different intervals:

annually	$\left(1+\frac{1}{1}\right)^{1} = 2$
semi-annually	$\left(1+\frac{1}{2}\right)^{2} = 2.25$
quarterly	$\left(1+\frac{1}{4}\right)^{4} = 2.44140625$
monthly	$\left(1+\frac{1}{12}\right)^{12} \approx 2.61303529022\ldots$
weekly	$\left(1+\frac{1}{52}\right)^{52} \approx 2.69259695444\ldots$
daily	$\left(1+\frac{1}{365}\right)^{365} \approx 2.71456748202\ldots$
hourly	$\left(1+\frac{1}{8760}\right)^{8760} \approx 2.71812669063\ldots$
every minute	$\left(1+\frac{1}{525\,600}\right)^{525600} \approx 2.7182792154\ldots$
every second	$\left(1+\frac{1}{31\,536\,000}\right)^{31536000} \approx 2.71828247254\ldots$

The number of times the interest is compounded is n. As n approaches infinity, the total investment approaches $2.71828…

The number 2.71828… is called **e** and it is the base rate for growth/decay of all continually growing processes.

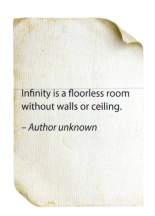

Infinity is a floorless room without walls or ceiling.

– Author unknown

> In general, all populations that grow continually are modelled by the population growth equation $P = P_0 e^{rt}$, where:
>
> P is the final population
>
> P_0 is the original population when $t = 0$
>
> r is the growth rate
>
> t is the time period.

Reflect and discuss 3

How could the formula $P = P_0 e^{rt}$ be used for problems involving exponential decay?

To calculate the growth rate or the time you can use logarithms with base e.

> If $y = e^x$ then the equivalent logarithmic statement is $\log_e y = x$, $y > 0$.
>
> The natural logarithm $\log_e y$ is written as: $\ln y$.

Your calculator should have an ln button, and an e^x button.

> The first mention of e, the natural logarithm, was noted in 1618 as an appendix to Napier's work on logarithms, but it wasn't officially named 'e' until Euler completed his work in 1748.
>
> The spiral galaxy on the cover of this book is a naturally occurring logarithmic spiral, where the shape does not alter but the size increases.

Example 6

A skydiver jumps off a cliff. Sensibly, the skydiver has remembered to wear a parachute. As he falls, his speed of descent is modelled by the equation $v(t) = 50(1 - e^{-0.2t})$, where his speed, $v(t)$, is in m/s and time t is in seconds.

Find:

a his initial speed

b the time it takes for him to reach a speed of 40 m/s.

a $v(0) = 50(1 - e^{-0.2 \times 0})$ — Initial speed occurs when $t = 0$.

$= 50(1 - 1) = 0$

Initial speed is 0 m/s.

b $40 = 50(1 - e^{-0.2t})$ — Substitute 40 for the speed and solve for t.

$\dfrac{40}{50} - 1 = -e^{-0.2t}$

$-0.2 = -e^{-0.2t}$ — Write the equivalent logarithmic statement.

$0.2 = e^{-0.2t}$

$\ln 0.2 = -0.2t$

$t = 8.05$ seconds (2 d.p.)

Example 7

In 2000, an island in the Caribbean had a population of 2600 people.

Assuming that the population was continually increasing at a rate of 1.68% per year, calculate:

a the predicted population in 2020

b the time taken for the population to double

c the time taken for the population in part **b** to return to 2600 if it began to decrease at a rate of 2% per year.

a $P = P_0 e^{rt}$ — Standard population equation.

$P = 2600 e^{0.0168 \times 20}$

$= 3638.3$ — Use your calculator.

The population in 2020 will be 3638 people.

b $P = P_0 e^{rt}$ — Double the population is 5200.

$5200 = 2600 e^{0.0168t}$

$2 = e^{0.0168t}$ — Rewrite as a logarithmic statement.

$\ln 2 = 0.0168 t$

$t = \dfrac{\ln 2}{0.0168} = 41.258\ldots$

The population will double in 42 years.

c $P = P_0 e^{rt}$ — For exponential decay, the rate is negative.

$2600 = 5200 e^{-0.02t}$

$0.5 = e^{-0.02t}$ — Rewrite as a logarithmic statement.

$\ln 0.5 = -0.02 t$

$t = \dfrac{\ln 0.5}{-0.02} = 34.657\ldots$

The population will return to 2600 people in 35 years.

Practice 4

1 Carbon-14 dating is a common method used to determine the age of fossils and bones.

One formula used to calculate the age t (in years) of an item is $P = e^{-0.000121 t}$, where P is the percentage of carbon-14 left in the item (written as a decimal).

A femur bone found in an Australian aboriginal burial site contains 11% of the carbon-14 found in a normal femur. Determine the femur's age.

E6.1 Ideal work for lumberjacks

2 Radioactive substances decay over time and are often described by their half-life (the time taken for half of the substance to decay/disappear). The formula used to describe this decay is $A = A_0 e^{bt}$, where A is the amount left after time t, A_0 is the original amount of the radioactive substance, and b is the decay constant.

Iodine-131 has a half-life of 8 days. It is a common substance used in diagnosing issues with the thyroid gland.

  a Find the decay constant b for iodine-131 if you begin with 3 grams.

  b Hence write down the formula for the exponential decay of iodine-131.

  c Find how many grams of iodine-131 will be present after 28 days if you start with 3 grams.

3 A biologist monitoring a fire ant infestation notices that the area infected by the ants can be modelled by $A = 1000e^{0.7n}$, where A is the area in hectares, and n is the number of weeks after the initial observation.

  a Find the initial population size.

  b Find the population after one week.

  c Calculate how long it will take the population to cover 50 000 hectares.

4 A group of 20 rabbits is introduced to a rabbit farm. After t years the number of rabbits N is modelled by the exponential equation $N = 20e^{0.6t}$.

  a Predict the number of rabbits after 3 years.

  b Determine how long it will take for the number of rabbits to reach 400.

  Give your answers to an appropriate degree of accuracy.

5 The model for the population P of wombats in a nature reserve is $P = 50e^{0.8t}$, where t is measured in years.

  a Write down how many wombats were introduced to the reserve.

  b Calculate how long it will take for the original population to quadruple.

Problem solving

6 The population of a small Dutch town was obtained for two consecutive years:

Year	2010	2011
Population	5101	5204

Assume that the population is modelled by the standard equation, and that $t = 0$ in 2010.

  a Calculate the rate of growth, r.

  b Calculate the predicted population in 2020.

  c Calculate the number of years until the population exceeds 10 000.

Reflect and discuss 4

- What do you think are the limitations of continuous growth and decay models?
- Why do you think ln is called a 'natural' logarithm?

Summary

An exponential equation has the form $a^x = b$:
- a is the base
- x is the exponent or power

For any two positive numbers a and b, there exists a third number c so that you can write

the exponential statement $a^c = b$

or the equivalent logarithmic statement
$\log_a b = c$; $a, b > 0$.

We say that c is 'the logarithm to base a of b', or 'the exponent of a that gives the answer b'.

In the logarithmic statement $\log_a b = c$, a is called the **base** of the logarithm and b is called the **argument** of the logarithm.

Change of base formula: $\log_a b = \dfrac{\log_c b}{\log_c a}$

Generally, all populations that grow continually are modelled by the population growth equation $P = P_0 e^{rt}$, where:

P is the final population

P_0 is the original population when $t = 0$

r is the growth rate

t is the time period.

If $y = e^x$ then the equivalent logarithmic statement is $\ln y = x$, $y > 0$.

The natural logarithm $\log_e y$ is written as $\ln y$.

The common logarithm $\log_{10} y$ is written as $\log y$.

Mixed practice

1 Write down the equivalent logarithmic statement for:

  a $7^x = 23$ **b** $10^x = 95$

  c $8^x = 6$ **d** $4^x = 47$

  e $12^x = 1200$

2 Write down an equivalent exponential statement for:

  a $\log_5 125 = 3$ **b** $\log_3\left(\dfrac{1}{9}\right) = -2$

  c $\log 1000 = 3$ **d** $\log_7 2401 = 4$

  e $\log_a m = n$

3 Find the value of x without a calculator:

  a $\log_2 16 = x$ **b** $\log_3 9 = x$

  c $\log_7\left(\dfrac{1}{7}\right) = x$ **d** $\log_6 1 = x$

  e $\log_3 3 = x$ **f** $\log_8 \sqrt{8} = x$

  g $\log_6 6\sqrt{6} = x$

4 Find the value of x to the nearest hundredth.

  a $\log_2 7 = x$ **b** $\log_4 12 = x$

  c $\log_5 312 = x$ **d** $\log_9 21 = x$

  e $\log 650 = x$ **f** $\ln 5 = x$

5 Find the value of x to 2 d.p.

  a $3^x = 20$ **b** $5^x = 20$

  c $e^x = 20$ **d** $3 \times 3^x = 15$

  e $2 \times 3^{2x+1} = 45$ **f** $3 \times e^{2x+1} = 30$

6 Find the value of x without using a calculator.

  a $2^{4x-1} = 8^{2x}$ **b** $5^{x+4} = 125^{3x-1}$

  c $3^{2x+3} = \left(\dfrac{1}{9}\right)^{2x-1}$

E6.1 Ideal work for lumberjacks

7 The table below shows the average movie ticket price in Canada from 1990 to 1994.

Year	Price of a movie ticket ($)
1990	4.00
1991	4.22
1992	4.45
1993	4.69
1994	4.95

a Find the exponential model for the average movie ticket price as a function of t, the number of years since 1990.

b Calculate the average price of a movie ticket in the year 2016.

c Find when the average price will reach $20.

8 A kettle of water is heated and then allowed to cool. The temperature can be modelled by the exponential equation $T = 100e^{-0.2t}$, where T is the temperature (in °C) and t is the time in minutes.

a Find the initial temperature.

b Determine the temperature after 3 minutes.

c Elizabeth can drink the water when the temperature is 40 °C. **Find** how long she must leave it to cool.

Review in context

Charles Francis Richter was an American seismologist and physicist who created the Richter magnitude scale, which quantifies the size of earthquakes by measuring their intensity. It is a logarithmic scale, and since 1935 it has been the standard measure of earthquake intensity.

1 One formula for modelling the magnitude of an earthquake using the Richter scale is:

$$M = \log_{10} \frac{I_c}{I_n}$$

where M is the magnitude, I_c is the intensity of the 'movement' of the earth from the earthquake, and I_n is the intensity of the 'movement' of the earth on a normal day-to-day basis.

a The intensity of the movement on a normal day in Oklahoma is 100 microns. Last December, the intensity of the movement of the earth from the earthquake was recorded as 250 000 microns. **Determine** the size of this earthquake on the Richter scale.

b Find the measurement of each earthquake on the Richter scale:

Earthquake	I_c	I_n
i Southeast Indian ridge	3 767 829 647	150
ii Near coast of Northern Chile	25 896 531	140
iii Admiralty Island, Papua New Guinea	45 487 563	160
iv Chiapas, Mexico	123 568 544	140
v Cuba region	86 532 658	110

c An earthquake in northern Peru was recorded as 4.8 on the Richter Scale. Usual movement in that area is 1450 microns. Using this information, **calculate** the intensity of the movement of the earth from the earthquake.

d An earthquake in Bermuda was recorded as 2.7 on the Richter Scale. The movement of the earth from the earthquake was calculated to be 2 145 000 microns. Using this information, **calculate** the usual movement.

2 To calculate the pH of a liquid you need to know the concentration of hydrogen ions (H^+) in moles per liter of the liquid.

The pH is then calculated using the logarithmic formula: $pH = -\log_{10}(H^+)$.

a HCl is a strong acid with a hydrogen ion concentration of 0.0015 moles per liter. **Find** the pH of the HCl solution.

b **Find** the hydrogen ion concentration of a liquid which has a pH of 9.24.

c A solution is said to be neutral (neither acid nor base) if its pH is 7. **Find** the concentration of hydrogen ions which would make a solution to be considered neutral.

3 The pOH of an aqueous solution measures the number of hydroxide ions (OH^-) in a solution, using the formula: $pOH = -\log_{10}(OH^-)$.

a **Find** the pOH of an aqueous solution that has a hydroxide ion concentration of 6.1×10^{-5} moles per liter.

b **Find** the hydroxide ion concentration in an aqueous solution that has a pOH of 2.3.

4 The decibel scale measures the intensity of sound D using the equation $D = 10 \log I$, where I is the intensity ratio of a sound. On the decibel scale, the purr of a cat measures $D = 10 \log 330 = 25.2$ (3 s.f.) decibels.

Sound intensity is measured on a traffic light system:

Green: no ear protection needed. 1–75 DB

Orange: ear protection recommended. 80–120 DB

Red: ear protection necessary. 120 + DB

Calculate the sound intensity of these noises and **determine** whether or not ear protection equipment is recommended or necessary.

a A chainsaw has an intensity ratio of 1.04×10^{11}.

b A flowing river has an intensity ratio of 3 100 000.

c A rocket launching has an intensity ratio of 8.2×10^{16}.

Reflect and discuss

How have you explored the statement of inquiry? Give specific examples.

Statement of Inquiry:

Generalizing changes in quantity helps establish relationships that can model duration, frequency and variability.

E7.1 Slices of pi

Global context: Orientation in space and time

Objectives

- Drawing and using the unit circle to find sines, cosines and tangents of angles
- Converting angles between degrees and radians
- Using the radian formula for the length of an arc
- Solving problems involving angular and linear displacement
- Drawing graphs of $\sin\theta$ and $\cos\theta$, where θ is in radians
- Describing transformed trigonometric functions of the form $f(x) = a\sin(bx - c) + d$

Inquiry questions

F
- What is a radian?
- What is the unit circle?
- How are the trigonometric functions related to the unit circle?

C
- Which parameter affects the horizontal translation of a sinusoidal function?
- What is the difference between phase shift and horizontal shift?

D
- Does the order matter when you perform transformations on sinusoidal functions?
- How do we define 'where' and 'when'?

RELATIONSHIPS

ATL Critical-thinking

Gather and organize relevant information to formulate an argument

Statement of Inquiry:

Generalizing and applying relationships between measurements in space can help define 'where' and 'when'.

GEOMETRY AND TRIGONOMETRY

You should already know how to:

• find exact values of trigonometric ratios of 30°, 60° and 45°	**1** Find the exact value of: **a** sin 30° **b** cos 45° **c** tan 60°
• draw the graphs of sine and cosine functions	**2** Sketch the graph of these functions for values of x between 0° and 360°. **a** sin x **b** cos x
• describe transformations on sinusoidal functions	**3 a** Find the amplitude and frequency of the graph of $y = 2 \sin 2x$. **b** Describe the transformations on the graph of $y = \cos x$ to get the graph of $y = 3 \cos x + 1$.
• find the length of an arc	**4 a** Find the circumference of this circle. **b** Find the length of arc AB.

F Radian measure and the unit circle

- What is a radian?
- What is the unit circle?
- How are the trigonometric functions related to the unit circle?

Until now, you have measured angles in degrees. Like distances, which can be measured in units as varied as kilometers, miles and light years, angles can also be measured in different units. Exploration 1 investigates radian measure for angles.

> Another unit of measurement for angle is the gradian, also known as the grad. One grad is equal to $\frac{1}{400}$ of one revolution. The French term centigrade means one-hundredth of a grad, but the term was also used as the name of the temperature scale. Hence, the term Celsius was adopted to replace Centigrade, in honour of Swedish astronomer Anders Celsius who proposed the scale.

You do not need to know how to work with gradians.

Objective B: Investigating patterns
ii. describe patterns as general rules consistent with findings

In Exploration 1, look for patterns in the central angle and length of the arc, and write a general rule to describe the pattern.

E7.1 Slices of pi   125

Exploration 1

ATL

1 A unit circle has a radius of 1 unit. Find the circumference of a unit circle in terms of π.

2 Copy and complete this table with the lengths of the arcs of the unit circle with different sized angles, in terms of π.

Central angle in degrees	Fraction of circle	Length of corresponding arc
360° (circumference)	1	2π
180°	$\frac{1}{2}$	
90°	$\frac{1}{4}$	$\frac{2\pi}{4} = \frac{\pi}{2}$
60°		
45°		
30°		

The lengths of the arcs are the **radian measures** of the angles in the first column. So 360° = 2π, 180° = π, and so on.

3 Suggest a formula to convert an angle of $a°$ to radians.

4 Consider a circle with radius r, as shown to the right. Copy and complete this table with the lengths of the arc s of this circle in terms of r and π:

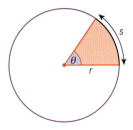

Central angle in degrees	Central angle in radians	Length of corresponding arc s
360° (circumference)	2π	$2\pi r$
180°		
90°	$\frac{\pi}{2}$	
60°		
45°		
30°		

5 Suggest a formula for the arc length s of a circle with radius r when the angle θ is in radians.

6 By considering your formula in step **5**, explain why radians are a measure without units.

7 Determine the central angle, in radians, of an arc equal in length to the radius of the circle.

8 Suggest a definition for the radian measure of an angle.

GEOMETRY AND TRIGONOMETRY

A **radian** is a unit for measuring angles.

1 radian is the angle at the center of an arc with length equal to the radius of the circle.

$\theta = \dfrac{s}{r}$ and $s = r\theta$

$360° = 2\pi$ radians

$1° = \dfrac{\pi}{180°}$ radians

Angle in radians = Angle in degrees $\times \dfrac{\pi}{180°}$

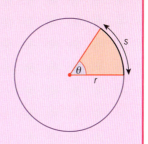

Practice 1

1 Convert these angles to radians.

    a 10°   **b** 15°   **c** 18°   **d** 36°   **e** 40°

    f 72°   **g** 126°   **h** 150°   **i** 225°   **j** 300°

2 Convert these angles to degrees.

    a $\dfrac{\pi}{6}$ **b** $\dfrac{2\pi}{3}$ **c** $\dfrac{3\pi}{4}$ **d** $\dfrac{3\pi}{5}$ **e** $\dfrac{7\pi}{8}$

    f $\dfrac{8\pi}{9}$ **g** $\dfrac{5\pi}{12}$ **h** $\dfrac{4\pi}{3}$ **i** $\dfrac{5\pi}{3}$ **j** $\dfrac{5\pi}{4}$

Problem solving

3 The area of a sector of a circle is given by $A = \dfrac{\pi r^2 \theta}{360°}$ where the angle θ is measured in degrees. Rewrite the formula for an angle θ measured in radians.

You can use radians to work out the distance a wheel has moved in a number of rotations. The formula $s = r\theta$ links the linear displacement s (distance travelled) and angular displacement θ (angle of rotation) for a wheel of radius r.

An angular displacement greater than 2π means that the wheel rotates more than one full turn of $360° = 2\pi$. For example, an angle of 10π is equivalent to 5 complete revolutions (turns).

Example 1

The second hand on a clock is 20 cm long. Find the linear displacement of the tip of the second hand (the distance the second hand travels) in 6 seconds.

Angular displacement for 1 second is: — Find the angular displacement each second (in radians).

$\dfrac{1}{60} \times 2\pi = \dfrac{\pi}{30}$

$\theta = 6 \times \dfrac{\pi}{30} = \dfrac{\pi}{5}$ —————— Find the angular displacement for 6 seconds.

▶ Continued on next page

E7.1 Slices of pi   127

$r = 20$ cm

$s = r\theta$

$= \dfrac{\pi}{5} \times 20$

$= 4\pi$ cm

$= 12.6$ cm (3 s.f.)

Practice 2

1 The hour hand on a clock is 15 cm long.

  a Find the linear displacement of the tip of the hour hand in 4 hours, to 3 s.f.

  b Find the linear displacement of the tip of the hour hand in 9 hours, to 3 s.f.

Problem solving

  c The linear displacement of the hour hand is 15π cm. Find how many hours have passed.

2 A bicycle odometer measures the linear distance travelled as the wheels rotate. The wheel of a road bike has diameter 62 cm.

  a Find how many meters the odometer measures when the wheel makes 20 rotations. Give your answer accurate to 3 s.f.

  b Find how many kilometers the odometer measures after 150 rotations. Give your answer accurate to 3 s.f.

  c The Tour de France is a 3500 km bike race. The rules specify that the diameter of the bike wheels used in the race must be between 55 cm and 70 cm. Find the minimum number of rotations a Tour de France bike wheel makes in the entire completed race, to the nearest hundred.

Problem solving

3 A wind turbine blade rotates through 90° in 3 seconds. During that time, the tip of the blade travels 18.85 m. Find the length of the blade, accurate to 2 d.p.

4 A horse-drawn carriage has a wheel of diameter 3 m.

  a Determine the angular displacement of the wheel in 15.25 rotations. Give your answer in radians.

  b Find the linear displacement of the wheel, to the nearest tenth of a meter.

The **unit circle** is a circle of radius 1 unit with its center at the origin of the graph.

The axes divide the circle into four **quadrants**.

The section at the top of page 129 shows the part of the unit circle in the first quadrant. OP is a radius of the circle, and thus has length 1.

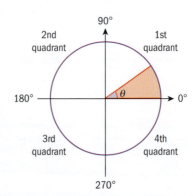

In the right-angled triangle OPQ:

$\sin\theta = \dfrac{PQ}{OP} = \dfrac{PQ}{1} = PQ$, so $PQ = \sin\theta$

$\cos\theta = \dfrac{OQ}{OP} = \dfrac{OQ}{1} = OQ$, so $OQ = \cos\theta$

$\tan\theta = \dfrac{PQ}{OQ} = \dfrac{y}{x}$

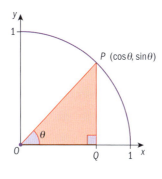

Since P has coordinates (x, y), the length of PQ is y and the length of OQ is x.

Therefore:

$\quad \sin\theta = y \qquad \cos\theta = x \qquad \tan\theta = \dfrac{y}{x}$

The coordinates of the point $P(x, y)$ in the 1st quadrant of the unit circle are therefore $(\cos\theta, \sin\theta)$.

Reflect and discuss 1

This diagram shows the full unit circle.

Point P_1 is in the 1st quadrant.

Point P_2 in the 2nd quadrant is the reflection of P_1 in the y-axis.

Point P_3 in the 3rd quadrant is the reflection of P_2 in the x-axis.

Point P_4 in the 4th quadrant is the reflection of P_1 in the x-axis.

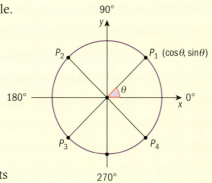

- What are the coordinates of points P_2, P_3 and P_4?
- What conclusions can you make about the signs of the sine and cosine ratios in the different quadrants?

Exploration 2

On graph paper, draw a quarter circle with radius 10 cm. Use your protractor to draw angles of 0°, 15°, 30°, 45°, 60°, 75° and 90° as shown.

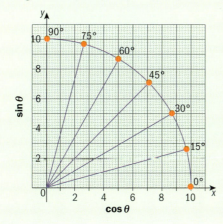

▶ Continued on next page

E7.1 Slices of pi

1 For each angle, read the values of $\cos\theta$ and $\sin\theta$ from the graph.

2 Copy and complete this table of values.

θ in degrees	θ in radians using π	$\sin\theta$	$\cos\theta$
0	0	0	1
15	$\frac{\pi}{12}$		
30			
...			
90			

3 Compare your values with the calculator values for $\sin 0°$, $\sin 15°$, $\sin 30°$, etc. Explain why they may be different.

4 Predict the values of sine and cosine for angles in the 2nd quadrant: $\theta = 105°, 120°, 135°, 150°, 165°$ and $180°$. Check with your calculator.

5 Repeat step 4 for the 3rd quadrant: $\theta = 195°, 210°, ..., 270°$ and the 4th quadrant: $\theta = 285°, 300°, ..., 360°$.

Reflect and discuss 2

- Use this unit circle to explain why:

$\sin\frac{\pi}{6} = \sin\frac{5\pi}{6}$

$\sin\frac{7\pi}{6} = -\sin\frac{\pi}{6}$

$\sin\frac{11\pi}{6} = -\sin\frac{\pi}{6}$

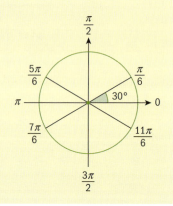

- Do your results confirm the conclusions you made in Exploration 2? Explain.

The trig functions secant, cosecant and cotangent can also be found on the unit circle. Pick a point P in the first quadrant and extend the line segment OP up and to the right, far enough for it to intersect the line $y = 1$ which is labelled RU here. Then draw the tangent line $x = 1$ from point S, upward to meet the line segment OU and label this point T.

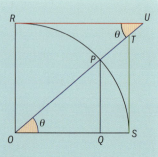

Then OT represents $\sec\theta$, OU is $\csc\theta$ and RU is $\cot\theta$.

Example 2

Write down these values without using a calculator.

a $\cos\dfrac{2\pi}{3}$ **b** $\sin\dfrac{\pi}{2}$ **c** $\tan\pi$

a $\dfrac{2\pi}{3} = 120°$ — Convert the radian measure to degrees so you can draw it easily.

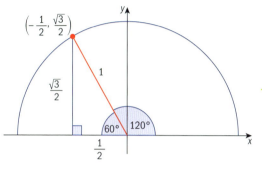

Draw the angle in the correct quadrant. Make a right-angled triangle with the x-axis and label the lengths of the sides.

$\cos\dfrac{2\pi}{3} = \cos 120° = -\dfrac{1}{2}$ — $\cos A = \dfrac{\text{adjacent}}{\text{hypotenuse}}$, or the x-coordinate.

b $\dfrac{\pi}{2} = 90°$ — Convert the radian measure to degrees.

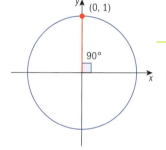

Draw the angle on the unit circle.

A point on the unit circle has coordinates $(\cos\theta, \sin\theta)$. So the value of $\sin\theta$ is the y-coordinate on the unit circle.

$\sin\dfrac{\pi}{2} = \sin 90° = 1$

c $\pi = 180°$ — Convert the radian measure to degrees.

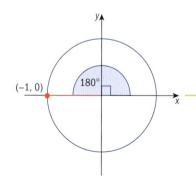

Draw the angle on the unit circle.

$\tan\pi = \dfrac{0}{-1} = 0$ — On the unit circle, $\tan\theta = \dfrac{y}{x}$

Reflect and discuss 3

- When you draw angles in the unit circle, why is the hypotenuse value always positive?
- If you drew a circle with radius = 2 for Example 2 part **a**, what would the other sides of the triangle be? What would the *x*-coordinate be? How does that compare to the value calculated in Example 2 part **a**?

To find trigonometric ratios of angles in the four quadrants, you can use:

- the special triangles and SOHCAHTOA

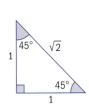 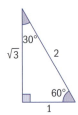

- similar special triangles where the hypotenuse is equal to one unit. This corresponds to the unit circle.

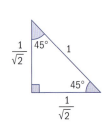 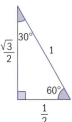

Practice 3

1 Write down these values.

  a $\sin \frac{\pi}{3}$ **b** $\cos \frac{\pi}{4}$ **c** $\tan \frac{2\pi}{3}$ **d** $\cos \frac{5\pi}{6}$ **e** $\cos \frac{3\pi}{2}$

  f $\tan 2\pi$ **g** $\sin \frac{5\pi}{4}$ **h** $\cos \frac{11\pi}{6}$ **i** $\tan \frac{3\pi}{4}$

Problem solving

2 Find two angles that have the given trigonometric ratio, $0 < A \leq 2\pi$.

  a $\tan A = -1$ **b** $\cos A = \frac{1}{2}$ **c** $\sin A = -\frac{\sqrt{2}}{2}$

  d $\cos A = 0$ **e** $\sin A = \frac{\sqrt{3}}{2}$

> If the domain is given in radians, give the answer in radians.

3 **i** Write down the coordinate values for each point.

  a $\left(\cos \frac{\pi}{6}, \sin \frac{\pi}{6}\right)$ **b** $\left(\cos \frac{2\pi}{3}, \sin \frac{2\pi}{3}\right)$

  c $\left(\cos \frac{5\pi}{4}, \sin \frac{5\pi}{4}\right)$ **d** $\left(\cos \frac{11\pi}{6}, \sin \frac{11\pi}{6}\right)$

  ii The unit circle has equation $x^2 + y^2 = 1$. Show that these four points are on the unit circle.

GEOMETRY AND TRIGONOMETRY

Reflect and discuss 4

- Can an angle measure more than 2π? Can an angle measure less than 0?
- How can you find the sine and cosine values of angles that measure less than 0 or more than 2π?

> To answer the questions in Example 3, you could also use the results from the unit circle you drew in Practice 3, question **3**.

Example 3

Write down these values without using a calculator.

a $\cos\left(\dfrac{11\pi}{4}\right)$ **b** $\sin\left(\dfrac{11\pi}{2}\right)$ **c** $\tan\left(-\dfrac{5\pi}{6}\right)$

a $\cos\left(\dfrac{11\pi}{4}\right) = \cos 495°$ — Convert the radian measure to degrees.

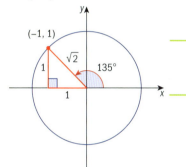

One rotation is 360°, so 495° is one rotation plus 135° more.

Draw an angle of 135°. Make a right-angled triangle with the x-axis and label the angles. Label the lengths of the sides in the special triangle with angles 90°, 45°, 45°.

$\cos\left(\dfrac{11\pi}{4}\right) = \cos 495° = -\dfrac{1}{\sqrt{2}}\left(\text{or } -\dfrac{\sqrt{2}}{2}\right)$ — $\cos A = \dfrac{\text{adjacent}}{\text{hypotenuse}}$

b $\sin\left(\dfrac{11\pi}{2}\right) = \sin 990°$

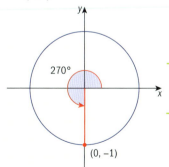

Two complete rotations is 720°. So 990° is 270° more. Draw an angle of 270°.

The value of $\sin\theta$ is the y-coordinate on the unit circle.

$\sin\left(\dfrac{11\pi}{2}\right) = \sin 990° = -1$

▶ Continued on next page

c $\tan\left(-\dfrac{5\pi}{6}\right) = \tan(-150°)$

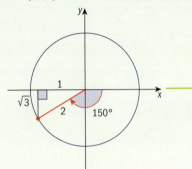

Negative angles start at zero degrees and rotate clockwise. Draw 150° clockwise from zero. Make a right-angled triangle with the x-axis. This is one of the special triangles.

$\tan\left(-\dfrac{5\pi}{6}\right) = \tan(-150°) = \dfrac{-1}{-\sqrt{3}} = \dfrac{\sqrt{3}}{3}$

Practice 4

1 Write down these values.

    a $\sin\left(\dfrac{8\pi}{3}\right)$ **b** $\cos\left(\dfrac{9\pi}{2}\right)$ **c** $\tan\left(\dfrac{7\pi}{6}\right)$

    d $\cos\left(-\dfrac{3\pi}{4}\right)$ **e** $\tan(-3\pi)$ **f** $\sin\left(-\dfrac{\pi}{6}\right)$

    g $\tan\left(\dfrac{13\pi}{4}\right)$ **h** $\sin\left(-\dfrac{7\pi}{4}\right)$ **i** $\cos\left(-\dfrac{\pi}{3}\right)$

2 Write down a negative angle (in radians) that has the given trigonometric ratio.

    a $\tan A = 1$ **b** $\cos A = -\dfrac{1}{2}$ **c** $\sin A = -1$

    d $\cos A = 0$ **e** $\sin A = -\dfrac{\sqrt{3}}{2}$

C Graphs of sinusoidal functions

- Which parameter affects the horizontal translation of a sinusoidal function?
- What is the difference between phase shift and horizontal shift?

Exploration 3

1 Graph the functions $f(x) = \sin(x)$ and $g(x) = \cos\left(x - \dfrac{\pi}{2}\right)$ on the same set of axes and write down the relationship between these two functions.

2 Describe the transformation on the graph of f that gives the graph of g.

The graph of $y = \cos\left(x - \dfrac{\pi}{2}\right)$ is a horizontal translation of the graph $y = \cos x$ by $\dfrac{\pi}{2}$ units in the positive x-direction.

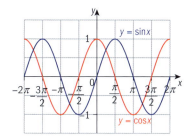 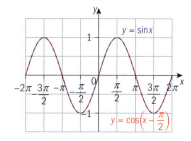

The graphs of $y = a \cos bx$ and $y = a \sin bx$ have amplitude a and frequency b. So the graph of $y = \cos\left(x - \dfrac{\pi}{2}\right)$ has amplitude 1, frequency $b = 1$, and is translated horizontally by $\dfrac{\pi}{2}$ units in the positive x-direction. For sinusoidal functions, when the parameter $b = 1$, the horizontal translation is called the **phase shift**.

The four parameters of a sinusoidal function $y = a\sin(bx - c) + d$ or $y = a\cos(bx - c) + d$ are interpreted graphically as:

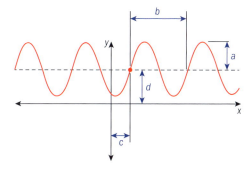

Exploration 4

1. Plot the graphs of $f(x) = \sin 2x$ and $g(x) = \sin\left(2x - \dfrac{\pi}{2}\right)$. Write down the horizontal translation on the graph of $f(x)$ that gives the graph of $g(x)$. Rewrite g in the form $g(x) = \sin 2(x - \square)$.

2. Plot the graphs of $f(x) = \cos \dfrac{x}{2}$ and $g(x) = \cos\left(\dfrac{x}{2} - \dfrac{\pi}{2}\right)$. Write down the horizontal translation on the graph of $f(x)$ that gives the graph of $g(x)$. Rewrite g in the form $g(x) = \cos \dfrac{1}{2}(x + \square)$.

3. Using your results from steps **1** and **2**:

   a convert $y = a \sin(bx - c) + d$ into an equivalent function of the form $y = a \sin b(x - \square) + d$.

   b convert $y = a \cos(bx - c) + d$ into an equivalent function of the form $y = a \cos b(x - \square) + d$.

The values of b and c both influence the horizontal translation of graphs of sinusoidal functions.

> For sinusoidal functions $y = a\sin(bx - c) + d$ and $y = a\cos(bx - c) + d$:
> - When $b = 1$, the parameter c is called the **phase shift**.
>   When $c > 0$, the shift is in the positive x-direction, to the right.
>   When $c < 0$, the shift is in the negative x-direction, to the left.
> - The **horizontal shift** is the actual horizontal translation of the graph, equal to $\frac{c}{b}$.
>   When $\frac{c}{b} > 0$, the shift is in the positive x-direction, to the right.
>   When $\frac{c}{b} < 0$, the shift is in the negative x-direction, to the left.

Example 4

State the amplitude, frequency, period, horizontal shift and vertical translation of $y = -2\sin(2x - \pi) - 4$. Draw the graph of the function using a window from 0 to 2π.

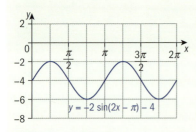

amplitude is $|a| = 2$ ———— In $y = a\sin(bx - c) + d$:
$|a|$ is the amplitude
frequency is $b = 2$ ———— b is the frequency

period is $\frac{2\pi}{b} = \frac{2\pi}{2} = \pi$ ———— $\frac{2\pi}{b}$ is the period.

horizontal shift is $\frac{c}{b} = \frac{\pi}{2}$ ———— $\frac{c}{b}$ is the horizontal shift.

vertical translation is $d = -4$ ———— d is the vertical translation.

Example 5

Without using a GDC, draw the graph of the function $y = \sin x$, from 0 to 2π. On the same set of axes, use transformations to draw the graph of $y = 1.5\sin(2x - \pi) - 2$.

amplitude = 1.5 ———— Find the amplitude, frequency, horizontal shift and vertical translation.
frequency = 2
horizontal shift = $\frac{\pi}{2}$
vertical translation = -2

▶ Continued on next page

GEOMETRY AND TRIGONOMETRY

$y = 1.5\sin(2x - \pi) - 2 = 1.5\sin 2\left(x - \frac{\pi}{2}\right) - 2$ ——— Rearrange so that x is 'on its own'. Work outward from x.

Step 1: Compare $y = \sin x$ with $y = \sin\left(x - \frac{\pi}{2}\right)$:

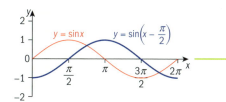

The graph of $y = \sin\left(x - \frac{\pi}{2}\right)$ is a horizontal translation of $\frac{\pi}{2}$ units in the positive x-direction of the original function $y = \sin x$.

Step 2: Compare $y = \sin\left(x - \frac{\pi}{2}\right)$ with $y = \sin 2\left(x - \frac{\pi}{2}\right)$:

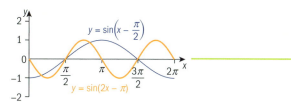

The graph of $y = \sin(2x - \pi)$ is a horizontal dilation of $y = \sin\left(x - \frac{\pi}{2}\right)$, scale factor $\frac{1}{2}$.
The frequency tells you the number of complete sine curve cycles in the period 2π.

Step 3: Compare $y = \sin 2\left(x - \frac{\pi}{2}\right)$ with $y = 1.5\sin 2\left(x - \frac{\pi}{2}\right)$:

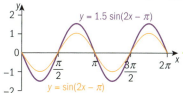

Redraw with amplitude 1.5.

Step 4: Compare $y = 1.5\sin 2\left(x - \frac{\pi}{2}\right)$ with $y = 1.5\sin 2\left(x - \frac{\pi}{2}\right) - 2$:

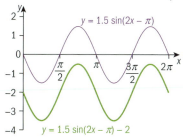

Redraw, translated 2 units down.

ATL Practice 5

1 State the amplitude, frequency, period, phase or horizontal shift, and vertical translation of:

   a $y = 3\cos 4x$

   b $y = \sin(3x - \pi) + 2$

   c $y = 0.5\sin\left(2x - \frac{\pi}{4}\right) + 7$

   d $y = 6\cos(3x + \pi) - 3$

   e $y = -\sin 3\left(x - \frac{\pi}{3}\right) + 2$

E7.1 Slices of pi   137

2 Without using a GDC, draw the graph of each function:

a $y = \cos\left(x - \frac{\pi}{2}\right) + 3$, from 0 to 2π

b $y = \cos(4x - 2\pi) - 1$, from 0 to π

c $y = 3\sin(0.5x) - 2$, from 0 to 6π

d $y = 0.5\sin(2x - \pi) - 2$, from 0 to 2π

Problem solving

3 The function $y = a\sin(bx - c) + d$ has amplitude = 2, period = π, horizontal shift = $\frac{\pi}{4}$ to the right, and vertical shift = -1. Write the equation of the function.

4 For each graph, find the equation of the sinusoidal functions in two equivalent forms: **i** as a transformation of $f(x) = \sin x$, and **ii** as a transformation of $f(x) = \cos x$. Verify that your two answers are equivalent using technology.

a
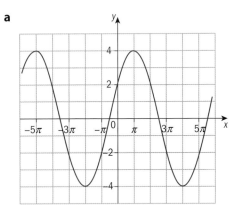

b
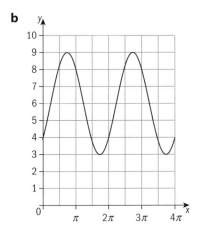

c
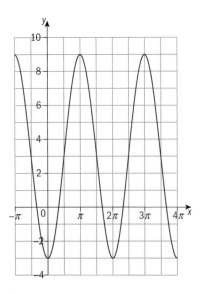

d
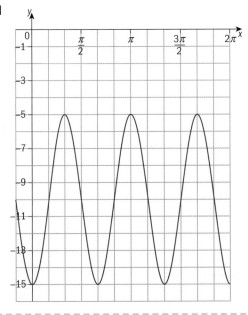

Reflect and discuss 5

- In Practice 5 question **4**, for which graphs was it easier to find the functions in terms of sine, and in which in terms of cosine? Explain.

Some periodic natural phenomena, such as tides, can be modelled using sinusoidal functions.

Example 6

In a seaside town, the depth d (in meters) of the water over the course of a day can be modelled by the sinusoidal function $d(t) = 1.6\sin(0.5t + 0.47) + 1.93$, where t is the number of hours after midnight.

a Draw the graph of the sinusoidal function.

b State the values of the amplitude, period, horizontal shift and vertical shift.

c Explain what the values you found in **b** represent.

a

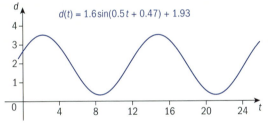

Draw the graph for 24 hours, for values of t from 0 to 24.

b amplitude = 1.6

Find the values from the function.

$$\text{period} = \frac{2\pi}{0.5} = 4\pi$$

$$\text{horizontal shift} = \frac{0.47}{0.5} = 0.94$$

vertical shift = 1.93

c The amplitude is the maximum rise and fall in depth over 24 hours.

Explain what these values mean in the context of the tides.

The period is 4π, or about 24 hours, and is the time taken for a full cycle of high and low tide.

The vertical shift is the mean depth over the 24 hours.

The horizontal shift means at midnight the water depth was already above the mean value.

Practice 6

1 A buoy bobs up and down in the water with the waves. The height h of the buoy in meters at time t (in seconds) can be modelled by the function $h(t) = 1.5 \sin\left(\dfrac{\pi}{5}t + \dfrac{\pi}{2}\right)$.

  a Draw the graph of the function.

  b State the values of the amplitude, period, horizontal shift and vertical shift. Explain what these represent in the context of the problem.

  c Use the model to find the height of the buoy after:

  i 2 seconds

  ii 5 seconds.

Problem solving

2 The gravitational force of the sun and moon cause changes in the heights of tides which can be modelled by the function $h(t) = 1.3 \sin\left(\dfrac{\pi}{6}(t+3)\right) + 0.67$, where t is the number of hours since midnight and h is the height in meters.

  a Draw a graph of the function.

  b State the amplitude, period, vertical and horizontal shifts, and interpret their values in this context.

  c Find the height of the tide at midnight.

  d Explain why predicting when high and low tides will occur is so important.

3 The height h in meters of a particular car on a giant Ferris wheel t minutes after the ride begins can be modelled by the function $h(t) = 86 \sin\left(\dfrac{2\pi}{30}(t - 7.5)\right) + 94$. The diameter of the Ferris wheel is 172 m.

  a Draw the graph of the function.

  b State the amplitude, period, vertical and horizontal shifts, and interpret their values in this context.

  c Find the height of the car after 10 minutes.

4 Another amusement park ride can also be modelled using a sinusoidal function. The ride starts 7 m above the ground. A car travels horizontally for 10 seconds, and then climbs and descends along a track. The entire ride lasts 32 seconds. The height h of the car in meters after t seconds can be modelled by the function $h(t) = 3\cos\left(\dfrac{\pi}{4}t - 3\pi\right) + 7$.

  a Draw a graph of this function for t from 10 to 32 seconds.

  b Interpret the meaning of each parameter in the function in this context.

5 The table shows the average monthly temperatures in Fahrenheit for a US city.

Jan	Feb	Mar	Apr	May	Jun	Jul	Aug	Sep	Oct	Nov	Dec
32°	35°	44°	53°	63°	73°	77°	76°	69°	57°	47°	37°

GEOMETRY AND TRIGONOMETRY

a Draw a graph of this data, using 1 to 12 on the *x*-axis to represent the months, and the temperatures on the *y*-axis.

b Use the data to find the amplitude, period, vertical and horizontal shifts, and interpret their meaning in this context.

c Write the sinusoidal function that models this problem.

D Order of transformations

- Does the order matter when you perform transformations on sinusoidal functions?
- How do we define 'where' and 'when'?

Reflect and discuss 6

- Write down the transformations used, in order, to draw the graph of $f(x) = 1.5\sin(2x - \pi) - 2$ in Example 5.
- Now draw the graph of $f(x) = 1.5\sin(2x - \pi) - 2$ by changing the order in which the transformations are used. Try a few different orders of transformations. Do you always get the same graph as in Example 5? Explain.

Exploration 5

1 a Describe the transformations on the graph of $y = \cos x$ that give the graph of $f(x) = 4\cos\left(x + \dfrac{\pi}{4}\right)$.

   b Describe the transformations on the graph of $y = \sin x$ that give the graph of $f(x) = \sin(x - \pi) - 3$.

   c Describe the transformations on the graph of $y = \cos x$ that give the graph of $f(x) = \cos\left(3x - \dfrac{\pi}{2}\right)$.

2 a Graph each function in step **1** by applying the two transformations in a different order.

   b State whether or not there is a difference in your two graphs.

3 Repeat steps **1** and **2** for these functions:

   $f(x) = -4\cos\left(x + \dfrac{\pi}{4}\right)$ $f(x) = 4\cos\left(-x + \dfrac{\pi}{4}\right)$ $f(x) = -4\cos\left(-x + \dfrac{\pi}{4}\right)$

4 Summarize your findings by indicating the order in which transformations need to be performed on the graph of $y = \sin x$ to give the graph of $f(x) = a\sin(bx - c) + d$.

Summary

On the unit circle, the coordinates of a point $P(x, y)$ on the terminal side of angle θ are $(\cos\theta, \sin\theta)$.

$\tan\theta = \dfrac{y}{x}$

A **radian** is a unit for measuring angles.

1 radian is the angle at the center of an arc with length equal to the radius of the circle.

$\theta = \dfrac{s}{r}$ and $s = r\theta$

$360° = 2\pi$ radians

$1° = \dfrac{\pi}{180°}$ radians

Angle in radians = Angle in degrees $\times \dfrac{\pi}{180°}$

For sinusoidal functions $y = a\sin(bx - c) + d$ and $y = a\cos(bx - c) + d$:

- When $b = 1$, the parameter c is called the **phase shift**.

  When $c > 0$, the shift is in the positive x-direction, to the right.

  When $c < 0$, the shift is in the negative x-direction, to the left.

- The **horizontal shift** is the actual horizontal translation of the graph, equal to $\dfrac{c}{b}$.

  When $\dfrac{c}{b} > 0$, the shift is in the positive x-direction, to the right.

  When $\dfrac{c}{b} < 0$, the shift is in the negative x-direction, to the left.

Mixed practice

1 Convert these angles from degrees to radians, in terms of π.

   a 12° **b** 240° **c** 288° **d** 480° **e** 125°

   f −225° **g** 810° **h** 336° **i** −80° **j** −700°

2 Convert these angles from radians to degrees.

   a $\dfrac{\pi}{12}$ **b** $\dfrac{2\pi}{5}$ **c** $\dfrac{3\pi}{2}$ **d** $\dfrac{8\pi}{15}$ **e** $-\dfrac{5\pi}{12}$

   f $\dfrac{7\pi}{5}$ **g** $-\dfrac{11\pi}{20}$ **h** $-\dfrac{\pi}{4}$ **i** $\dfrac{23\pi}{30}$ **j** $-\dfrac{17\pi}{60}$

3 Without using a calculator, **write down**:

   a $\sin\left(\dfrac{\pi}{2}\right)$ **b** $\cos\left(\dfrac{\pi}{3}\right)$ **c** $\sin\left(\dfrac{3\pi}{4}\right)$ **d** $\tan\left(-\dfrac{\pi}{6}\right)$

   e $\cos\left(\dfrac{7\pi}{6}\right)$ **f** $\tan\left(\dfrac{13\pi}{6}\right)$ **g** $\sin\left(\dfrac{3\pi}{2}\right)$ **h** $\cos\left(\dfrac{7\pi}{4}\right)$

   i $\tan(4\pi)$ **j** $\sin\left(-\dfrac{5\pi}{6}\right)$

Problem solving

4 One car on a Ferris wheel rotates through an angle of 120° in 11 seconds. During that time, the car travels a distance of 10.47 m. **Find** the radius of the Ferris wheel to the nearest cm.

5 State the amplitude (a), frequency (f), period (p), phase shift or horizontal shift (hs), and vertical shift (vs) of:

   a $y = 2\sin(\pi x)$ **b** $y = 5\cos(4x + 2\pi)$

   c $y = 0.2\sin 6\left(x + \dfrac{\pi}{3}\right) - 1$ **d** $y = -3\cos\left(\dfrac{\pi}{2}x - \dfrac{3\pi}{2}\right) + 8$

   e $y = -\sin(6x - 12) - 1$

6 Without using a GDC, **draw** the graphs of:

   a $y = 3\cos(2x) - 3$, from 0 to 2π

   b $y = \sin\left(x - \dfrac{\pi}{2}\right) - 4$, from 0 to 2π

   c $y = 2\cos\left(0.5x + \dfrac{\pi}{4}\right) - 3$, from 0 to 2π

Problem solving

7 Find the equation that defines each function.

   a Points shown on graph: $\left(-\dfrac{\pi}{12}, 0\right)$, $\left(\dfrac{7\pi}{12}, 0\right)$, $\left(\dfrac{23\pi}{12}, 0\right)$, $\left(\dfrac{31\pi}{12}, 0\right)$

b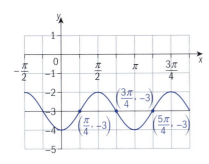

8 The diameter of a car tire is given in inches.

  a Explain why you need to know the diameter of the tire to calculate the distance a car wheel travels.

  b Find the distance travelled, in whole meters, in 1050 revolutions by a:

  i 16″ tire   **ii** 17″ tire   **iii** 20″ tire

  > Convert inches to cm to obtain metric distances. 1 inch = 2.54 cm.

  c Find the radian measure of 0.6 revolutions of the tire.

  d Hence, find the distance travelled, in meters, in 0.6 revolutions by a:

  i 15″ tire   **ii** 18″ tire   **iii** 22″ tire

  e Find the exact distance travelled, in meters, in 2147.3 revolutions by a:

  i 14″ tire   **ii** 15″ tire   **iii** 20″ tire

  f Explain how the radian measure of the number of revolutions helps you find the distance travelled.

Problem solving

9 The seal population on an iceberg increases and decreases annually. The population can be modelled using a trigonometric function. In early January, the population is 3000 seals. In early July, after all the baby seals are born, there are 5000 seals. The population reduces to 3000 seals again by the following January.

  a Draw a graph of the trigonometric function to represent the population of seals at any time of the year.

  b Write down the trigonometric equation that represents this function.

  c Find the seal population at the end of May.

  d Find the seal population at the beginning of November.

  e Find in which month(s) there are 4500 seals.

Review in context

1 For science class, a sundial with a diameter of 12 m is going to be made on a school playground. A surveyor's wheel with radius 20 cm is used to measure the length of arcs around the edge of the circle of the sundial. **Find** the central angle (in radians) of the arc when the surveyor's wheel makes:

  a 2 rotations   **b** 5 rotations
  c 3.5 rotations   **d** 6.25 rotations

2 The Romans were the first to use catapults as a means of increasing the distance from which they could effectively neutralize enemies. A catapult launches a stone into the air whose trajectory (until it lands on the ground) can be modelled by the function: $t(x) = a \sin(bx - c)$, where $c = 0$.

  a i Write down the simplified function $t(x)$.

  ii State what the variable x represents.

  iii State what the parameter a represents. **Suggest** how the catapult could influence its value.

  iv State what the parameter b represents. **Explain** how it determines the stone's point of impact.

  v The stone starts travelling when it is launched and stops as soon as it hits the ground. **State** the domain and range of $t(x)$ in terms of a and b.

  b State what the parameter c represents. **Suggest** what this says about the catapult's position.

c Keeping in mind the trajectory of the stone as explained in part **a iv**, **state** the domain and range of $t(x)$ in terms of a, b and c.

d Suggest why the vertical displacement of the catapult's position is ignored in this question.

> Does the sine curve accurately model a trajectory if the starting point of the stone is 10 m above the ground?

e A first stone is launched, and its trajectory is shown in this graph:

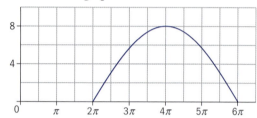

i State the domain and range of the stone's trajectory.

ii Using your GDC or otherwise, **find** the values of a, b and c and **write down** the equation of this stone's trajectory.

f A second stone's trajectory is modelled by the function $f(x) = 5\sin(0.5x - \pi)$.

i State the coordinates of the starting position of the stone at the moment it is launched.

ii State the coordinates of the position where it lands on the ground.

iii State the stone's maximum height, and where it reaches this height.

iv Hence, state the domain and range of this stone's actual trajectory.

Astronomers can increase both resolution and sensitivity of new telescopes by either building very large individual telescopes or by combining the signals from an array of ever-greater numbers of smaller telescopes spread over long distances. This is called an interferometer.

The Square Kilometre Array (SKA) is the next-generation interferometer for radio astronomy, due to start operating in the early 2020s. It will consist of hundreds of radio telescopes spread over hundreds of kilometers in southern Africa and Australia. With an instrument of this size, astronomers will have access to a radio telescope on a scale that is orders of magnitude better than any current facility, allowing them to revolutionize our understanding of the universe.

Reflect and discuss

How have you explored the statement of inquiry? Give specific examples.

Statement of Inquiry:

Generalizing and applying relationships between measurements in space can help define 'where' and 'when'.

E8.1 Making it all add up

Global context: Scientific and technical innovation

Objectives
- Finding the sum of an arithmetic series
- Finding the sum of a finite geometric series
- Finding the sum of an infinite geometric series, where appropriate

Inquiry questions

F
- What is a series?
- How can rewriting a series make it easier to sum?

C
- What are the similarities and differences between real-life patterns that require arithmetic and geometric series?

D
- Can you evaluate a series that goes on forever?
- What are the risks of making generalizations?

ATL Communication

Understand and use mathematical notation

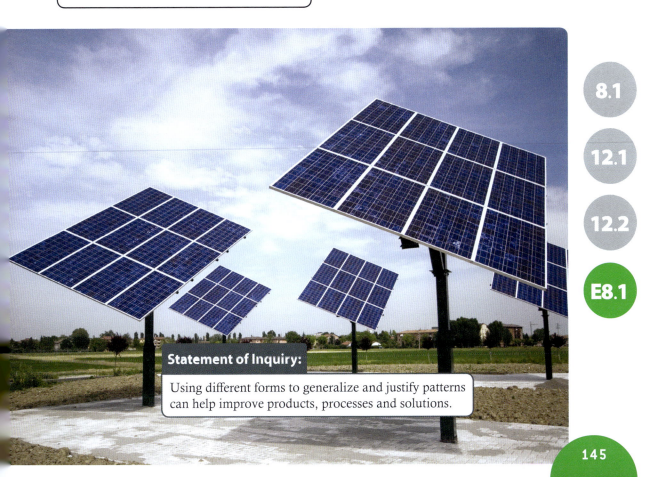

Statement of Inquiry:

Using different forms to generalize and justify patterns can help improve products, processes and solutions.

You should already know how to:

• work with arithmetic sequences	1 Find the nth term and the 20th term of each arithmetic sequence. **a** 5, 9, 13, 17, … **b** 25, 22.5, 20, 17.5, … **c** 42.7, 43.9, 45.1, …
• work with geometric sequences	2 Find the next two terms and the nth term of these geometric sequences. **a** 15, 30, 60, 120, 240, … **b** 100, 25, 6.25, … **c** 162, −54, 18, −2, …
• generate the triangular numbers	3 Write down: **a** the first 5 triangular numbers **b** the general formula for the triangular numbers.
• generate a sequence from an explicit formula	4 Generate the first five terms of each sequence: **a** $u_n = 3n - 1, n \in \mathbb{Z}^+, n \geq 1$ **b** $u_n = 3 \times 2^n, n \in \mathbb{Z}^+, n \geq 1$

F Series

- What is a series?
- How can rewriting a series make it easier to sum?

A **sequence** is an ordered list of numbers. A **series** is the sum of the terms of a sequence. The sequence and the series below have a finite number of terms.

1, 7, 4, 11, 2, −4 is a sequence, even though there is no pattern in it.

1 + 7 + 4 + 11 + 2 + −4 is a series, because it is the sum of the terms of a sequence. In this case, the sum is 21.

> In MYP Mathematics Standard 8.1 you saw that sequences do not have to follow a pattern.

> An **arithmetic series** is the sum of the terms of an arithmetic sequence.
> A **geometric series** is the sum of the terms of a geometric sequence.

Exploration 1

1 Find the value of each series.
   a 1 + 3 + 5 + 7 + 9 + 11
   b 7 + 7 + 7 + 7 + 7
   c 1 + 4 + 9 + 16 + 25 + 36
   d 7 + 14 + 28 + 56 + 112 + 224

▶ Continued on next page

ALGEBRA

2 Look at the patterns in the series in step **1**.

   a Which are arithmetic series? Which are geometric series?

   b One is neither geometric nor arithmetic. Suggest a suitable description of it.

3 The terms of a series are given by the formula $u_n = 4n - 2$. Find the value of the series $u_1 + u_2 + u_3 + u_4 + u_5$.

4 The terms of a series are given by the formula $u_n = 3 \times 4^n$. Find the value of the series $u_1 + u_2 + u_3 + u_4$.

> S_n means the sum of the first n terms of a series.
>
> For example, $S_4 = u_1 + u_2 + u_3 + u_4$.

ATL

5 Given that $u_n = n^2 - 2n$, find the value of S_6, the sum of the first 6 terms of the series.

Some types of series have special formulae that help you find the sum easily. In Practice 6 of MYP Standard 8.1 you saw that the general formula for the triangular numbers is:

$$S_n = \frac{n(n+1)}{2}$$

The nth triangular number is the number of dots needed to make a triangular pattern with n rows:

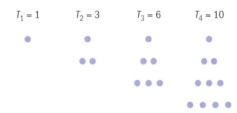

Since each triangle is made of rows which get longer by one every time, the number of dots in the nth triangle can be found by adding $1 + 2 + 3 + 4 + \ldots + n$, so the nth triangular number is the same as the sum of the first n integers.

Using the formula, the sum of the first five positive integers is $S_5 = \frac{5(5+1)}{2} = \frac{5 \times 6}{2} = 15$, which you can easily check by adding $1 + 2 + 3 + 4 + 5$.

Example 1

ATL

Find the total of the series $47 + 48 + 49 + \ldots + 82$.

$47 + 48 + 49 + \ldots + 82$

$= S_{82} - S_{46}$

$= \dfrac{82 \times 83}{2} - \dfrac{46 \times 47}{2} = 2322$

Sum the series $1 + 2 + \ldots + 82$, and from it subtract the sum of the series $1 + 2 + \ldots + 46$:

$S_{82} = 1 + 2 + 3 + 4 + \ldots + 46 + 47 + 48 + 49 + \ldots + 82$

$-S_{46} = 1 + 2 + 3 + 4 + \ldots + 46$

$S_{82} - S_{46} = 47 + 48 + 49 + \ldots + 82$

E8.1 Making it all add up

ATL The sum S of the first n consecutive positive integers is $S_n = \frac{n(n+1)}{2}$.
You can use $S_n = \frac{n(n+1)}{2}$ to find the total of any series of consecutive integers.

You can also use $S_n = \frac{n(n+1)}{2}$ to find the sum of any series of integers with a common factor, as shown in Example 2.

Example 2

Find the value of the series $5 + 10 + 15 + \ldots + 725$.

$5 + 10 + 15 + \ldots + 725$ — The common factor is 5.

$= 5(1 + 2 + 3 + \ldots + 145)$

$= 5 \times \frac{145 \times 146}{2}$

$= 52\,925$

Practice 1

1 Find the total of the series $1 + 2 + 3 + 4 + \ldots + 30$.
2 Find the value of $25 + 26 + 27 + \ldots + 53$.
3 Find the value of $4 + 8 + 12 + 16 + \ldots + 128$.
4 Find the value of $180 + 182 + 184 + 186 + 188 + 190 + \ldots + 324$.

In **3** and **4**, take out a common factor.

Problem solving

5 **a** Find the value of the series $2 + 4 + 6 + 8 + \ldots + 60$.
  b Hence find the value of the series $1 + 3 + 5 + 7 + 9 + \ldots + 59$.
6 **a** Find the value of $1 + 3 + 5 + 7 + 9 + \ldots + 101$.
  b Find the value of $-2 + -4 + -6 + -8 + \ldots + -100$.
  c Hence find the value of $1 - 2 + 3 - 4 + \ldots - 100 + 101$.
7 The terms of a series are given by the formula $u_n = 3n + 7$.
  Find the value of S_5.
8 The sum of the first n squared numbers $1 + 4 + 9 + \ldots + n^2 = \frac{n(n+1)(2n+1)}{6}$.
  a Use this formula to confirm your answer to $1 + 4 + 9 + 16 + 25 + 36$ in Exploration 1 step **1c**.
  b Find the value of $1 + 4 + 9 + 16 + \ldots + 100$.
  c Find the value of $1 + 4 + 9 + \ldots + 400$.
  d Hence find the value of $121 + 144 + 169 + \ldots + 400$.

Problem solving

9 Find the value of $1 - 4 + 9 - 16 + \ldots + 361$.

Exploration 2

1 a Copy and complete the calculations to find the value of:

$$40 + 44 + 48 + 52 + 56 + 60 + 64 + 68 + 72$$

$$S = 40 + 44 + 48 + 52 + 56 + 60 + 64 + 68 + 72$$
$$+ S = 72 + 68 + \_ + \_ + \_ + \_ + \_ + \_ + 40$$
$$2S = 112 + \_ + \_ + \_ + \_ + \_ + \_ + \_$$

b Hence find the value of $2S$.

c Deduce the value of S.

2 Use a similar method to find the value of:

a $57 + 63 + 69 + 75 + 81 + 87 + 93 + 99 + 105 + 111$

b $1.7 + 2.9 + 4.1 + 5.3 + 6.5 + 7.7 + 8.9 + 10.1$

c $30 + 21 + 12 + 3 + -6 + -15 + -24$

Check your answers using a calculator.

> The terms in the sum in the second line are just the reverse of the order in the first line.

Reflect and discuss 1

- Do you need to write out all the numbers in the series twice to find $2S$?
- Without writing out all the terms, explain how you could find $2S$ and then S for this series with seven terms: $5 + 9 + \ldots + 29$

This method is sometimes called Gauss's method, as it is the method that 8-year-old Karl Gauss is said to have used when his teacher asked him to calculate the sum of all the numbers from 1 to 100. Gauss was not the first to discover it, and his teacher probably already knew it – it wouldn't have made sense for him to set the task if he had to work out the total himself slowly.

Practice 2

1 Find the value of these arithmetic series:

a $23 + 30 + 37 + 44 + 51 + 58 + 65 + 72 + 79$

b $30 + 28 + 26 + 24 + 22 + 20 + 18 + 16 + 14$

c $24 + 28.4 + 32.8 + 37.2 + 41.6 + 46 + 50.4 + 54.8 + 59.2 + 63.6$

2 An arithmetic series begins $6 + 19 + \ldots$ and has 40 terms.

  a Find the 40th term.

  b Hence find the sum of the series.

3 Consider the arithmetic series $S = 15 + 19 + \ldots + 199$.

  a Write down the value of the first term and the common difference.

  b Hence find the number of terms in the series.

  c Find the sum of the series.

Problem solving

4 Find the sum of the arithmetic series $41 + 48 + \ldots + 104$.

5 An arithmetic series begins $29 + 25 + 21 + \ldots$
Find the maximum value that the series could have.

Exploration 3

1 Use the method from Exploration 2 to find the sum of the arithmetic series $u_1 + u_2 + \ldots + u_n$.

$$\begin{aligned} S_n &= u_1 + u_2 + \ldots + u_{n-1} + u_n \\ + S_n &= u_n + u_{n-1} + \ldots + u_2 + u_1 \\ \hline 2S_n &= (u_1 + u_n) + (u_2 + u_{n-1}) + \ldots + (u_{n-1} + u_2) + (u_n + u_1) \end{aligned}$$

  a Explain clearly why $(u_1 + u_n)$ and $(u_2 + u_{n-1})$ are equal in value.

  b Hence explain why $2S_n = n(u_1 + u_n)$.

  c Write down a formula for S_n in terms of n, u_1 and u_n.

2 Use the formula for the nth term $u_n = a + (n-1)d$ to write down a formula for S_n for an arithmetic series with n terms, first term a and common difference d. Give your formula in terms of a, d and n only.

> You might find it useful to use the general formula for the nth term: $u_n = a + (n-1)d$.

Reflect and discuss 2

How have you used the concept of generalization when moving from Exploration 2 to Exploration 3?

For an arithmetic series with first term $u_1 = a$ and common difference d, the formula for the sum of the first n terms is given by:

$S_n = \dfrac{n}{2}(u_1 + u_n)$, where u_n is the nth term, or $S_n = \dfrac{n}{2}(2a + (n-1)d)$

Learn both of these two formulas. Examples 3 and 4 show they are both useful for different types of problems.

Example 3

Find the value of the arithmetic series with 11 terms that begins 15 + 17 + ...

$a = 15, d = 2$ and $n = 11$

$S_n = \dfrac{n}{2}(2a+(n-1)d)$ — Use this formula when you know a, d and n.

$S_{11} = \dfrac{11}{2}(30+10\times 2)$

$= 275$

Example 4

Find the value of the arithmetic series with 9 terms: first term 22, last term 78.

$u_1 = 22, u_9 = 78, n = 9$

$S_n = \dfrac{n}{2}(u_1 + u_n)$ — Use this formula when you know the first and last terms, and n.

$S_9 = \dfrac{9}{2}(22+78)$

$= 450$

Practice 3

1 Find the total of these arithmetic series.

  a 5 + 8 + 11 + 14 + 17 + 20 + 23 + 26 + 29 + 32

  b 1.4 + 2 + 2.6 + 3.2 + 3.8 + 4.4 + 5 + 5.6 + 6.2 + 6.8 + 7.4 + 8

  c 15 + 29 + 43 + ... + 141

  d 57 + 52 + 47 + ... + −3

2 Find the sum of:

  a an arithmetic series with 15 terms, first term 20 and common difference 5

  b an arithmetic series with 18 terms, first term 6 and last term 43

  c an arithmetic series with 11 terms, first term 9 and second term 15

  d an arithmetic series with 32 terms, second term 16 and fourth term 22

  e an arithmetic series with common difference 12, first term 39 and last term 291.

3 Find the value of the arithmetic series with 12 terms, where the first term is 18 and the last term is 102.

4 Find the value of the arithmetic series that begins 17, 19.2, 21.4, ... and which has 11 terms.

Problem solving

5 Find the sum of the arithmetic series with 15 terms, where the eighth term is 42 and the eleventh term is 57.

6 An arithmetic sequence has terms $x_1 + x_2 + x_3 + \ldots + x_{55} = 726$.

Find the value of x_{28}.

Exploration 4

You have seen that you can sum an arithmetic series by first writing it backwards to find $2S$. What happens when you try to use the same approach to find the sum of a *geometric* sequence?

1 Try to use Gauss's method to find the value of the series:

$S = 1 + 3 + 9 + 27 + 81 + 243 + 729 + 2187 + 6561$

Describe the problems with the method.

2 Copy and complete the values of $3S$:

$S = 1 + 3 + 9 + 27 + 81 + 243 + 729 + 2187 + 6561$
$3S = 3 + 9 + 27 + + + + + 19683$

3 Copy and complete the subtraction:

$3S = 3 + 9 + 27 + + + + + 19683$
$- \; S = 1 + 3 + 9 + 27 + 81 + 243 + 729 + 2187 + 6561$

$2S = -1 + 0 + 0 + \ldots$

4 Hence find the value of $2S$, and deduce the value of S.

5 Explain why it was useful to multiply by 3. Suggest what you should multiply this series by, and check that your suggestion works:

$S = 2 + 8 + 32 + 128 + 512 + 2048 + 8192$

6 Explain why the pattern behind geometric sequences means that this approach will always work when trying to sum a geometric sequence.

Practice 4

1 Use the method from Exploration 4 to find the value of these geometric series.

   a $14 + 42 + 126 + 378 + 1134 + 3402 + 10\,206 + 30\,618 + 91\,854 + 275\,562$

   b $2.5 + 15 + 90 + 540 + 3240 + 19\,440$

   c $1 + 1.3 + 1.69 + 2.197 + 2.8561 + 3.71293 + 4.826809 + 6.2748517$

   d $625 - 500 + 400 - 320 + 256 - 204.8 + 163.84$

2 Find the value of these geometric series.

   a $6 + 48 + 384 + 3072 + 24\,576 + 196\,608 + 1\,572\,864$

   b $2500 + 1500 + 900 + 540 + 324 + 194.4 + 116.64$

   c $160 + 240 + 360 + 540 + 810 + 1215 + 1822.5$

3 A geometric series begins 2 + 6 + 18 + 54 + ….

  a Write down an expression for the nth term of the series.

  b Hence find the value of S_8, the sum of the first 8 terms of the series.

Exploration 5

A geometric series has first term a and common ratio r.

1. Write down expressions in terms of a and r for u_1, u_2, u_3 and u_n (the first, second, third and nth terms).

2. Let S_n be the sum of the first n terms.
   Apply the method in Exploration 4 to show that $(r - 1)S_n = a(r^n - 1)$.

3. Hence write down a formula for S_n in terms of a, r and n.

4. Use your formula to find the sums of:

   a a geometric series with first term 3, common ratio 7 and eight terms

   b a geometric series with nine terms, first term 12 and second term 18

   c a geometric series with seven terms, second term 168 and third term 504.

5. For the geometric series:

   $1000 + 800 + 640 + 512 + 409.6 + 327.68 + 262.144$

   a write down the value of the first term and the common ratio

   b show that the sum formula gives: $S_n = 1000\left(\dfrac{-0.7902848}{-0.2}\right)$

   c explain where the negative numbers have come from.

6. Show clearly that the sum formula can be rearranged to give:

   $S_n = a\left(\dfrac{1-r^n}{1-r}\right)$

   State when the fraction in this formula will have a positive numerator and a positive denominator.

7. Use either formula above to find the value of:

   a a geometric series with first term 16, seven terms in total, and a common ratio of 0.4

   b a geometric series with ten terms whose first term is 20 and whose second term is −10

   c a geometric series with twenty terms whose fourth term is 7 and seventh term is 56.

You saw in MYP Standard Mathematics 12.2 that geometric sequences behave very differently depending on whether or not $-1 < r < 1$. The same is true for geometric series, and you will explore this in the debatable section. For now, the biggest difference is that it is easier to use $S_n = a\left(\dfrac{1-r^n}{1-r}\right)$ when $-1 < r < 1$.

> For a geometric series with first term a and common ratio r:
> $$S_n = a\left(\dfrac{r^n - 1}{r - 1}\right) = a\left(\dfrac{1 - r^n}{1 - r}\right)$$
> The first version is easier to use when $r > 1$ or $r < -1$.
> The second version is easier to use when $-1 < r < 1$.

Reflect and discuss 3

- What happens if you use the formula for a geometric series with common ratio 1?
- How would you find the sum of a geometric series for which the common ratio is 1?

Example 5

Find the sum of the first 13 terms of the geometric series which begins $10 + 5 + 2.5 + \ldots$

$a = 10$, $r = \dfrac{5}{10} = 0.5$ and $n = 13$

$S_n = a\left(\dfrac{1 - r^n}{1 - r}\right)$ ——— $-1 < r < 1$

$S_{13} = 10\left(\dfrac{1 - 0.5^{13}}{1 - 0.5}\right)$

$= 19.9975585938$

$= 20.0$ (3 s.f.) ——— Round to a sensible degree of accuracy.

Practice 5

1 Choose the most suitable formula to find the value of:

   a a geometric series with first term 16, seven terms and common ratio 0.4

   b a geometric series with ten terms, first term 20 and second term -10

   c a geometric series with twenty terms, fourth term 7 and seventh term 56.

ALGEBRA

2 Find the total of each geometric series.

  a $5 + 25 + 125 + 625 + 3125 + 15\,625$

  b $7^3 + 7^4 + 7^5 + 7^6 + 7^7 + 7^8 + 7^9 + 7^{10}$

  c $96 + 144 + 216 + 324 + 486 + 729$

3 Find the value of the geometric series with ten terms, first term 24 and second term 96.

4 Find the value of the geometric series that begins $18 + 24 + \ldots$ and has eight terms.

5 Find the value of the geometric series with first term 10, common ratio 9 and last term 430 467 210.

6 A geometric series has first term 14 and third term 56.

  a Show that there are two possible values for the common ratio.

  b Hence find two possible sums of the first seven terms of the series.

7 A geometric series has first term 15 and second term −10. Find the sum of the first eight terms.

> **Tip**
>
> In **8**, if you know how to use logarithms (E7.3), they might help you to find the number of terms in this series. If not, you can find the number of terms using trial and error and your GDC.

Problem solving

8 A geometric series is formed by summing the integer powers of 2:

$S_n = 1 + 2 + 4 + \ldots$

Determine the least number of terms so that $S_n > 1\,000\,000\,000$.

C Series in real-life

- What are the similarities and differences between real-life patterns that require arithmetic and geometric series?

There are many real-life situations where problems can be solved using arithmetic or geometric series. In each worked example, consider why it is appropriate to use an arithmetic or geometric sequence.

> Arithmetic and geometric sequences in real-life contexts are explored in MYP Mathematics Standard 12.2.

Example 6

> A company launches a new music streaming service. Customers pay \$4 per month to subscribe. In the first month, 1200 customers sign up. For the rest of the year, another 500 customers join every month.
>
> **a** Explain why the monthly income forms an arithmetic sequence.
>
> **b** Find the income in the first month.
>
> **c** Assuming no customers leave the service, find the total income from the service in the first year.

a The monthly income grows by a constant amount, so this forms an arithmetic series.

> Each month the income increases by $500 \times \$4 = \$2000$.

b $1200 \times 4 = \$4800$

▶ Continued on next page

E8.1 Making it all add up   **155**

c $a = 4800$, $d = \$2000$ and $n = 12$

$S_n = \dfrac{n}{2}(2a + (n-1)d)$ ———— Use this formula when the last term is unknown.

$S_{12} = \dfrac{12}{2}(2 \times 4800 + 11 \times 2000)$

$= \$189\,600$ ———— Remember to write the units.

Example 7

A social networking site gains new members very quickly. Each month, the number of new members is 1.5 times the number in the previous month.

In the first month of 2016 the site gains 15 000 new members.

a Explain why the number of new members per month forms a geometric sequence.

b Find the number of new members in the 2nd and 3rd months of 2016.

c Find the total number of new members that will be gained in 2016 if this growth continues.

d Explain why the growth cannot continue in this way indefinitely.

a The number of new members forms a geometric sequence because it is increasing by a constant scale factor. ———— The common ratio is 1.5.

b $r = 1.5$, $a = 15\,000$

Number of new members in 2nd month = $15\,000 \times 1.5 = 22\,500$.
Number of new members in 3rd month = $22\,500 \times 1.5 = 33\,750$.

c $S_n = a\left(\dfrac{r^n - 1}{r - 1}\right)$ ———— Use this formula because $r > 1$.

$S_{12} = 15\,000\left(\dfrac{1.5^{12} - 1}{1.5 - 1}\right)$

$= 3\,862\,390.137\ldots$

$= 3\,860\,000$ (3 s.f.)

d If growth continued in this way, after 36 months there would be 22 billion new members, which is much greater than the current world population (7 billion). ———— Give an example to justify your explanation.

Examples 6 and 7 asked you to find the value of a series, given some information about a scenario — the first term and the common ratio or common difference.

The next example shows you how to form an equation and work backwards when you know the value of a series.

ALGEBRA

Example 8

Your aunt gives you 10 AUD on your first birthday, 20 AUD on your second birthday and so on, increasing by the same amount every year.

a Find the total amount given for your first 13 birthdays.

b On which birthday will you have been given 1530 AUD in total?

a $S_n = \dfrac{n}{2}(u_1 + u_n)$ ——— The series increases by a common difference, so it is arithmetic.

$S_{13} = \dfrac{13}{2}(10 + 130) = 910$ AUD

b $a = 10, d = 10, S_n = 1530$

$S_n = \dfrac{n}{2}(2a + (n-1)d)$

$1530 = \dfrac{n}{2}(20 + 10(n-1))$

$3060 = n(10n + 10)$

$10n^2 + 10n - 3060 = 0$ ——— Rearrange and solve, by factorizing, using the quadratic formula or your GDC.

$n^2 + n - 306 = 0$

$(n + 18)(n - 17) = 0$

$\Rightarrow n = 17$ or $n = -18$

Since n is a number of years, $n = -18$ does not make sense. ——— Interpret your solution in the context of the problem.

On your 17th birthday, the total amount given is 1530 AUD.

> There are over 20 officially recognized Dollar currencies. The Australian Dollar – AUD, just written using the dollar sign ($) in Australia – replaced the Australian Pound in 1966.

Practice 6

Problem solving

1 A pyramid scheme has three Level 1 investors. Each person recruits four Level 2 investors. Each Level 2 investor recruits four Level 3 investors, and so on.

   a Show that there are 48 Level 3 investors.

   b Write down a formula for the number of Level n investors.

   c Find the number of investors needed to fill eight levels.

   d Show that over 1 000 000 000 investors would be needed to fill 15 levels.

> A pyramid scheme is a fraudulent 'investment' scheme. Some members may make large profits but most will lose their money. Pyramid schemes are illegal in many countries.

E8.1 Making it all add up

2 Competitors in a beanbag race run on a straight track. Beanbags are placed at various distances from the starting line: 20 m, 24 m, 28 m and so on. The final beanbag is placed 40 m from the starting line.

In each problem, consider whether your answer makes sense in the given context. Give your answers to an appropriate degree of accuracy.

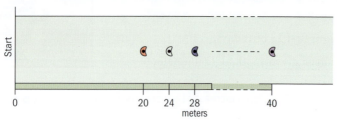

Competitors run to the first beanbag, pick it up and then return it to the starting line. They repeat this until all the beanbags have been returned to the starting line.

  a Find the number of beanbags that each competitor must collect.

  b Find the total distance that each competitor runs.

3 A ball is kicked into the air from the ground. It reaches a height of 6 m before it falls and hits the ground. It then bounces up again, several times.

Each bounce is $\frac{2}{3}$ the height of the previous bounce.

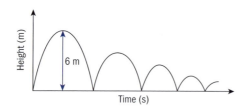

  a In the first bounce, the ball travels 12 m (6 m up and 6 m down). Write down the total distance it travels in the second bounce.

  b Find the total distance it has travelled as it returns to the ground for the sixth time.

4 European standard paper sizes are A0, A1, A2 and so on.
The area of A1 paper is half the area of A0 paper.
The area of A2 paper is half the area of A1 paper.
This relationship continues to the smallest official size, A10, which has dimensions 26 by 37 mm.

  a Find the area of a sheet of A10 paper.

  b Find the total area covered by one sheet of each size from A0 to A10.

5 A set of building blocks contains blocks of 9 different lengths.
The first block is 50.8 mm long (2 inches), the second is 76.2 mm long (3 inches), and so on, up to the last block, which is 254 mm long (10 inches).
Find the total length of all the blocks combined, giving your answer in mm.

ALGEBRA

6 In a fitness program, Pedro does 20 sit-ups on the first day, and increases this by 2 sit-ups every day.

  a Show that he does 30 sit-ups on the sixth day.

  b Find the day that he first does 50 sit-ups.

  c Determine the day when the total number of sit-ups is 1100.

7 Hannah runs 5 km in 30 minutes. Her goal is to decrease her time by 3% every time she runs. Find the total amount of time she would run for in the first 20 runs.

Reflect and discuss 4

How can you tell when a real-life situation generates either an arithmetic or geometric series?

D Series and fractions

- Can you evaluate a series that goes on forever?
- What are the risks of making generalizations?

In the arithmetic sequence $u_n = 8n - 2$, for $n \geq 1$, there is no limit to the size that n could take. No matter how large a value of n you choose, someone else can always name a larger value of n, and hence find a larger value for u_n. To describe this situation mathematically, you say: 'as n tends to infinity (gets larger and larger, without limit), u_n also tends to infinity (also gets larger and larger, without limit).'

Reflect and discuss 5

- Explain why this sequence could go on forever: 4, 6, 8, 10, 12, 14, …
- Explain what happens if you try to find the value of this series:

  $4 + 6 + 8 + 10 + 12 + 14 + \ldots$

ATL

Exploration 6

1 In all these sequences $n \geq 1$. For each, determine what happens to the terms of the sequence as n tends to infinity (becomes infinitely large).

  $a_n = 3n - 2$ $b_n = 9 - 0.5n$ $c_n = 3n^2 - 7n$ $d_n = \sqrt{2n - 1}$

  $e_n = 6 \times 4^{n-1}$ $f_n = 5 \times 0.2^n$ $g_n = 3 \times 10^{-n}$ $h_n = \dfrac{4}{(1+n)^n}$

2 For each geometric sequence in step **1**, find the value of the common ratio, r. Describe how the value of r affects the behavior of the sequence.

▶ Continued on next page

> **Tip**
>
> You could use your GDC or a spreadsheet.

E8.1 Making it all add up   159

3 a Show that the areas of the squares and rectangles in this diagram form a geometric sequence with first term 1.

b Explain why the area of the whole diagram represents the sum of a geometric series.

c Use the diagram to determine the sum of the series $1 + \frac{1}{2} + \frac{1}{4} + \frac{1}{8} + \ldots$

4 Explain how you could use the diagram below to illustrate that $3 + \frac{3}{4} + \frac{3}{16} + \frac{3}{64} + \ldots = 4$.

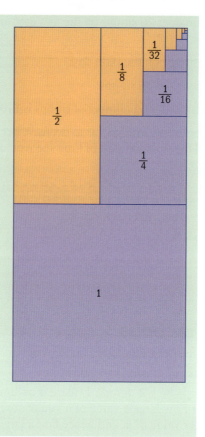

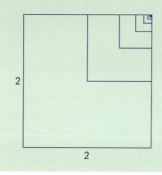

Exploration 6 shows that you can find the sum of an infinite decreasing geometric series. It is an interesting notion that you can add up an infinite number of terms even in a finite amount of time!

> In a decreasing geometric series, each term is smaller than the previous one.

Example 9

Find the value of the geometric series:
$S = 10 + 8 + 6.4 + 5.12 + \ldots$

Common ratio $= \frac{8}{10} = 0.8$

$S = 10 + 8 + 6.4 + 5.12 + \ldots$
$-0.8S = 8 + 6.4 + 5.12 + \ldots$
$0.2S = 10$

$S = \frac{10}{0.2} = 50$

Multiply by the common ratio and subtract.

You may have used a similar method to convert a recurring decimal into a fraction. Example 10 shows how to do this.

ALGEBRA

Example 10

Express $0.\dot{2}\dot{7}$ as a fraction.

Let $x = 0.\dot{2}\dot{7}$

Then $100x = 27.\dot{2}\dot{7}$ ——— Multiply by 100 so the repeating part will cancel out when you subtract.

$100x = 27.27272727\ldots$

$-\quad x = 0.27272727\ldots$

$99x = 27$

$x = \dfrac{27}{99} = \dfrac{3}{11}$

Practice 7

1 Find the value of each geometric series.

  a $S = 1000 + 500 + 250 + 125 + \ldots$

  b $S = 10 + 9 + 8.1 + 7.29 + \ldots$

  c $S = 20 + 12 + 7.2 + 4.32 + \ldots$

2 An infinite geometric series has first term 36 and second term 12. Find the sum of the series.

3 An infinite geometric series has second term 25 and third term 5. Find the sum of the series.

4 By considering each as a geometric series, express as a fraction:

  a $0.2222\ldots$ **b** $0.\dot{3}\dot{7}$ **c** $0.\dot{2}0\dot{7}$ **d** $0.\dot{1}4\dot{8} + 0.\dot{4}$

Problem solving

5 Express $x = 0.\dot{5}\dot{5}$ as a fraction by writing it as $x = 0.55 + 0.0055 + 0.000055 + \ldots$

Examining the formula for the sum of the terms of a geometric series shows why it is sometimes possible to find the sum of an infinite geometric series.

$S_n = a\left(\dfrac{r^n - 1}{r - 1}\right)$ when $r > 1$ or $r < -1$.

If $r > 1$ or $r < -1$ then as n gets very large, r^n gets very large. Therefore, the numerator of the fraction $\left(\dfrac{r^n - 1}{r - 1}\right)$ gets very large, hence S_n gets very large.

$S_n = a\left(\dfrac{1 - r^n}{1 - r}\right)$ when $-1 < r < 1$.

If $-1 < r < 1$ then as n gets very large, r^n gets closer and closer to 0, so the numerator of the fraction $\left(\dfrac{1 - r^n}{1 - r}\right)$ gets closer and closer to 1. Hence, S_n gets closer to $\dfrac{1}{1 - r}$.

E8.1 Making it all add up

> You can find the value of an infinite geometric series with first term a and common ratio r when $-1 < r < 1$.
>
> The sum to infinity is $S_\infty = \dfrac{a}{1-r}$ for $|r| < 1$.

Example 11

A geometric series has third term 6, fourth term 4. Find its sum to infinity.

$r = u_4 \div u_3 = \dfrac{4}{6} = \dfrac{2}{3}$

$|r| < 1$, so the series has a finite sum. ——— *First show that $|r| < 1$.*

$u_3 = ar^2$

so $a = u_3 \div r^2 = 13.5$ ——— *Find a.*

$S_\infty = \dfrac{a}{1-r} = \dfrac{13.5}{1 - \frac{2}{3}} = 40.5$

Practice 8

1 Determine whether or not each series has a finite sum.

   a $10 + 7 + 4.9 + 3.43 + \ldots$ **b** $10 + 7 + 4 + 1 + \ldots$

   c $2 + 4 + 8 + 16 + \ldots + 16384$ **d** $2 + 4 + 6 + 8 + \ldots + 16384$

   e $1 + 0.1 + 0.01 + 0.001 + 0.0001 + \ldots$ **f** $1 - 0.2 + 0.04 - 0.008 + \ldots$

   g $3 + 6 + 9 + 12 + \ldots$ **h** $3 + 6 + 12 + \ldots$

2 A geometric series has first term 4 and common ratio $\dfrac{1}{3}$.

   a Write down the first three terms.

   b Find the sum of the first five terms.

   c Find its sum to infinity.

> I am incapable of conceiving infinity, and yet I do not accept finity.
> – Simone de Beauvoir

3 An infinite geometric series has first term 18 and second term 15. Show that it has a finite sum, and find the sum of the series.

4 An infinite geometric series has second term 24 and fourth term 1.5.

   a Show that there are two possible values for the common ratio.

   b Show that one value of the common ratio gives the series a sum of 128, then find the other possible value of the sum to infinity.

ALGEBRA

Problem solving

5 A geometric series has first term 6 and a sum to infinity of 60. Find its common ratio.

6 A geometric series has first term x and second term $x - 3$. Its sum to infinity is 75. Find the value of x.

7 A geometric series has first term x and second term $x + 4$. Its sum to infinity is -2.25. Find the value of x.

Activity

The Koch Snowflake is a geometric pattern called a fractal. It is formed by starting with an equilateral triangle (Stage 1) and then replacing each line with four smaller lines, like this:

Objective B: Investigating patterns
ii. describe patterns as general rules consistent with findings

Make a mathematical observation about the patterns; for example, a rule linking the Stage number to the perimeter, or some conclusion about the eventual area of the shape.

Investigate the geometric properties of the Koch Snowflake by answering some or all of these questions:

Explore how the Stage number relates to:

- the number of line segments in the diagram
- the length of each of the line segments
- the total perimeter
- the number of triangles in the diagram
- the area of each triangle in the diagram
- the total area

You could investigate other fractals, such as the *Gosper Island* or *Sierpinski Triangle*, which you can search for online.

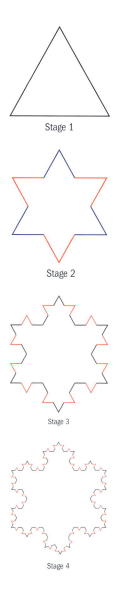

Stage 1

Stage 2

Stage 3

Stage 4

E8.1 Making it all add up

Summary

An **arithmetic series** is the sum of the terms of an arithmetic sequence.

A **geometric series** is the sum of the terms of a geometric sequence.

S_n means the sum of the first n terms of a series.

The sum of the first n consecutive positive integers $= \frac{n(n+1)}{2}$.

For an arithmetic series with first term $u_1 = a$ and common difference d, the sum of the first n terms,

$S_n = \frac{n}{2}(u_1 + u_n)$, where u_n is the nth term, or

$S_n = \frac{n}{2}(2a + (n-1)d)$

For a geometric series with first term a and common ratio r:

$S_n = a\left(\frac{r^n - 1}{r - 1}\right)$ when $r > 1$ or $r < -1$

$S_n = a\left(\frac{1 - r^n}{1 - r}\right)$ when $|r| < 1$

You can find the sum of an infinite geometric series with first term a and common ratio r when $-1 < r < 1$.

The sum to infinity is $S_\infty = \frac{a}{1-r}$ for $|r| < 1$

Mixed practice

1 **Find** the sum of each series:

   a $1 + 3 + 5 + \ldots + 197$

   b $34 + 31 + 28 + 25 + \ldots + -2$

   c $1 + 4 + 16 + \ldots + 1024$

   d $50\,000 + 10\,000 + 2000 + \ldots + 0.64$

2 An arithmetic series begins $15 + 19 + 23 + \ldots$ and has 32 terms. **Find** its value.

3 **Find** the value of the arithmetic series that begins $18 + 17.2 + 16.4 + \ldots + 2$.

4 A geometric series begins $10 + 8 + \ldots$ and has 15 terms. **Find** the sum of the series.

5 A geometric series begins $8 + 12 + \ldots$ and has eight terms. **Find** the total of the series.

6 An arithmetic series has first term 5 and common difference 11. **Find** the sum of the first 14 terms.

Problem solving

7 A geometric series has first term 8 and common ratio 1.25. **Find** the sum of the first eight terms.

8 A geometric series has third term 96 and fourth term 128. **Find** the sum of the first seven terms.

9 An arithmetic series has eighth term 48. The sum of the first ten terms is 405. **Find** the common difference.

10 **Find** a fraction equivalent to:

   a $0.\dot{2}$ **b** $0.\dot{1}\dot{8}$ **c** $0.\dot{1}2\dot{3}$

11 An arithmetic series has second term 9 and fifth term 30. **Find** the sum of the first 104 terms.

12 An arithmetic series has first term 11, and the sum of the first fifteen terms is 585. **Find** the common difference.

13 An arithmetic series has fourth term 40, and the sum of the first 10 terms is 520. **Find** the first term and common difference.

14 A geometric series has common ratio 1.5, and the sum of the first six terms is 1995. **Find** the value of its first term.

15 An infinite geometric series has first term 10, and a sum to infinity of 15. **Find** the value of the common ratio.

16 An arithmetic series has first term a and common difference d. The first five terms have sum 240 and the first ten terms have sum 630. **Find** the values of a and d.

17 An arithmetic series has first term $2x + 1$ and common difference x. The sum of the first seven terms is 182. **Find** x.

18 A geometric series has first term x and second term $x - 3$. The sum to infinity is $\frac{100}{3}$. **Find** the value of x.

19 I receive \$10 from my uncle on my 10th birthday, \$11 on my 11th birthday, \$12 on my 12th birthday and so on. **Find** the total amount given to me by my uncle from my 10th to 20th birthdays (inclusive).

20 In January I received 11 emails. In February I received 15 emails. In March I received 19 emails. I predict that this will continue as an arithmetic sequence. **Find** the total number of emails that I would receive in the whole year.

21 A4 paper has an area of 625 cm^2. A5 paper has half the area of A4 paper, A6 paper has half the area of A5 paper and this pattern continues on to A10 paper. I have one sheet each of A4, A5, A6, …, A10 paper. **Find** the total area.

Review in context

1 A small business repairs mobile phones. The managers predict 10 phones will be repaired in the first week, 12 phones in the second week, 14 in the third week and so on, increasing capacity by two phones every week as the business grows by word of mouth.

  a **Write down** an expression for the number of phones repaired in the nth week.
  Explain why the number of phones repaired each week forms an arithmetic sequence.

  b **Find** the number of phones repaired in the 10th week.

  c **Find** the total number of phones repaired in the first 10 weeks.

2 A movie producer expects that 1 000 000 people will see her new film in its first week in the cinemas. Past experience tells her that in each subsequent week, the number of people who see the film will be 80% of the week before. **Use** her modelling assumptions to **estimate** the number of people who will see the film in the first eight weeks. Give your answer to an appropriate degree of accuracy.
Explain why your answer is only an estimate.

3 An internet company prints and sells T-shirts. In their first month they sell 100 shirts. In the second month they sell 120, and in the third month they sell 144.

  a **Show that** these numbers form a geometric sequence.

  b The owners assume that the sales will continue to follow a geometric sequence. **Determine** the number of shirts they will sell in total in their first 12 months if this assumption is correct.

4 A network of computers includes a server and a large number of terminals which can communicate directly with each other. A network engineer wants to test all the possible connections on the network. If the network had only one terminal, there would just be one connection to test: Server-T_1. Adding a second terminal adds two more connections: Server-T_2 and T_1-T_2. Adding a third terminal adds three more: Server-T_3, T_1-T_3 and T_2-T_3.

  a **Explain** why adding the nth terminal to the network will add another n connections.

  b **Hence explain** why the number of connections added when the nth terminal is added forms an arithmetic sequence.

  c **Find** the total number of possible connections in the network when there are 50 terminals.

  d **Determine** the number of terminals that would lead to 2080 possible connections to test.

5 A blog gains 150 000 followers in its first month, 120 000 in its second month, and 96 000 in its third month.

  a **Show that** the number of new followers per month in the first three months forms a geometric sequence.

  b Assuming that this pattern continues, **find** the number of followers that the blog would gain in its first six months.

  c Assuming that the number of new followers per month continues to fall in the same way, **find** the maximum number of followers the website should expect to have in total.

6 A new subscription music streaming service gains 1000 subscribers in its first month, 2500 in its second month, 4000 in its third month, and so on for the whole first year, increasing by a constant amount each month.

  a **Find** the total number of subscribers it would gain in its first year.

  b **Write down** the number of new subscribers that it would gain in its 13th month if it continues to follow this model.

After the 13th month, the number of new subscribers starts to decrease on a monthly basis, with each month attracting only 75% of the number of new subscriptions as the month before.

  c Assuming that members do not leave the service, **find** the overall maximum number of subscribers that the service can expect if this trend continues.

7 A parcel delivery service starts with warehouses in two UK cities: Aberdeen and Birmingham. They connect the warehouses with a daily service to transport packages.
They subsequently open a third warehouse, in Cambridge, and start running two new services: Aberdeen–Cambridge and Birmingham–Cambridge.

  a **Write down** the number of additional routes that would be required if they were to open a fourth warehouse in Doncaster, and then connect it to all previous depots.

  b They open additional depots in Edinburgh and Falmouth, as well as the one in Doncaster. **Find** the total number of connections needed for their entire network if each warehouse is to be connected directly to each other warehouse.

  c **Determine** how many depots could be built before the number of connections needed would exceed 200.

Reflect and discuss

How have you explored the statement of inquiry? Give specific examples.

Statement of Inquiry:

Using different forms to generalize and justify patterns can help improve products, processes and solutions.

E9.1 Another dimension

Global context: Personal and cultural expression

Objectives
- Understanding and using coordinates in three dimensions
- Using the distance, section and midpoint formulae in three dimensions
- Using Pythagoras and trigonometry in three dimensions
- Finding the angle between a line and a plane

Inquiry questions

F
- How can you describe space in three dimensions?
- How can you find distances and angles in 3D solids?

C
- How does your understanding of 2D space extend to 3D space?
- How can you divide a line segment in a given ratio?

D
- Can a 4th dimension exist?
- Is human activity random or calculated?

RELATIONSHIPS

| ATL | Critical-thinking |

Analyse complex concepts and projects into their constituent parts and synthesize them to create new understanding

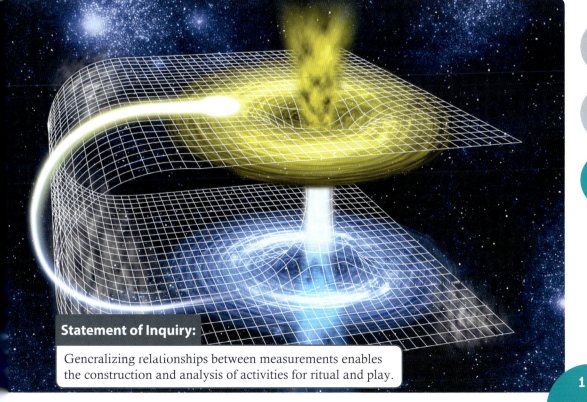

Statement of Inquiry:
Generalizing relationships between measurements enables the construction and analysis of activities for ritual and play.

7.3

9.1

E9.1

You should already know how to:

• find the distance between two points in two dimensions	**1** Find the distance between the two points: **a** (4, 5) and (7, 9) **b** (5, 6) and (−2, 30)
• find the midpoint of two points in two dimensions	**2** Find the midpoint of the points: **a** (3, 7) and (7, 9) **b** (2, 6) and (−2, 9)
• use Pythagoras' theorem in right-angled triangles	**3** Find the length of side x, giving your answer in exact (surd) form. (triangle with sides 11, 8, and x)
• use trigonometry to find angles in right-angled triangles	**4** Find angle y, to the nearest degree. (triangle with 6 cm, 4.5 cm, angle y)

F Describing space

- How can you describe space in three dimensions?
- How can you find distances and angles in 3D solids?

You can describe space in two dimensions using coordinates, but the real world has three dimensions. Two dimensions are very useful for describing flat surfaces – for example, in a computer program to display something on a computer screen. But they are not useful for giving instructions in 3D space, say when creating a model of a new building or using a 3D printer.

Reflect and discuss 1

- To describe a cuboid, or an object shaped like a cuboid, such as a cereal box, what measurements would you give?

GEOMETRY AND TRIGONOMETRY

To fully describe objects such as cuboids in three dimensions, you use three perpendicular measurements: width, depth and height.

You can describe positions in 3D space using x, y and z coordinates. The x-axis, y-axis and z-axis are perpendicular to each other.

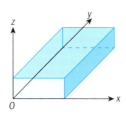

> **Tip**
> The convention is to draw the x- and y-axes as if on the horizontal base of a cuboid, and the z-axis pointing upward.

A flat surface in 3D space is called a **plane**. The plane containing the x- and y-axes is the **x-y plane**, and the planes including the other pairs of axes are named similarly.

The cuboid below is drawn on a 3D coordinate grid.

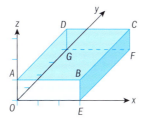

From the origin O to point A is 0 units in the x-direction, 0 units in the y-direction, and 1 unit in the z-direction. The coordinates of A are $(0, 0, 1)$.

To find the point with coordinates $(3, 4, 1)$ start at the origin and move:

  3 units in the positive x-direction to point E

  4 units in the positive y-direction to point F

  1 unit in the positive z-direction, to point C.

> **Tip**
> Some GDCs and dynamic graphing software can draw 3D graphs. Practise plotting some points.

Example 1

> Two points have coordinates $C(3, 4, 7)$ and $D(3, 4, -18)$.
>
> Calculate the distance CD.

C has the same x and y coordinates as D, so it is vertically above D.

$7 - (-18) = 25$ — The distance between C and D is the difference between their z coordinates.

$CD = 25$ units

> **Objective C:** Communicating
> **iii.** move between different forms of mathematical representation
>
> *With some questions in Practice 1 you will need to translate written coordinates into a diagram. Sometimes it might help to draw a diagram, even if the question does not specifically mention it.*

E9.1 Another dimension

Practice 1

1 Write the (*x*, *y*, *z*) coordinates for each lettered vertex in these cuboids.

a 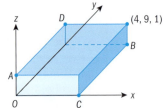 **b**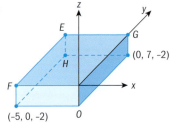

2 Find the distance between these pairs of points.

  a (1, 1, 4) and (1, 1, 11)
  b (3, 2, 0) and (9, 2, 0)
  c (−4, −2, 11) and (−4, 11, 11)

> The **space diagonal** of a cuboid is the straight line joining one vertex to the opposite vertex, passing through the center of the cuboid.
>
> A **face diagonal** is a straight line joining one vertex to another, passing through one of the faces of the shape.
>
>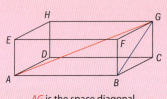
>
> *AG* is the space diagonal
> *BG* is the face diagonal

Exploration 1

This box measures 2 cm by 3 cm by 6 cm.

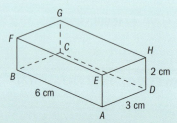

1 Sketch the triangle *BCA*. Use Pythagoras' theorem to find the length of *CA*. Give your answer in exact (surd) form.

2 Sketch the triangle *GCA*. Label the lengths of *GC* and *CA*. What is the size of ∠*ACG*? Explain how you know.

3 Use Pythagoras' theorem to find the length of *GA* to 1 decimal place.

4 Find the length *BH*. What do you notice about *GA* and *BH*?

5 Another box is 1 cm deep, 4 cm wide and 8 cm long. Using the same method as you did to find *GA*, find the length of this box's space diagonal.

6 A cuboid measures *x* cm by *y* cm by *z* cm. Find, in terms of *x* and *y*, the length of one of its face diagonals.

▶ Continued on next page

GEOMETRY AND TRIGONOMETRY

Hence show that the space diagonal has length $\sqrt{x^2 + y^2 + z^2}$.

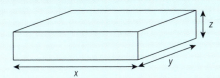

In a cuboid with dimensions x, y and z, the space diagonal has length $\sqrt{x^2 + y^2 + z^2}$.

The space diagonal is also called the **volume diagonal**.

Reflect and discuss 2

- How many different face diagonals does a cuboid have?
- How many space diagonals does a cuboid have? Are they all equal?
- In the first box shown in Exploration 1, explain why the space diagonal GA is the shortest distance between G and A.

Practice 2

1 Find the length of the space diagonal of these cuboids:

  a a cereal box measuring 9 cm by 12 cm by 20 cm

  b a room with floor measuring 4 m by 7 m, which is 4 m tall

  c a pencil box that is 2 cm by 7 cm by 26 cm.

2 Find the length of the space diagonal of each cuboid, giving your answer to a suitable degree of accuracy:

  a a cardboard box measuring 50 cm by 15 cm by 22 cm

  b a smartphone whose dimensions are 123.8 mm by 58.6 mm by 7.6 mm

  c a storage box measuring 13 inches by 13 inches by 18 inches.

3 Find the length of the space diagonal of a cube whose sides are 5 cm. Give your answer in an exact form.

Problem solving

4 A pencil just fits diagonally into a cuboid shaped box with dimensions 16 cm by 4 cm by 2 cm. Find the length of the pencil correct to 3 s.f.

Sets of three integers which can form the sides of a right-angled triangle are known as Pythagorean triples. Examples include {3, 4, 5} and {5, 12, 13}, because $3^2 + 4^2 = 5^2$ and $5^2 + 12^2 = 13^2$.

If a cuboid has three integer sides and also an integer space diagonal, the four numbers are called a Pythagorean quadruple. Examples of these include {1, 2, 2, 3} and {2, 14, 23, 27}, because $1^2 + 2^2 + 2^2 = 3^2$ and $2^2 + 14^2 + 23^2 = 27^2$.

E9.1 Another dimension

Exploration 2

1 Points $C(4, 7, 11)$ and $D(19, 17, 17)$ are two vertices of a cuboid of width w, height h, and depth d as shown.

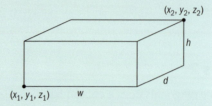

   a Write down the values of w, h and d.

   b Hence find the distance CD.

2 Points $P_1(x_1, y_1, z_1)$ and $P_2(x_2, y_2, z_2)$ are two vertices of a cuboid of width w, height h, and depth d.

   a $w = x_2 - x_1$. Write similar expressions for h and d.

   b Hence show that the distance between P_1 and P_2 is $\sqrt{(x_2-x_1)^2 + (y_2-y_1)^2 + (z_2-z_1)^2}$.

The distance formula

The distance between two points $P_1(x_1, y_1, z_1)$ and $P_2(x_2, y_2, z_2)$ is:
$\sqrt{(x_2-x_1)^2 + (y_2-y_1)^2 + (z_2-z_1)^2}$.

Reflect and discuss 3

- In two dimensions, the distance between (x_1, y_1) and (x_2, y_2) is given by $\sqrt{(x_2-x_1)^2 + (y_2-y_1)^2}$.
  How does this relate to the distance formula in three dimensions?

Example 2

Find the distance between points $(5, 7, -3)$ and $(-4, 8, -2)$.
Give your answer correct to 3 significant figures.

$d = \sqrt{(x_2-x_1)^2 + (y_2-y_1)^2 + (z_2-z_1)^2}$

$= \sqrt{(-4-5)^2 + (8-7)^2 + (-2-(-3))^2}$ — Substitute the values into the formula. Be careful with the negative numbers.

$= \sqrt{81+1+1} = \sqrt{83}$

$= 9.11$ (3 s.f.)

GEOMETRY AND TRIGONOMETRY

Practice 3

1 Find the distance between each pair of points, giving your answer in exact form.
   - **a** (0, 0, 0) and (2, 3, 6)
   - **b** (5, 6, 8) and (6, 8, 10)
   - **c** (23, 11, 18) and (17, 17, 11)
   - **d** (−4, 3, −2) and (0, 7, −9)
   - **e** (−3, 11, −16) and (33, 3, −19)
   - **f** (5.5, 11, −4) and (−4.5, 1, 13.5)

2 **a** Find the distance OA, where O is the origin and A is the point (4, 6, 12).

   b Use a similar method to find the distance from the origin to $B(3, 14, 18)$.

Problem solving

3 Three points have coordinates $O(0, 0, 0)$, $A(3, 6, 6)$ and $B(4, 4, 7)$.
   Show that OAB forms an isosceles triangle.

4 Three points have coordinates $P(0, 9, 1)$, $Q(1, 5, 9)$ and $R(5, 1, 2)$.
   Show that PQR forms an isosceles triangle. Find the area of triangle PQR.

In this cuboid, $ABCD$ is in a horizontal plane. The angle between the diagonal CE and the plane $ABCD$ is labelled θ.

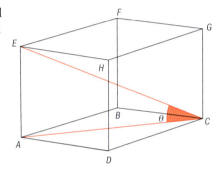

Example 3

Find the angle between the diagonal CE and the plane $ABCD$ in this cuboid. Give your answer to the nearest degree.

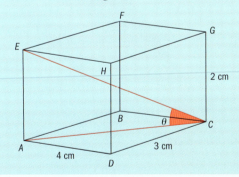

$AC^2 = 3^2 + 4^2 = 25$ ———————————— Find the length AC.
$\Rightarrow AC = 5$ cm

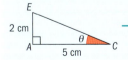
———————————— Sketch the triangle ACE.

$\tan \theta = \dfrac{2}{5}$ ———————————— Use the tan ratio.

$\theta = \tan^{-1}\left(\dfrac{2}{5}\right) = 21.801\ldots°$

$\theta \approx 22°$

E9.1 Another dimension

Practice 4

1 Find the angle between the diagonal *FD* and the plane *ABCD* in this cuboid. Give your answer to the nearest degree.

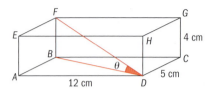

2 a In cuboid *PQRSTUVW*, show that the length of *PR* is $\sqrt{149}$ cm.

b Find the angle between *VP* and the plane *PQRS* correct to the nearest degree.

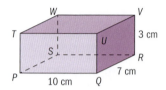

3 a In cuboid *ABCDEFGH*, find the angle between the diagonal *BH* and the plane *ABCD* to 1 d.p.

b Hence find the angle between *BH* and the plane *EFHG*.

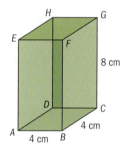

4 In this triangular prism, *BCE* and *AFD* are right-angled triangles. The other faces are rectangles.

  a Find the length *BC*, correct to 3 s.f.

  b Find the length *DB*, correct to 3 s.f.

  c Find the angle that *DB* makes with the plane *CDEF*, correct to the nearest degree.

  d Find angle *BCE*, the angle between planes *ABCD* and *CDEF*, correct to the nearest degree.

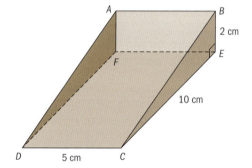

5 *ABCDE* is a square-based right pyramid, with *AB* = 8 cm. The point *E* is 8 cm directly above the point *M*, which is the midpoint of the base.

  a Find *BD*.
  b Hence find *BM*.
  c Find *EB*.
  d Find the angle between *EB* and the base *ABCD*. Give your answer to the nearest degree.

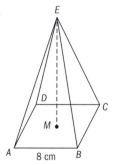

E9 Space

GEOMETRY AND TRIGONOMETRY

Problem solving

6 The volume of a cuboid is 440 cm³. The base measures 5 cm by 8 cm. To the nearest decimal place, what is the length of the cuboid's longest face diagonal?

7 Micol holds the string of a kite 1 m above ground level. He stands 6 m due east and 4 m due north of Rachel, who is holding the kite flat on the ground. As soon as Rachel lets go of the kite it rises 11 m vertically upward.

Calculate the length of the kite string, assuming that it makes a straight line from Micol to the kite.

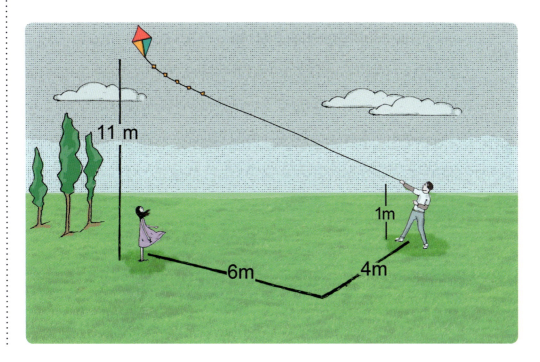

C Dividing space

- How does your understanding of 2D space extend to 3D space?
- How can you divide a line segment in a given ratio?

Exploration 3

1 The midpoint of the line segment joining (x_1, y_1) and (x_2, y_2) is given by $\left(\dfrac{x_1+x_2}{2}, \dfrac{y_1+y_2}{2}\right)$.

Conjecture a similar formula for the midpoint between $P_1(x_1, y_1, z_1)$ and $P_2(x_2, y_2, z_2)$.

2 Use your formula to find M, the midpoint between $A(7, 11, 14)$ and $B(19, 19, 38)$.

▶ Continued on next page

E9.1 Another dimension

3. Use the distance formula to verify that $AM = BM$.

4. Explain why knowing that $AM = BM$ is not enough to show that M is the midpoint of AB.

5. Find the distance AB.

6. Hence explain how you can be certain that M is the midpoint of AB.

7. Prove that your formula from step **1** gives the midpoint of segment P_1P_2.

Example 4

Find M, the midpoint of $A(3, -2, 9)$ and $B(1, 8, -3)$.

$M = \left(\dfrac{3+1}{2}, \dfrac{-2+8}{2}, \dfrac{9+(-3)}{2}\right)$ — Find the mean of the x, y and z coordinate values in A and B.

$= (2, 3, 3)$

The midpoint M of the line segment joining $P_1(x_1, y_1, z_1)$ and $P_2(x_2, y_2, z_2)$ is given by $M = \left(\dfrac{x_1+x_2}{2}, \dfrac{y_1+y_2}{2}, \dfrac{z_1+z_2}{2}\right)$.

Example 5

P, Q and R have coordinates $(p, 3, 8)$, $(1.5, q, 9)$, and $(7, -2, r)$ respectively. Q is the midpoint of PR. Find p, q and r.

$(1.5, q, 9) = \left(\dfrac{x_1+x_2}{2}, \dfrac{y_1+y_2}{2}, \dfrac{z_1+z_2}{2}\right) = \left(\dfrac{p+7}{2}, \dfrac{3+(-2)}{2}, \dfrac{8+r}{2}\right)$ — Use the midpoint formula.

Therefore:

$\dfrac{p+7}{2} = 1.5 \Rightarrow p = -4$ — Form equations using the x, y and z coordinate values. Solve for p, q and r.

$q = \dfrac{3-2}{2} \Rightarrow q = 0.5$

$9 = \dfrac{8+r}{2} \Rightarrow r = 10$

Practice 5

1. Find the midpoint of each pair of points:
   a. $A(0, 0, 0)$ and $B(4, 8, 16)$
   b. $C(1, 1, 1)$ and $D(25, 11, 7)$
   c. $E(-2, -6, -1)$ and $F(-4, 4, 23)$
   d. $G(1, 0, 8)$ and $H(8, 10, 8)$
   e. $I(-11, 32, 12)$ and $J(-4, -7, 5)$
   f. $K(1.7, -0.4, 11.8)$ and $L(2.3, 4.1, -12.1)$

2 Two points have coordinates $A(1, -1, 3)$ and $M(11, 3, 0.5)$.
M is the midpoint of AB. Find the coordinates of B.

3 Three points have coordinates $P(a-2, b, 2c)$, $Q(a+2, b+3, 3b)$ and $R(4a+3, 3b, -3c)$. Q is the midpoint of PR. Find a, b and c.

Problem solving

4 Points A, B, C and D have coordinates $A(1, 3, 11)$, $B(-2, 11, -1)$, $C(7, 11, 7)$ and $D(10, 3, 19)$. Show that the midpoint of AC and midpoint of BD are the same. Describe the quadrilateral $ABCD$.

5 Points P, Q and R have coordinates $P(a, b, c)$, $Q(d, e, f)$ and $R(g, h, i)$.
Q is the midpoint of PR. The letters a to i represent the integers from 1 to 9 inclusive, and each integer is used exactly once.
Find a possible assignment of integers 1 to 9 to the variables a to i such that both of these mathematical statements are satisfied:

$a < c < g < e$

$b < h$

The midpoint formula finds the point that divides a line segment into two equal parts, thus in the ratio 1 : 1. How would you find a point that divided a line segment into different proportions, for example the ratio 1 : 2 or the ratio 2 : 3? Generalizing, how would you find the point that divided a line segment into the ratio $m : n$?

ATL

Exploration 4

1 Two points in the x-y plane have coordinates $P_1(x_1, y_1)$ and $P_2(x_2, y_2)$.
P_3 lies on P_1P_2 and divides it in the ratio $m : n$.

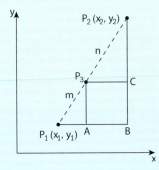

a Explain why triangles AP_1P_3, BP_1P_2, and CP_3P_2 are all similar.

b Find the scale factor that transforms BP_1P_2 to AP_1P_3.

c Hence show that $P_1A = \dfrac{m}{m+n}(x_2 - x_1)$ and derive a similar expression for AP_3.

d Use your expressions for P_1A and AP_3 to show that the coordinates of P_3 are given by $\left(\dfrac{nx_1 + mx_2}{m+n}, \dfrac{ny_1 + my_2}{m+n}\right)$. This is called the **section formula**.

▶ Continued on next page

> **Tip**
> When you write that P_3 divides P_1P_2 in the ratio $m : n$, the order matters – so the m section is closer to P_1 and the n section is closer to P_2.

2 Points A and C have coordinates $A(4, -2)$ and $C(10, 16)$.

  a Use the section formula to find B, which divides AC in the ratio $2:1$.

  b Verify that $AB = 2BC$.

3 Verify that when $m = n = 1$, the section formula in two dimensions matches the midpoint formula.

4 Two points have coordinates $P_1(x_1, y_1, z_1)$ and $P_2(x_2, y_2, z_2)$ in three dimensions. Suggest a suitable formula for the coordinates of the point P_3, which lies on P_1P_2 and divides it in the ratio $m:n$.

5 Use your formula to find the coordinates of T, the point that divides UV in the ratio $3:2$ where U and V have coordinates $(1, -5, -8)$ and $(16, 30, -3)$ respectively. Use the distance formula to verify that $2UT = 3VT$.

The section formula

If $P_1(x_1, y_1, z_1)$ and $P_2(x_2, y_2, z_2)$ where P_3 divides P_1P_2 in the ratio $m:n$, then P_3 has coordinates $\left(\dfrac{nx_1 + mx_2}{m+n}, \dfrac{ny_1 + my_2}{m+n}, \dfrac{nz_1 + mz_2}{m+n}\right)$.

Example 6

Point T divides the line segment SU in the ratio $1:4$. The coordinates of S and U are $(4, -3, -8)$ and $(24, -18, 21)$ respectively. Find the coordinates of T.

$T = \left(\dfrac{nx_1 + mx_2}{m+n}, \dfrac{ny_1 + my_2}{m+n}, \dfrac{nz_1 + mz_2}{m+n}\right)$

$= \left(\dfrac{4 \times 4 + 1 \times 24}{1+4}, \dfrac{4 \times (-3) + 1 \times -18}{1+4}, \dfrac{4 \times (-8) + 1 \times 21}{1+4}\right)$ — Substitute the values into the formula.

$= (8, -6, -2.2)$

Practice 6

1 Use the section formula to divide each line segment in the given ratio.

  a Divide $A(5, 11)$ and $B(25, 27)$ in the ratio $1:3$.

  b Divide $C(-4, 6)$ and $D(5, 0)$ in the ratio $1:2$.

  c Divide $E(11, 4)$ and $F(1, 24)$ in the ratio $2:3$.

2 Use the section formula to divide each line segment in the given ratio.

  a Divide $A(1, 3, -5)$ and $B(25, -9, 4)$ in the ratio $1:2$.

  b Divide $C(3, -2, 6)$ and $D(15, 6, 0)$ in the ratio $3:1$.

  c Divide $E(-2, -8, 3)$ and $F(12, -1, 24)$ in the ratio $4:3$.

3 Find the coordinates of a point three fifths of the way from the point $(4, 6, 7)$ to the point $(24, -4, 42)$.

Problem solving

4 Three points have coordinates $A(a, 11, b+1)$, $B(11, 36, 3b+1)$ and $P(2, 21, 2b)$. P divides AB in the ratio $m : n$. Find the ratio $m : n$ in its simplest form and hence find the values of a and b.

D Higher dimensional space

- Can a 4th dimension exist?
- Is human activity random or calculated?

In two dimensions, the distance formula is $\sqrt{(x_2 - x_1)^2 + (y_2 - y_1)^2}$, and in three dimensions it is $\sqrt{(x_2 - x_1)^2 + (y_2 - y_1)^2 + (z_2 - z_1)^2}$.

Similarly, in two dimensions the midpoint formula is $\left(\dfrac{x_1 + x_2}{2}, \dfrac{y_1 + y_2}{2}\right)$, and in three dimensions it is $\left(\dfrac{x_1 + x_2}{2}, \dfrac{y_1 + y_2}{2}, \dfrac{z_1 + z_2}{2}\right)$. There is a clear link between the formulae in two and three dimensions.

What if you extended these formulas to contain another letter, w?

Exploration 5

Imagine you could draw points in four dimensions, with coordinates of the form (w, x, y, z).

1. Use your knowledge of the distance formula in two and three dimensions to conjecture a formula for the distance between (w_1, x_1, y_1, z_1) and (w_2, x_2, y_2, z_2).

2. Use your formula to find the distance between each pair of points:
   a. $(0, 0, 0, 0)$ and $(1, 4, 4, 4)$
   b. $(2, -4, 3, 8)$ and $(7, 0, -5, 0)$
   c. $(1, 3, 0, 2)$ and $(7, -1, 12, 15)$

3. Use your knowledge of the midpoint formula in two and three dimensions to conjecture a formula for the midpoint between (w_1, x_1, y_1, z_1) and (w_2, x_2, y_2, z_2).

4. Use your formula to find M, the midpoint between $A(7, 4, 1, -3)$ and $B(-3, -2, 3, -1)$.

5. Use your distance formula to find the distances AM, BM and AB.

6. Explain how your answers to step **5** support the claim that M is the midpoint of A and B.

What could it mean to consider geometry in more than three dimensions? In his book *Flatland: A Romance of Many Dimensions*, Edwin Abbott explores life in a world whose inhabitants believe they occupy a 2D space, but then one day discover that they are actually part of a 3D universe. Similarly, could we have misunderstood the space we live in? Could there be other geometric dimensions which we cannot perceive?

Some problems in theoretical physics have been better understood by imagining that we inhabit a space with more than three dimensions – sometimes using as many as 12 to solve a problem. For the time being, we can't know whether this makes sense, but maybe as time goes on scientists will develop a greater understanding of the idea.

Reflect and discuss 4

- Is it possible to find and describe familiar geometric shapes in 4D?
- Explain what is meant by the term *isosceles triangle*.
- Show that the three points $T(1, 6, 5, 8)$, $U(5, 2, 4, 4)$ and $V(1, 1, 0, 0)$ form an isosceles triangle.
- How could you identify a square in four dimensions? Is it sufficient simply to have four equal sides?

One test that you can apply is that for the points A, B, C and D to form a square, all four sides must be equal in length, diagonals AC and BD must have equal midpoints, and AC and BD must also be equal in length.

- Show that the points $A(0, 0, 0, 0)$, $B(6, 4, 2, 13)$, $C(2, 10, 15, 11)$ and $D(-4, 6, 13, -2)$ form a square.
- Do you think it is possible that we could be living in a world with dimensions we can't see?

Summary

- In a cuboid with dimensions x, y and z, the space diagonal has length $\sqrt{x^2 + y^2 + z^2}$.
- A flat surface in 3D space is called a **plane**. The plane containing the x- and y-axes is the **x-y plane**, and the planes including the other pairs of axes are named similarly.
- **The distance formula**
  The distance between $P_1(x_1, y_1, z_1)$ and $P_2(x_2, y_2, z_2)$ is $\sqrt{(x_2 - x_1)^2 + (y_2 - y_1)^2 + (z_2 - z_1)^2}$

- The midpoint M of the line segment joining $P_1(x_1, y_1, z_1)$ and $P_2(x_2, y_2, z_2)$ is given by
  $M\left(\frac{x_1 + x_2}{2}, \frac{y_1 + y_2}{2}, \frac{z_1 + z_2}{2}\right)$.
- **The section formula**
  If $P_1(x_1, y_1, z_1)$ and $P_2(x_2, y_2, z_2)$ where P_3 divides P_1P_2 in the ratio $m : n$, then P_3 has coordinates $\left(\frac{nx_1 + mx_2}{m+n}, \frac{ny_1 + my_2}{m+n}, \frac{nz_1 + mz_2}{m+n}\right)$

Mixed practice

1 **Find** the length of the space diagonal of a shoebox measuring 50 cm by 15 cm by 22 cm. Give your answer to a reasonable degree of accuracy.

Problem solving

2 A cuboid shaped storage box measures 13 inches by 13 inches by 18 inches. **Determine** whether a 20 inch drumstick would fit into the box.

3 **Find** the midpoint of, and distance between, these pairs of points:

  a (1, 1, 3) and (3, −2, 9)

  b (5, 5, −2) and (6, 1, −10)

  c (13, −2, 11) and (9, 2, 4)

GEOMETRY AND TRIGONOMETRY

4 A straight path runs directly from points with coordinates $B(0, 3, 2)$ and $C(15, 6, 8)$, where the axes are measured in meters.

   a **Find** the length of the path and the coordinates of its midpoint.

   b **Find** the difference in height between the two ends of the path.

5 A single strand of a spider web runs in a straight line from one corner of a room at the floor to the opposite corner at the ceiling. The room is shaped like a cuboid 4 meters wide by 6 meters long and 3 meters tall.

   a **Find** the length of the strand of web.

   b **Find** the angle between the strand of web and the floor.

6 Square-based pyramids in Egypt were built as tombs for queens and pharaohs. The most famous are the three pyramids at Giza.

   a Each of the sides of the base of the Great Pyramid of Giza measures 230.4 m, and the height measures 145.5 m. **Find** the angle between the base of the pyramid and any of the sloping sides. Give your answer accurate to two decimal places.

   b The middle pyramid at Giza has base sides of 216 m and a height of 143 m. **Find** the angle between the base of the pyramid and any of the sloping sides. Give your answer accurate to two decimal places.

   c **Determine** if these two pyramids are mathematically similar. **Justify** your answer.

7 Two points have coordinates $A(13, -2, 7)$ and $B(34, 19, 42)$. **Find** the point that divides AB in the ratio 2 : 5.

Problem solving

8 Two points have coordinates $U(-5, 11, 17)$ and $V(7, -1, -7)$. A divides UV in the ratio 1 : 2. B divides UV in the ratio 1 : 3. **Determine** which of A or B is closer to the origin.

9 Two points have coordinates $A(3, 5, 8)$ and $B(-3, -1, z)$. **Find** all possible values for z given that A and B are 11 units apart.

10 **Show that** the points $A(-13, 2, 4)$, $B(11, 6, -12)$ and $C(5, -4, 3)$ form an isosceles triangle. **Find** its area.

Review in context

1 Laser tag can be played indoors or outdoors, with players trying to 'tag' each other with lasers. One of the largest arenas is in Arizona, USA. To plan it, the designers used a 3D coordinate system to simulate the space, with one unit representing one foot. With the origin O as one corner of the rectangular prism, the furthest corner from the origin was at $C(150, 120, 20)$.

   a In the arena, a refuge tower is located so that its top is $\frac{3}{4}$ of the way along the line joining O and C. **Find** the coordinates of the top of the tower.

   b The company that makes the lasers says that lasers will reliably hit targets up to 175 feet away. **Determine** whether the lasers are appropriate for use within this arena.

   c **Find** the longest distance your laser could travel within the structure if you stood in the middle of the arena floor.

2 When creating video games, programmers use 3D space to create the game and transformational geometry to move objects and characters. In one game, a programmer is creating two tunnels. Tunnel EF has end coordinates $E(22, 17, 16)$ and $F(215, 28, 18)$, while tunnel GH runs from end coordinate $G(28, 19, 17)$ to $H(213, 30, 19)$.

   a **Find** the difference between the lengths of the two tunnels.

   b There is a sign at the midpoint of the tunnel entrances F and H. **Find** the coordinates of the sign.

   c An access corridor runs from the midpoint of one tunnel to the midpoint of the other tunnel. **Find** the length of this corridor.

E9.1 Another dimension

3 A new roller coaster ride is designed to have a long straight tunnel that passes through a natural hill. The roller coaster gains speed as it passes through the tunnel before turning sharply as it emerges. The ride control center is situated at (0, 0, 0), and on a scale of 1 unit to 1 m, the tunnel enters the hill at $A(15, 25, 26)$ and leaves the hill at $B(75, 50, -14)$.

  a **Find** the total change in height from one end of this tunnel to the other.

  b **Find** the total length of this tunnel in meters.

  ### Problem solving

  c If the tunnel is created by boring a cylindrical hole with a cross-sectional area of 9 m², **estimate** the volume of earth to be excavated. **Explain** why your answer is an estimate.

  d A refuge point is to be built in the tunnel, two thirds of the way from A to B. **Find** the coordinates of this refuge point.

4 A skier travels down a straight ski slope. With units measured in meters, her starting coordinates relative to the ski lift are (4, 7, −2). The coordinates of the point where she finishes are (−208, 355, −78).

  a **Find** the shortest distance between her starting and finishing points.

  b **Suggest** reasons why the distance she travels will be greater than this.

  c Given that her average speed is 20 ms⁻¹, **find** the minimum time needed for her to ski down the slope.

Another skier travels the same route down the slope, heading in a straight line from the starting point to the finish. Three-quarters of the way down the slope, he loses control and drops one of his ski poles.

  d **Find** the coordinates of the point where he drops the ski pole.

The male skier leaves his skis at a café at point (−240, 370, −75) and then walks back up to the slope to find his ski pole.

  e **Find** the distance from the café to his ski pole.

  f **Find** the angle of elevation of his ski pole from the café.

Reflect and discuss

How have you explored the statement of inquiry? Give specific examples.

Statement of Inquiry:

Generalizing relationships between measurements enables the construction and analysis of activities for ritual and play.

Mapping the world

Global context: Orientation in space and time

Objectives
- Finding the area of a non-right-angled triangle
- Using the sine rule
- Using the cosine rule
- Finding the area of a segment of a circle
- Using the sine and cosine rules to solve problems involving bearings

Inquiry questions

- How can you find the area of a non-right-angled triangle?
- How does the area formula help you find the area of other shapes?

- How is the sine rule related to the area of a triangle?
- How is the cosine rule related to the Pythagorean theorem?

- Which is more useful, the sine rule or the cosine rule?
- How do we define 'where' and 'when'?

RELATIONSHIPS

ATL	Transfer

Apply skills and knowledge in unfamiliar situations

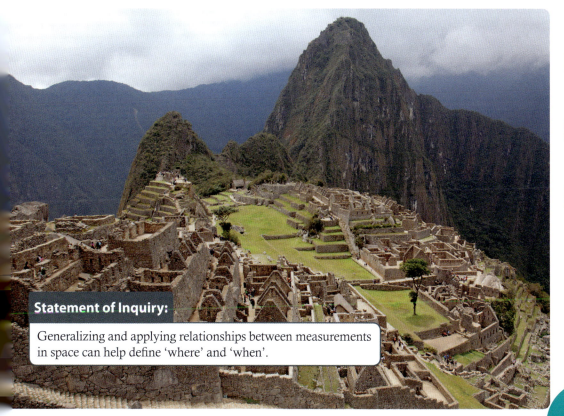

Statement of Inquiry:
Generalizing and applying relationships between measurements in space can help define 'where' and 'when'.

You should already know how to:

• find the area of a triangle given two perpendicular measurements	**1** Find the area of each triangle. **a** 3 cm, 7 cm **b** 6 cm, 14 cm
• use trigonometry in right-angled triangles	**2** Find x in each triangle. **a** 8 cm, 62°, x **b** 7 cm, 5 cm, x
• find the area of a sector of a circle	**3** Find the area of this sector. 115°, 6 cm
• use radians	**4** Find x in each sector. **a** Area = x cm², 1.7 rad, 5 cm **b** 14 cm, x rad, 8 cm

 ## Area of a triangle

- How can you find the area of a non-right-angled triangle?
- How does the area formula help you find the area of other shapes?

When working with triangles, the convention is to label vertices and the interior angle at each vertex with a capital letter, and the side opposite that vertex with the same letter in lower case, like this:

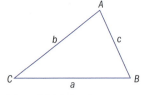

> A **convention** is a practice which, although usually not essential, makes things easier if everybody follows it. For example, in algebra, the convention is to write variables in alphabetical order.

Practice 1

1 Here are triangles *ABC*, *DEF*, *LMN* and *UVW*. Copy each triangle and label the remaining vertices and sides.

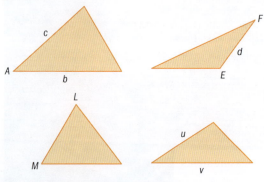

GEOMETRY AND TRIGONOMETRY

Reflect and discuss 1

Here is Paco's calculation of the area of this triangle:

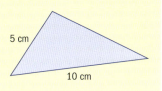

$$\text{Area} = \tfrac{1}{2} \times \text{base} \times \text{height} = \tfrac{1}{2} \times 5 \times 10 = 25 \text{ cm}^2$$

- What mistake has he made?
- Will the true answer be greater or less than 25 cm²? How do you know?
- What other information might help you to calculate the true area?

It would be better if Paco had remembered the area formula as

Area = $\tfrac{1}{2}$ × base × *perpendicular* height.

In mathematics, 'height' means 'perpendicular height'.

Exploration 1

1 A triangle has sides 13 cm and 15 cm.

  a Use trigonometry to work out the length of the dashed line, which meets the 15 cm side at right angles.

  b Hence find the area of the triangle.

2 Triangle *ABC* has sides *BC* = 7 cm, *AC* = 22 cm and ∠*ACB* = 42°.

  a Sketch triangle *ABC*.

  b Hence find the area of the triangle in cm².

3 A triangle has sides *a*, *b* and *c*:

  By considering the length of the line through *A* perpendicular to *BC*, find a formula for the area of the triangle in terms of the lengths *a* and *b* and the size of angle *C*.

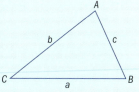

4 Use your formula from step **3** to show that when *BC* = 10 cm, *AC* = 5 cm and ∠*ACB* = 30°, the area of triangle *ABC* is 12.5 cm².

In Exploration 1 you found the area of a triangle by drawing an altitude and finding its length and then deriving a formula. It is usually easier to use the formula to solve problems than to draw in extra lines.

Area of a triangle = $\tfrac{1}{2} ab \sin C$

Tip

An **altitude** is a line perpendicular to one side of a triangle that passes through the opposite vertex. The height of a triangle is one of its three altitudes.

E9.2 Mapping the world

Reflect and discuss 2

- What measures do you need to use this new area formula?
- Where is the angle in relation to the sides?
- How would you change the formula for a triangle that is not labelled *ABC*?

Practice 2

1 Use the formula to find the area of each triangle. Give your answers to 3 s.f.

a 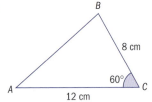 b 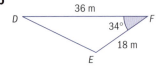 c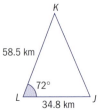

2 Find the area of each triangle to the nearest square centimeter.

a 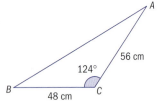 b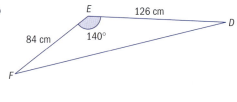

3 In triangle *PQR*, *PQ* = 34 cm, *PR* = 67 cm and $\angle RPQ = 120°$. Sketch the triangle and hence find its area correct to 3 s.f.

4 Here is a triangular tile.

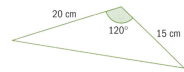

a Find the area of the tile.

b A tiler divides 10 m² by the area of one tile and rounds up to the nearest whole number.

  i Explain what this calculation represents.
  ii Explain why it is likely to be an underestimate.

Problem solving

5 In triangle *DEF*, *d* = 12 cm, *e* = 18 cm, $\angle E = 49°$ and $\angle D = 30°$. Sketch the triangle and hence find its area correct to 3 s.f.

6 A large regular tetrahedron has four faces, each of which is an equilateral triangle of side 8 m. Maria is going to paint all four faces of the tetrahedron. The paint thickness on each surface will be 1 mm. Paint is sold in 1.5 liter tins. Determine how many tins of paint Maria should buy.

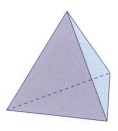

Reflect and discuss 3

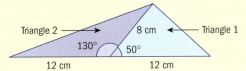

- Find the area of Triangle 1.
- Explain why Triangle 2 has the same area as Triangle 1.
- How does this relate to the trigonometry of the unit circle?

Exploration 2

In this Exploration, use the formula Area $=\frac{1}{2}ab\sin C$ to help you find the areas of some other shapes. Give answers correct to 3 significant figures.

1 A parallelogram has sides 20 cm and 45 cm. Its acute angle is 75°.

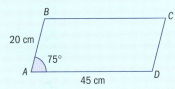

   a Copy the diagram and draw the line segment BD.
   b Explain why triangles ABD and BCD are identical.
   c Find the area of triangle ABD.
   d Hence find the area of the parallelogram.

2 An equilateral triangle has side length a.

   a Write down the exact value of sin 60°.
   b Hence find a formula for the area of an equilateral triangle with side length a. Use your formula to find the area of an equilateral triangle with side length 4 cm.

3 Explain how a regular hexagon can be divided into identical equilateral triangles. Hence show that the area of a regular hexagon with side length s is given by $\frac{3\sqrt{3}}{2}s^2$.

4 A company logo is a sector OAB of a circle with radius 4 cm and center O, a portion of which is shaded.

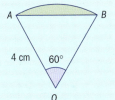

   a Find the area of the sector OAB.
   b Explain why $\triangle OAB$ is equilateral.
   c Hence find the area of $\triangle OAB$.
   d Find the area of the shaded segment.

The area formula for a triangle can be used with other results to find the areas of familiar shapes. The key results are given below, but it is more important to learn how to divide shapes into simple parts than to memorize these formulae.

Area of a parallelogram

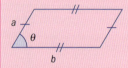

Area = $ab \sin \theta$

Area of an equilateral triangle

Area = $\frac{\sqrt{3}}{4}s^2$

Area of a segment

Degrees: Area = $\frac{\pi r^2 \theta}{360} - \frac{1}{2} r^2 \sin \theta$

Radians: Area = $\frac{1}{2} r^2 \theta - \frac{1}{2} r^2 \sin \theta = \frac{1}{2} r^2 (\theta - \sin \theta)$

Practice 3

Give answers correct to 3 significant figures unless otherwise directed.

1 Find the shaded areas. The central angle is given in degrees.

a 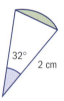   b 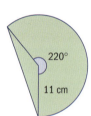   c

> In **1b**, there are two parts to the total area involved.

2 Find the shaded areas. The central angle is given in radians.

a   b   c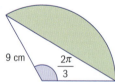

Problem solving

3 Find the area of each shape.

a 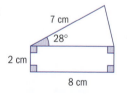   b 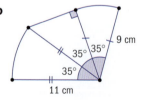   c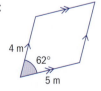

4 A segment of a circle has radius 7 cm and area 2 cm².

Find, correct to two decimal places, the central angle of the segment in radians.

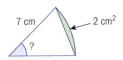

GEOMETRY AND TRIGONOMETRY

  The sine and cosine rules

- How is the sine rule related to the area of a triangle?
- How is the cosine rule related to the Pythagorean theorem?

Exploration 3

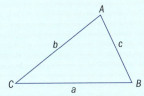

You have already shown that this triangle has area $= \frac{1}{2}ab \sin C$.

1. Explain why the triangle's area can also be found using $\frac{1}{2}ac \sin B$.
2. Hence show that $\dfrac{b}{\sin B} = \dfrac{c}{\sin C}$.
3. Show similarly that $\dfrac{a}{\sin A} = \dfrac{b}{\sin B}$.
4. Use your formula from step **3** to show that $a \approx 10.6$ cm in the triangle below.

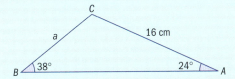

5. Show that $\dfrac{\sin A}{a} = \dfrac{\sin B}{b} = \dfrac{\sin C}{c}$.
6. Hence find the size of the acute angle at B in this triangle:

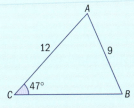

The sine rule

For a triangle ABC:

$$\dfrac{\sin A}{a} = \dfrac{\sin B}{b} = \dfrac{\sin C}{c}$$

and

$$\dfrac{a}{\sin A} = \dfrac{b}{\sin B} = \dfrac{c}{\sin C}$$

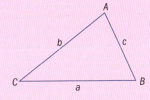

E9.2 Mapping the world

Example 1

Find the value of a, giving your answer correct to 3 significant figures.

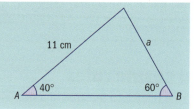

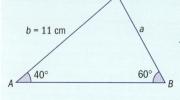

Label the triangle.

$$\frac{a}{\sin A} = \frac{b}{\sin B}$$

To find a missing side, use the sine rule with sides 'on top'.

$$\frac{a}{\sin 40°} = \frac{11}{\sin 60°}$$

$$\Rightarrow a = \frac{11 \sin 40°}{\sin 60°} = 8.16 \text{ cm}$$

Example 2

Find the size of $\angle B$ correct to the nearest degree.

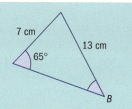

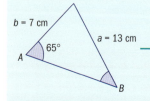

Label the triangle.

$$\frac{\sin A}{a} = \frac{\sin B}{b}$$

To find a missing angle, use the sine rule with angles 'on top'.

$$\frac{\sin 65°}{13} = \frac{\sin B}{7}$$

$\sin B = \frac{7 \sin 65°}{13} = 0.488\ldots$

$B = \arcsin(0.488\ldots)$

$ = 29°$ (to the nearest degree)

Practice 4

1 Find the length of the sides marked with letters, correct to 3 s.f.

a 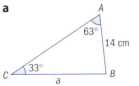 b 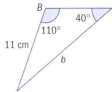 c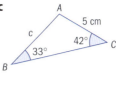

2 For each triangle, find the size of the unknown angle, to the nearest degree.

a 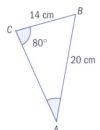 b 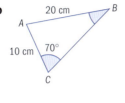 c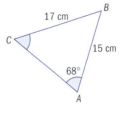

3 Find the value of x for each triangle. Give each answer correct to 1 d.p.

a 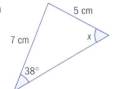 b 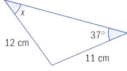 c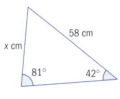

4 In triangle XYZ, $x = 16$ cm, $\angle Y = 44°$ and $\angle Z = 72°$. Find the length of side y.

Problem solving

5 A yacht crosses the starting line of a race at C and sails on a bearing of 026°. After sailing 2.6 km, it rounds a buoy B and sails on a bearing of 335° until it reaches another buoy A which is due north of C.

   a Sketch a diagram to show the path followed by the yacht.

   b Calculate the size of $\angle CBA$ and find the size of $\angle CAB$.

   c Calculate the total distance sailed by the yacht correct to 3 s.f.

6 From point X, a ship sights a lighthouse L on a bearing of 350°. The ship is moving on a bearing of 020° at 10 km/h. Thirty minutes later it is at point Y and sights the lighthouse again, this time on a bearing of 325°.

   a Sketch a triangle to illustrate this information.

   b Determine the size of all three angles in the triangle.

   c Calculate the distance from point Y to the lighthouse. Give your answer in km, accurate to 3 significant figures.

Reflect and discuss 4

Explain why you cannot use the sine rule to find length x in this triangle.

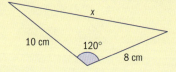

Exploration 4

1. **a** Copy triangle ABC, with the altitude through C meeting AB at right angles at point P.

   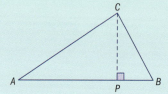

   Label the sides a, b and c.

   b Label length AP as x. Hence label length PB in terms of c and x.

   c Use Pythagoras' theorem to write two expressions for CP. Hence show that $b^2 - x^2 = a^2 - (c-x)^2$.

   Expand the brackets and simplify this equation as far as possible.

   d Use trigonometry to express x in terms of A and b.

   e Hence show that $a^2 = b^2 + c^2 - 2bc \cos A$.

2. Explain why it is possible to rewrite the formula from step **1e** so that it uses angle B as $b^2 = a^2 + c^2 - 2ac \cos B$.

   Rewrite the formula so that it uses angle C.

The cosine rule

For a triangle ABC:

$$a^2 = b^2 + c^2 - 2bc \cos A$$

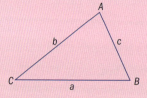

Reflect and discuss 5

- How does the information required to use the cosine rule compare to the information required to use the formula: Area $= \frac{1}{2} ab \sin C$?
- How would you modify the formula if the triangle is labelled MNP and you wanted to use the cosine rule to find the measure of side p?
- How is the cosine rule related to the Pythagorean theorem? Explain.

GEOMETRY AND TRIGONOMETRY

Example 3

Find the length of side a correct to 3 significant figures.

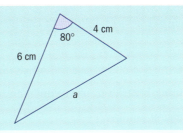

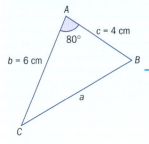

Label sides and vertices.

$a^2 = b^2 + c^2 - 2bc \cos A$ — Use the cosine rule.

$ = 6^2 + 4^2 - 2 \times 6 \times 4 \times \cos 80°$

$ = 43.665...$

$\Rightarrow a = 6.61$ cm (3 s.f.) — Check that your answer seems sensible.

Example 4

Find the value of x correct to the nearest degree.

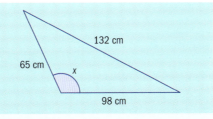

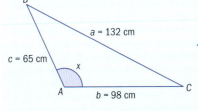

Label sides and vertices.

$a^2 = b^2 + c^2 - 2bc \cos A$ — Use the cosine rule.

$132^2 = 98^2 + 65^2 - 2 \times 98 \times 65 \times \cos x$

$17\,424 = 9604 + 4225 - 12\,740 \cos x$ — Rearrange.

$12\,740 \cos x = -3595$

$\cos x = -0.28218...$

$x = \arccos(-0.28218...)$

$ = 106°$ (to the nearest degree)

E9.2 Mapping the world

Reflect and discuss 6

- Show that the cosine rule $a^2 = b^2 + c^2 - 2bc \cos A$ can be rearranged to $\cos A = \dfrac{b^2 + c^2 - a^2}{2bc}$.
- Which version do you think is easier to use to find an unknown side?
- Which version do you think is easier to use to find an unknown angle?
- Does it matter which version you use?

Practice 5

1 Use the cosine rule to find the lengths of the marked sides correct to 1 d.p.

a 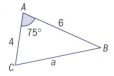 **b** 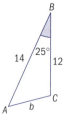 **c**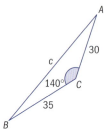

2 Use the cosine rule to find the size of the marked angles correct to the nearest degree.

a 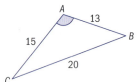 **b** **c**

3 Find these side lengths and angles, accurate to 3 s.f.

  a $\triangle PQR$, $p = 14$, $q = 11$, $R = 98°$. Find side r.
  b $\triangle DEF$, $d = 6$, $e = 12$, $f = 9$. Find angle E.
  c $\triangle XYZ$, $x = 7.2$, $z = 5.4$, $Y = 121°$. Find side y.
  d $\triangle BCD$, $b = 25$, $c = 31$, $d = 34$. Find angle D.

4 Two boats leave the same harbor. One travels on a bearing of 060° for 13 km, the other on a bearing of 135° for 8 km.

  a Draw a diagram illustrating their positions.
  b Find the angle between their courses.
  c Find the distance between the two boats.

Problem solving

5 A triangle has sides of length 4 cm, 6 cm and 7 cm.

  a Find the size of the largest angle in the triangle.
  b Find the size of the smallest angle in the triangle.

GEOMETRY AND TRIGONOMETRY

 ## Applying the sine and cosine rules

- Which is more useful, the sine rule or the cosine rule?
- How do we define 'where' and 'when'?

You can use the sine and cosine rules to find missing lengths and angles in triangles. To decide which rule to use, look at the information you have and the information you are trying to find.

Reflect and discuss 7

Referring to the triangle here:

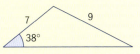

- Which measurements can you calculate? What rule will you use? Explain your selection.
- What information needs to be given in order to be able to use the sine rule?
- What information leads you to use the cosine rule?

By examining the sine rule and cosine rule carefully, you can work out when you should use each one. For the sine rule, you only ever use two of the three parts, so only two are shown here.

Cosine rule: $a^2 = b^2 + c^2 - 2bc \cos A$ **Sine rule:** $\dfrac{a}{\sin A} = \dfrac{b}{\sin B}$

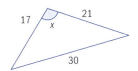   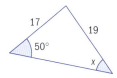

The cosine rule involves three sides and one angle.

This triangle problem involves three sides and one angle (the unknown), so use the cosine rule.

The sine rule involves two sides and two angles.

This triangle problem involves two sides and two angles (one is the unknown), so use the sine rule.

Tip

If an angle and the side opposite are known, then the sine rule can be used. If not, then use the cosine rule.

Example 5

Find the size of the angle at A.

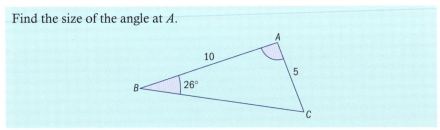

$\dfrac{\sin B}{b} = \dfrac{\sin C}{c}$ ────── The problem involves two sides and two angles, so use the sine rule.

$\dfrac{\sin 26°}{5} = \dfrac{\sin C}{10}$

$C = \arcsin(2 \sin 26°)$ ────── Rearrange to make C the subject.

$= 61.25$

$\Rightarrow A = 180° - 26° - 61.25°$ ────── The angles in a triangle add up to 180°.

$= 92.7°$ (3 s.f.)

E9.2 Mapping the world

Practice 6

1 In each triangle, determine whether the cosine rule or sine rule should be used to find the value of x. Explain how you have made your decision.

a   b   c

d 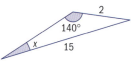 e   f

2 In each triangle, find the value of x, giving your answer correct to 3 s.f.

a 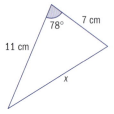 b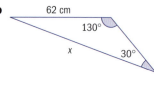

c 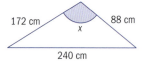 d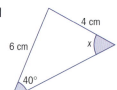

e 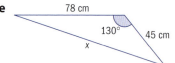 f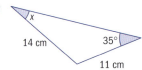

3 Use the sine rule or cosine rule to find the marked sides and angles. Give answers correct to 3 significant figures.

a   b 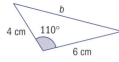 c

d   e 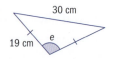 f

196 E9 Space

Problem solving

4 Find the area of this triangle.

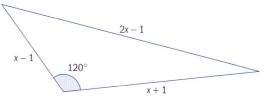

5 Circles of radius 10 cm and 6 cm overlap as shown:

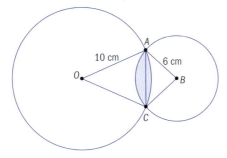

The circles meet at points A and C, and $\angle AOC = 0.8$ rad.

a Find the length of the straight line segment joining A to C.

b Hence find $\angle ABC$ in radians.

c Hence find the shaded area.

One handy and beneficial use of trigonometry is to help you calculate measurements in space that you can't measure directly.

Objective D: Applying mathematics in real-life contexts
iv. justify the degree of accuracy of a solution

In the Activity, comment on the accuracy of the original measurements and the effect that inaccuracy in the original measurements would have on your calculations.

Activity

1 Telephone cables from a house are connected to a telegraph pole C at the roadside, but C is not yet connected to the network.

▶ Continued on next page

E9.2 Mapping the world

On the other side of the road, two other telegraph poles A and B are 30 meters apart.

An engineer uses a theodolite to measure ∠CAB = 27° and ∠ABC = 68°.

  a Copy and complete this aerial view of the street, adding the size of angle B and the length AB.

  b Find the size of angle C.

  c Hence calculate the lengths of BC and AC. Explain why the telegraph pole at C should be connected to pole B.

  d The engineer discovers an error in the measurements. The angle at A is actually 23°.

     i   Calculate the true length of BC. Justify your degree of accuracy.

     ii Find the percentage error in your original calculated value of BC.

     iii Suggest three reasons why the engineer should bring more cable than the calculated value to connect B to C.

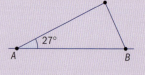

A theodolite is a tool used to measure angles when making maps or surveying construction sites.

Practice 7

1 From point P, a ship sails 5.5 km to point Q on a bearing of 039°. It then sails 4.1 km to point R on a bearing of 072°.

  a Show this information on a diagram.

  b Calculate the size of angle PQR.

  c Find the distance from R to P.

2 Two boats leave a harbor: one travels on a bearing of 050° for 14 km, the other on a bearing of 145° for 18 km.

  a Draw a diagram illustrating their positions.

  b Find the angle between their courses.

  c Find the distance between the two boats.

Problem solving

3 Peter, Alex and Mary are sea-kayaking. Peter is 430 m from Alex on a bearing of 113°. Mary is on a bearing of 210° and a distance of 310 m from Alex. Find the distance between Peter and Mary.

4 I am standing at point O, near a major road. Along the road there are markers every 100 m. I can see three consecutive markers: P, Q and R. I am 20 m away from marker P. The bearing of P from my current position is 340° and the bearing of Q from my current position is 030°.

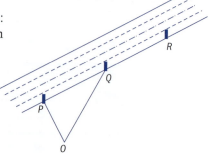

  a Find the angle POQ.

  b Find the bearing of Q from P.

  c Find my distance from Q.

  d Find my distance from R.

GEOMETRY AND TRIGONOMETRY

Summary

When working with a triangle ABC, side a is opposite vertex A and so on.

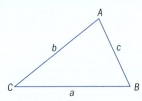

Area of triangle $= \frac{1}{2}ab \sin C$

The sine rule

$$\frac{\sin A}{a} = \frac{\sin B}{b} = \frac{\sin C}{c}$$

and

$$\frac{a}{\sin A} = \frac{b}{\sin B} = \frac{c}{\sin C}$$

The cosine rule

$$a^2 = b^2 + c^2 - 2bc \cos A$$

Mixed practice

1 Find the shaded areas correct to 3 s.f.

a 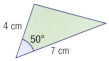   b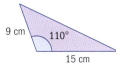

c 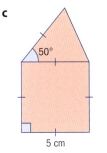   d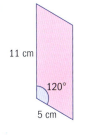

e 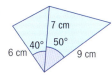   f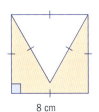

g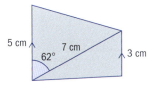

2 Find the areas of these segments correct to 3 s.f.

a 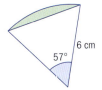   b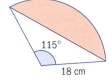

3 Find the areas of these segments, whose central angles are given in radians. Give your answers correct to 3 s.f.

a   b

4 Find the labelled side correct to 3 s.f.

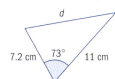

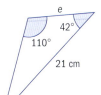

5 Find the labelled angle in degrees, correct to 3 s.f.

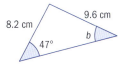

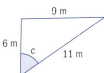   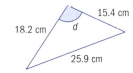

E9.2 Mapping the world

6 Find the labelled angle in radians, correct to 3 s.f.

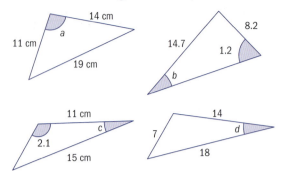

7 Find, in degrees, the size of the smallest angle in a triangle with sides 4 cm, 9 cm and 11 cm.

8 Two friends start walking from the same spot. One walks 50 meters due north. The other walks 35 meters on a bearing of 057°.

Find the distance between their final positions, correct to 3 s.f.

Problem solving

9 Find the perimeter of this triangle, correct to 3 s.f., given that $0 < x < 90°$.

ATL Review in context

Many maps are created by **triangulation**; this is when a network of points are joined across a landscape. Surveyors measure distances and bearings to a high degree of accuracy and use them to form the skeleton that underpins the map, as illustrated here in this triangulation of the Hawaiian island of Oahu.

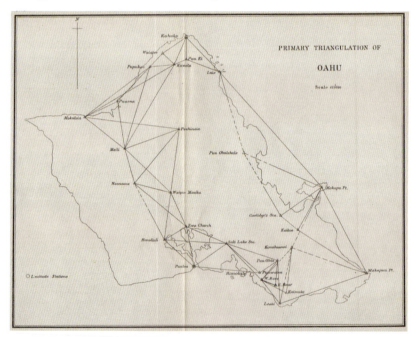

Use the map on the next page to help you answer these questions. Give your answers correct to a sensible degree of accuracy.

1 It is 9.5 km from Rolle to Saint-Prex on a bearing of 076°. Thonon-les-Bains lies on a bearing of 130° from Rolle, and on a bearing of 170° from Saint-Prex.

  a Using the letters R, S and T to denote the three towns, **sketch** triangle RST to show their positions relative to each other.

  b Find and label the size of angles $\angle RST$, $\angle STR$ and $\angle TRS$.

  c Hence find the distances from Rolle to Saint-Prex and from Rolle to Thonon-les-Bains.

  d When standing in Saint-Prex, there is an angle of 100° between Lausanne and Thonon-les-Bains. Lausanne lies on a bearing of 038° from Thonon-les-Bains. Add this information to your diagram.

  e Hence find the distance from Lausanne to Saint-Prex and Thonon-les-Bains.

GEOMETRY AND TRIGONOMETRY

The map below shows some key towns on the banks of Lac Leman, which lies on the French-Swiss border and is known as Lake Geneva in most English-speaking countries. Because of the presence of the lake itself, and the mountains that surround it, finding point-to-point distance measurements between places on the shores of the lake is difficult.

Problem solving

2 The distance from Evian-les-Bains to Saint-Gingolph is measured to be 16.9 kilometers. Viewed from Lausanne, there is an angle of 54° between Evian-les-Bains and Saint-Gingolph. Viewed from Evian-les-Bains, there is an angle of 66° between Lausanne and Saint-Gingolph. Evian-les-Bains lies on a bearing of 200° from Lausanne.

  a **Find** the bearing of Saint-Gingolph from Lausanne.

  b **Find** the distance from Lausanne to Evian-les-Bains.

3 The distance from Vevey to Montreux is 6.5 km. The distance from Montreux to Saint-Gingolph is 9.6 km. The distance from Vevey to Saint-Gingolph is 8.6 km on a bearing of 200°.

  a **Find** the angle between Montreux and Saint-Gingolph as viewed from Vevey.

  b **Hence find** the bearing of Montreux from Vevey.

  c **Find** the bearing of Saint-Gingolph from Montreux.

4 Nyon is 7.0 km from Chens-sur-Leman. Coppet is 6.1 km from Chens-sur-Leman. As viewed from Chens-sur-Leman, there is an angle of 84° between Nyon and Coppet.
Find the distance from Nyon to Coppet.

Reflect and discuss

How have you explored the statement of inquiry? Give specific examples.

Statement of Inquiry:

Generalizing and applying relationships between measurements in space can help define 'where' and 'when'.

E9.2 Mapping the world

E10.1 Time for a change

Global context: Orientation in space and time

RELATIONSHIPS

Objectives
- Drawing graphs of logarithmic functions
- Finding the inverse of an exponential function
- Justifying algebraically and graphically that a logarithmic function and the corresponding exponential function are mutual inverses
- Identifying and applying function transformations on graphs of logarithmic functions

Inquiry questions

F
- What does a logarithmic function look like?
- How are logarithmic functions related to exponential functions?

C
- What are the properties of logarithmic functions?
- How are the properties of logarithmic functions related to those of exponential functions?

D
- How do you transform logarithmic functions?
- Is change measurable and predictable?

ATL	Critical-thinking

Draw reasonable conclusions and generalizations

E6.1

E10.1

E12.1

Statement of Inquiry:
Generalizing changes in quantity helps establish relationships that can model duration, frequency and variability.

202

NUMBER

You should already know how to:

• rewrite exponential statements as logarithmic statements and vice versa	**1** Write $y = 5^x$ as a logarithmic statement. **2** Write $y = \log_7 x$ as an exponential statement.
• find the inverse of a function	**3** Find the inverse of the functions: **a** $f(x) = 3x + 2$ **b** $g(x) = x^2 - 1$
• describe transformations on exponential functions	**4** Describe the transformations on the graph of $y = 3^x$ that give the graph of $y = 2 \times 3^x - 1$.
• justify algebraically and graphically that two functions are mutual inverses	**5** Justify algebraically and graphically that the function $f(x) = \dfrac{5}{x}$ is its own inverse.

F From exponential to logarithmic functions

- What does a logarithmic function look like?
- How are logarithmic functions related to exponential functions?

Objective B: Investigating patterns
ii. describe patterns as general rules consistent with findings

In Exploration 1, you will need to find the correct rules for the patterns you generate in order to solve the given problem.

Exploration 1

1 Fold an A4 size piece of paper in half. You now have two layers. Assume the thickness of the paper is 0.05 mm. The total thickness of the folded paper is 0.1 mm.

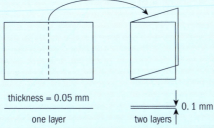

A4 size paper is part of the international standard used in most countries around the world. In the USA and Canada, 'Letter' size is the standard, measuring 8.5 by 11.0 inches.

2 Fold the paper in half again, and write down the number of layers and the total thickness of the folded paper. Repeat once more.

3 Find a function f for the number of layers of the folded paper in terms of the number of folds x.

4 Write down a function g for the total thickness of the folded paper in terms of the number of folds x.

▶ Continued on next page

E10.1 Time for a change

5 Use function g to find the total thickness of the folded paper after:

   a 10 folds   b 20 folds   c 30 folds

6 Use function f to determine how many folds give:

   a 10 layers   b 20 layers   c 30 layers

7 Use function g to determine how many folds give a total thickness of:

   a 10 cm   b 1 m   c 10 m   d 1 km

8 The average distance from Earth to the Moon is 384 400 km. Determine the number of folds needed to give a total thickness of 384 400 km.

Reflect and discuss 1

- The world record for folding an A4 sheet of paper in half is seven times. It is mathematically impossible to fold a sheet of A4 paper more than seven times. Why do you think this is? Think about the relationship of the area of the paper to the number of folds of the paper.

- Do you think it is possible to fold a larger piece of paper more than seven times? Explain.

- In 2002, a 16-year-old American student, Britney Gallivan, demonstrated that a roll of paper 1200 m long could be folded in half 12 times. She developed the formula $L = \frac{\pi t}{6}(2^n + 4)(2^n - 2)$, where L is the length of the paper, t is its thickness, and n is the number of folds. Use her formula to find the thickness of the paper she used.

- Britney Gallivan held the world record for paper folding until a group of students at a school in Massachusetts showed that paper can be folded 13 times. This paper was 0.058 mm thick. Use the formula to find the length of the paper that was folded 13 times.

- For paper 0.058 mm thick, how long would a sheet of paper need to be in order to fold it 14 times?

Exploration 1 shows that the inverse of an exponential function is a logarithmic function. For example, for $y = 2^x$, the equivalent logarithmic statement is $\log_2 y = x$.

Interchanging the x and y in the logarithmic expression $y = 2^x$ gives the inverse function $y = \log_2 x$.

Graphing $y = 2^x$ and $y = \log_2 x$ together verifies that they are mutual inverses, since their graphs are reflections of each other in the line $y = x$.

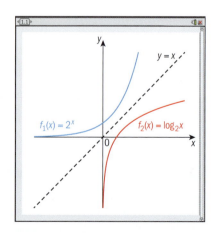

To find the inverse of an exponential function $y = a^x$, use the equivalent logarithmic statement $\log_a y = x$, then interchange the x and y to obtain $y = \log_a x$. The domain of the exponential function $y = a^x$ is $\mathbb{R}$, and its range is $\mathbb{R}^+$. So the domain of its inverse $y = \log_a x$ is $\mathbb{R}^+$, but the range is $\mathbb{R} - \{0\}$, since $\log_a x \neq 0$.

> The inverse of an exponential function is the corresponding logarithmic function. The inverse of $f(x) = a^x$ is $f^{-1}(x) = \log_a x$ and vice versa.

Practice 1

1 State the inverse function of $f(x)$.

   a $f(x) = 4^x$
   b $f(x) = \log_6 x$
   c $f(x) = \ln x$
   d $f(x) = 3\log_5 x$
   e $f(x) = \dfrac{3\ln x}{4}$
   f $f(x) = \dfrac{2}{3}(11^x)$

Problem solving

2 The population of seagulls P in a coastal village can be modelled by the exponential equation $P(x) = 15 \times 1.15^x$, where x is in months.

   a Draw the graph of P (on your GDC) and state P_0, the initial seagull population.
   b State the inverse function.
   c Draw the graph of the inverse function on the same axes.
   d Using the graphs, find how long it would take for the population to reach 20 seagulls.

C Properties of logarithmic functions

- What are the properties of logarithmic functions?
- How are the properties of logarithmic functions related to those of exponential functions?

ATL

Exploration 2

1 Graph the function $y = \log_a x$ and add a slider for the parameter a. Experiment with different values of $a > 1$.

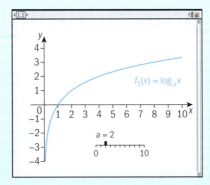

2 a State what happens to the value of y as a increases.

   b State the domain and range of the function.
   c State any asymptotes of the function.
   d State whether the function is concave up or concave down.

▶ Continued on next page

E10.1 Time for a change

e Does the *x*-intercept remain the same for all values of *a*? Explain.

  f Determine if the function has a *y*-intercept. Explain.

  g Explain why *x* cannot be less than or equal to 0.

3 Graph the function $y = \log_a x$ for different values of *a*, where $0 < a < 1$.

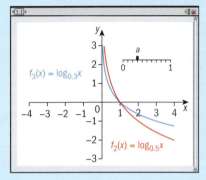

  a State what happens to the value of *y* as *a* increases in the interval (0, 1).

  b Repeat steps **2 b–g** for this graph.

4 Explain how graphs of logarithmic functions with different bases are alike, and how they are different.

5 Summarize your results in a table like this:

	$y = a^x$	$y = \log_a x$
domain		
range		
equation of asymptote		
intercept(s)		

Properties of graphs of logarithmic functions

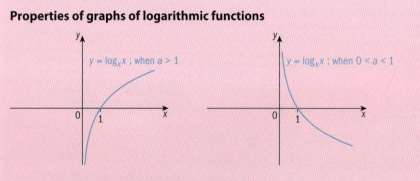

- A logarithmic function is of the form $y = \log_a x$, where $a \in \mathbb{R}^+, a \neq 1$. Its domain is $\mathbb{R}^+$ and its range is $\mathbb{R} - \{0\}$.

- All graphs of logarithmic functions of the form $y = \log_a x$ have *x*-intercept (1, 0) and vertical asymptote $x = 0$.

▶ Continued on next page

- The shape of the graph is determined by the base a. For $a > 1$, the graph is concave down and increasing. For $0 < a < 1$, the graph is concave up and decreasing.
- The value of a controls how rapidly the graph is increasing or decreasing.

Example 1

Find the inverse of $f(x) = 2^{x-3}$ and verify your result graphically. State the domain and range of the function and its inverse.

$y = 2^{x-3}$ ——————————————————— Interchange the x and y variables.

$x = 2^{y-3}$

$\log_2 x = y - 3$ ——————————————————— Rewrite as a logarithmic statement.

$y = \log_2 x + 3$

$f^{-1}(x) = \log_2 x + 3$.

Graphically: ——————————————————— Graph f and f^{-1} and confirm that f and f^{-1} are reflections of each other in $y = x$.

Since f and f^{-1} are reflections of each other in the line $y = x$, they are mutual inverses.

Domain of $f = \mathbb{R}$, range of $f = \mathbb{R}^+$.

Domain of $f^{-1} = \mathbb{R}^+$; range of $f^{-1} = \mathbb{R}$.

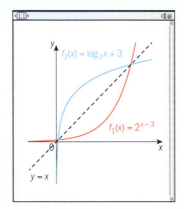

Reflections and their physical properties are well known for just about every type of wave you can think of. Geologists study seismic waves; echoes and sonic effects are studied in acoustics; and of course, reflections are observed with many types of electromagnetic wave, including visible light, which often produces beautiful and interesting images. Perhaps not coincidentally, reflection can also mean the act of peaceful meditation.

Objective C: Communicating
iii. move between different forms of mathematical representation

In Exploration 3 you will move between graphs, tables and equations, and different forms of logarithmic functions.

Exploration 3

1 Here is the graph of a logarithmic function $y = \log_a x$, with its table of values. Explain how you can find a using the table of values.

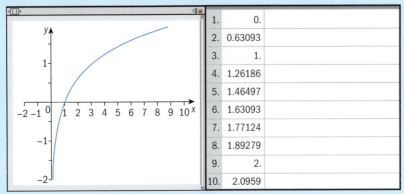

1.	0.
2.	0.63093
3.	1.
4.	1.26186
5.	1.46497
6.	1.63093
7.	1.77124
8.	1.89279
9.	2.
10.	2.0959

2 Copy the graph from step **1**. Explain how you can draw the inverse of the function. Draw the graph of its inverse function.

3 Identify the coordinates on the graph of the function and its inverse that allow you to find a in both.

4 Use what you learned in steps **1–3** to find a in this graph of $y = \log_a x$.

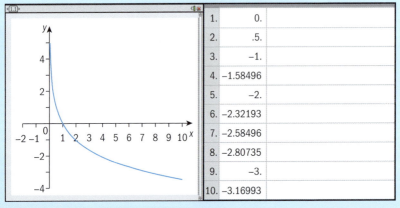

1.	0.
2.	.5.
3.	−1.
4.	−1.58496
5.	−2.
6.	−2.32193
7.	−2.58496
8.	−2.80735
9.	−3.
10.	−3.16993

5 Approximate the value of a from this graph and table of values for $y = \log_a x$. State what you think the exact value of a might be. Explain.

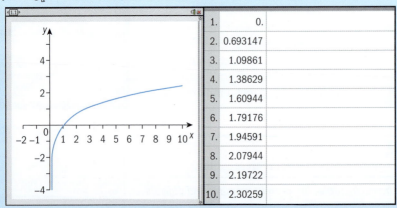

1.	0.
2.	0.693147
3.	1.09861
4.	1.38629
5.	1.60944
6.	1.79176
7.	1.94591
8.	2.07944
9.	2.19722
10.	2.30259

6 Explain an easy way to determine the unknown base of a logarithmic function $y = \log_a x$ from its graph or table of values.

The graph of $y = \log_a x$ passes through the point $(a, 1)$.

Practice 2

1 Match each graph to its function.

   a $f(x) = \log_2 x$ **b** $f(x) = \log_5 x$ **c** $f(x) = \log_{0.3} x$

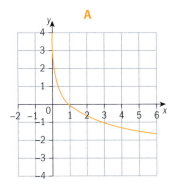
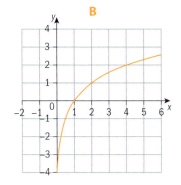
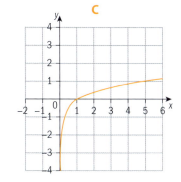

2 The table shows the height in meters of a silver maple tree, measured at the end of each year. The height data can be approximately modelled by the function $y = \log_a x$.

Age of tree x (years)	1	2	3	4	5	6	7	8	9	10
Height y (meters)	0.1	3.5	6	7.5	9	10	10.5	11.5	12	12.5

  a Plot these values on a graph and estimate the value of a.

  b Find the inverse of this function, and state its domain and range.

  c Confirm graphically that the function in **b** is the inverse of the function you found in **a**.

3 Each graph below represents a function of the form $f(x) = \log_a x$. For each graph:

    i state the natural domain and the range of the function

    ii state the base of the logarithmic function.

  a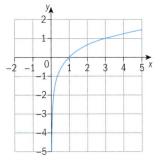
  b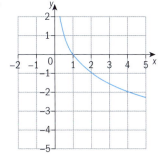
  c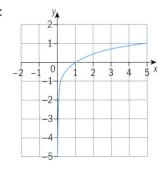

E10.1 Time for a change

4 For each function below:

   i state the natural domain and range of the function

   ii find the inverse of the function, and state any domain restrictions on the function in order for the inverse to be a function

   iii confirm graphically that the function you found is the inverse of the original function.

   a $f(x) = e^{x+2}$ **b** $f(x) = 3\log(x+2)$ **c** $f(x) = 4\log(1+2x)$

   d $f(x) = 2 - \log(3x - 2)$ **e** $f(x) = \ln(2x) - 4$

Problem solving

5 Explain how the change of base property of logarithms ensures that every logarithmic function of the form $y = \log_a x$ can be expressed in the form $y = b \ln x$.

6 For each description below, write down a function with the given characteristics.

   a A logarithmic function, passing through (1, 0) and (4, 2), domain = $\mathbb{R}^+$.

   b An exponential function whose inverse is $f^{-1}(x) = 2\log_5 x + 7$.

   c A logarithmic function, passing through (5, −1), (1, 0) and $\left(\dfrac{1}{25}, 2\right)$.

D Transformations of logarithmic functions

- How do you transform logarithmic functions?
- Is change measurable and predictable?

Exploration 4

1 State the effect of each transformation on the graph of $f(x)$.

   a $f(-x)$ **b** $-f(x)$ **c** $af(x)$ **d** $f(ax)$

   e $f(x - h)$ **f** $f(x) + k$ **g** $f(x - h) + k$

2 Using a graphing tool, verify that the transformations in step **1** have the same effect on the graph of $y = \log_4 x$. Draw a graph of $f(x)$ and the transformed graph to demonstrate what you have discovered. Summarize your results in a table.

3 Confirm your results with another function of the form $y = \log_a x$.

4 For each of the transformations in step **1**, indicate which property it affects (domain, range, x-intercept, asymptote) and how.

Transformations of exponential functions

Reflection:

For the logarithmic function $f(x) = \log_a x$:

- the graph of $y = -f(x)$ is the reflection of $y = f(x)$ in the x-axis.
- the graph of $y = f(-x)$ is the reflection of $y = f(x)$ in the y-axis.

Translation:

For the logarithmic function $f(x) = \log_a x$:

- $y = f(x) + k$ translates $y = f(x)$ by k units in the y-direction.

  When $k > 0$, the graph moves in the positive y-direction (up).

  When $k < 0$, the graph moves in the negative y-direction (down).

- $y = f(x - h)$ translates $y = f(x)$ by h units in the x-direction.

  When $h > 0$, the graph moves in the positive x-direction (to the right).

  When $h < 0$, the graph moves in the negative x-direction (to the left).

Dilation:

For the logarithmic function $f(x) = \log_b x$:

- $y = af(x)$ is a vertical dilation of $f(x)$, scale factor a, parallel to the y-axis.

- $y = f(ax)$ is a horizontal dilation of $f(x)$, scale factor $\frac{1}{a}$, parallel to the x-axis.

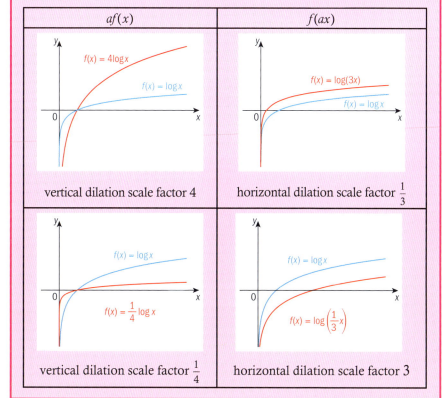

Example 2

Describe the transformations on $f(x) = \log_3 x$ to get $g(x) = 3\log_3(x+4)$.

$f(x) = \log_3 x$ to $f_1(x) = \log_3(x+4)$ is a translation of 4 units in the negative x-direction. —— Work outward from the variable.

$f_1(x) = \log_3(x+4)$ to $g(x) = 3\log_3(x+4)$ is a vertical dilation, scale factor 3.

Practice 3

1 Describe the transformations applied to $f(x)$ to get $g(x)$.

  a $f(x) = \log_4 x$; $g(x) = \log_4(x-7)+1$
  b $f(x) = \log_8 x$; $g(x) = 2\log_8 x - 3$
  c $f(x) = \log_2 x$; $g(x) = \log_2(6x) - 4$
  d $f(x) = \log_5 x$; $g(x) = -4\log_5(x+2)+8$
  e $f(x) = 2\log_3(x+1) - 5$; $g(x) = 6\log_3(x-4)+2$

:**Problem solving**

  f $f(x) = \log_7 x$; $g(x) = \frac{2}{3}\log_7(3x+12) - 10$

2 Describe the transformations applied to $f(x) = \ln x$ to give:

  a $g(x) = \ln(x-3)+2$
  b $h(x) = 0.5\ln x - 1$
  c $j(x) = \ln(2x) - 1$
  d $k(x) = 2\ln x + 1$
  e $m(x) = -3\ln(x+1)$
  f $n(x) = \ln(0.5x)+2$
  g $p(x) = -2\ln(5x) - 3$
  h $q(x) = \frac{1}{5}\ln(2x-6)+4$

3 Each graph shows an original function $f(x)$, labelled A, and transformed function, labelled B. For each one, describe the transformation(s). Hence write down the equation of the transformed function in terms of f.

a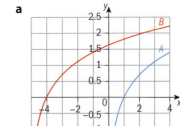

b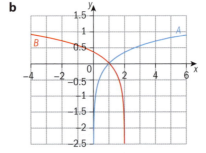

c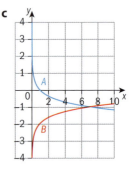

:**Problem solving**

4 Draw the graphs of $f(x) = \ln x$ and $g(x) = \ln\left[2\left(x - \frac{1}{2}\right)\right]$ on the same set of axes.

Describe the transformations on the graph of f that give the graph of g.

5 Draw the graphs of $f(x) = \log_5 x$ and $h(x) = -\log_5[5(x+2)]$ on the same set of axes. Describe the transformations on the graph of f that give the graph of h.

Summary

A logarithmic function is of the form $y = \log_a x$, where $a \in \mathbb{R}^+$, $a \neq 1$. Its domain is $\mathbb{R}^+$ and its range is $\mathbb{R} - \{0\}$, since x cannot take the value 0.

The inverse of the inverse of an exponential function is the corresponding logarithmic function.

The inverse of $f(x) = a^x$ is $f^{-1}(x) = \log_a x$ and vice versa.

Graphs of logarithmic functions

- All graphs of logarithmic functions of the form $y = \log_a x$ have x-intercept $(1, 0)$ and vertical asymptote $x = 0$.
- The shape of the graph is determined by the base a. For $a > 1$, the graph is concave down and increasing. For $0 < a < 1$, the graph is concave up and decreasing.
- The value of a controls how rapidly the graph is increasing or decreasing.
- The graph of $y = \log_a x$ passes through the point $(a, 1)$.

Transformations of logarithmic functions

Reflection:

For the logarithmic function $f(x) = \log_b x$:

- the graph of $y = -f(x)$ is the reflection of $y = f(x)$ in the x-axis.
- the graph of $y = f(-x)$ is the reflection of $y = f(x)$ in the y-axis.

Translation:

For the logarithmic function $f(x) = \log_b x$:

- $y = f(x) + k$ translates $y = f(x)$ by k units in the y-direction.

  When $k > 0$, the graph moves in the positive y-direction (up).

  When $k < 0$, the graph moves in the negative y-direction (down).

- $y = f(x - h)$ translates $y = f(x)$ by h units in the x-direction.

  When $h > 0$, the graph moves in the positive x-direction (to the right).

  When $h < 0$, the graph moves in the negative x-direction (to the left).

Dilation:

For the logarithmic function $f(x) = \log_a x$:

- $y = bf(x)$ is a vertical dilation of $f(x)$, scale factor b, parallel to the y-axis.
- $y = f(bx)$ is a horizontal dilation of $f(x)$, scale factor $\frac{1}{b}$ parallel to the x-axis.

Mixed practice

1 Match the function with its graph:

   a $y = \log_5 x$ **b** $y = \log_5(-x)$ **c** $y = \log_{0.5} x$

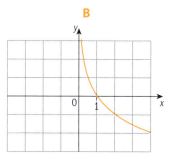
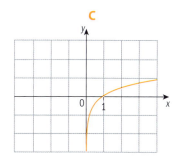

E10.1 Time for a change

2 While training for a race, Joseph tracks his average speed for a 10 km run. The data, shown in the table below, can be approximately modelled by the function $s = \log_a w + 14$, where s = speed and w = week of training.

Week	Speed (km/h)
1	14
2	14.88
3	15.39
4	15.76
5	16.04
6	16.27
7	16.47
8	16.64
9	16.79
10	16.92

a Explain why a logarithmic model is a good fit for this kind of data.

b Plot these values on a graph and estimate the value of a.

c Find the inverse of this function. **State** the relationship that the inverse function represents.

3 For each function in parts **a** to **e**:

  i state the natural domain and range of the function

  ii find the inverse of the given function, and state any domain restrictions on the original function in order for the inverse to be a function

  iii confirm graphically that the function you found is the inverse of the original function.

  a $y = \ln(x - 3)$

  b $y = 5(7^{0.2x}) + 2$

  c $y = \frac{1}{2}\log_2(4x - 8) - 5$

  d $y = \frac{\log_2(x+1)}{\log_2 3}$

  e $y = 4^{1-x}$

4 Describe the transformation(s) necessary to change the graph of $f(x) = \log_a x$ into the graph of the given function, assuming the base remains the same. Where more than one transformation is needed, make sure you write them in the correct order.

a $y = \log_5 x + 2$ **b** $y = \log_7(x + 2)$

c $y = \ln(x - 2) - 3$ **d** $y = \ln(-x) + 2$

e $y = \log\left(\frac{x}{4}\right)$ **f** $y = \ln(2 - x)$

g $y = -2\log_3(-x)$ **h** $y = 2\log_8(x - 2) - 1$

i $y = -2\ln(1 - x) + 2$

Problem solving

5 Sketch the graphs of $f(x) = \ln x$ and $g(x) = -7\ln(3x - 12) + 1$ on the same set of axes. **Describe** the transformations on the graph of f to give the graph of g.

6 Sketch the graphs of $f(x) = \log x$ and $g(x) = \frac{1}{4}\log(2 - x)$ on the same set of axes. **Describe** the transformations on the graph of f to give the graph of g.

7 The data in the table here follow an exponential model of the form $y = a^x + k$.

x	0	$\frac{1}{2}$	1	$\frac{3}{2}$
y	4	5	7	11

a Find the inverse of this exponential function.

b Describe the different ways you can find this inverse function.

Review in context

In E6.1 you explored the logarithmic models of some natural phenomena. The questions here explore the graphs of these phenomena by considering the transformations on their graphs.

1 pH is calculated using the logarithmic formula
pH = $-\log_{10}(H^+)$.

Using transformations of functions, **describe** the difference in graphs of this model and the standard logarithmic model $y = \log x$.

2 The formula for modelling the magnitude M of an earthquake using the Richter scale is $M = \log 10 \frac{I_c}{I_n}$, where I_c is the intensity of the 'movement' of the earth from the earthquake and I_n is the the intensity of the 'movement' of the earth on a normal day-to-day basis.

a A series of earthquakes were recorded in Russia, which followed the standard model. A series of earthquakes were recorded in Mexico which were 4 times this intensity.

Describe the graph of these earthquakes using transformations of functions.

b In Australia, the magnitude of the quakes is known to be double the magnitude of the Russian quakes. **Describe** how the graph of this model would differ from the standard model using transformations of functions.

3 The amount of energy released during an earthquake can be modelled by the function $r = 0.67 \log E - 7.6$, where r is the number on the Richter scale, and E is the energy released by the earthquake in ergs.

a By comparing this model to the standard logarithmic function $f(x) = \log x$, **describe**, using transformations of functions:

i the effect of the scale factor 0.67 on the graph of $f(x)$, and the name of this transformation

ii the effect of the parameter -7.6, and the name of this transformation.

b The range for this model is the number on the Richter scale, $0 \le r \le 10$.

Determine the domain of this model.

c **Explain** why the graph of this model has no y-intercept. **State** the kind of transformation on the graph that *would* give a y-intercept.

4 Height above sea level and atmospheric air pressure can be modelled by the function

$$h = -26400 \ln\left(\frac{p}{2120}\right)$$

where h is measured in feet, and p in pounds per square foot. **Describe** this model as a transformation of the graph of $y = \ln x$.

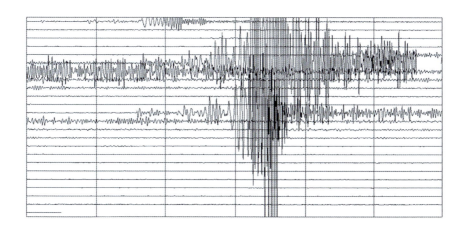

Reflect and discuss

How have you explored the statement of inquiry? Give specific examples.

Statement of Inquiry:

Generalizing changes in quantity helps establish relationships that can model duration, frequency and variability.

E10.1 Time for a change

E10.2 Meet the transformers

Global context: Scientific and technical innovation

Objectives
- Graphing rational functions of the form $f(x) = \dfrac{a}{x-h} + k$
- Finding asymptotes of graphs of rational functions
- Transforming rational functions using translations, reflections and dilations
- Identifying transformations of graphs
- Finding the inverse of a rational function

Inquiry questions

F
- What is a rational function?
- How are asymptotes linked to domain and range?

C
- How can understanding transformations of rational functions help you graph them?

D
- Can a function be its own inverse?
- Does science solve problems or create them?

ATL Critical-thinking

Revise understanding based on new information and evidence

Statement of Inquiry:

Representing change and equivalence in a variety of forms has helped humans apply our understanding of scientific principles.

10.2

11.3

E10.2

216

ALGEBRA

You should already know how to:

• find the domain and range of a function	1 State the domain and range of these functions.
	  a $f(x) = 2x + 5$ b $f(x) = x^2$
	  c $f(x) = \dfrac{2}{1-x}$

• write statements of proportionality	2 Write the statement of proportionality for:
	  a y is proportional to x
	  b y is inversely proportional to x.

• transform functions	3 State the combination of transformations applied to $f(x)$ to get $g(x)$: $f(x) = x^2$, $g(x) = -2(x+3)^2 + 5$
• find an inverse function algebraically	4 Find the inverse relation of each function algebraically, and state whether or not the inverse relation is also a function. a $f(x) = \dfrac{3}{4}x + 2$ b $f(x) = (x-2)^2$

F Introducing rational functions

- What is a rational function?
- How are asymptotes linked to domain and range?

> A **rational function** is any function which can be expressed as a quotient $\dfrac{f(x)}{g(x)}$ where $f(x)$ and $g(x)$ are functions of x, and $g(x) \neq 0$.

> The function $f(x) = \dfrac{1}{x}$, $x \neq 0$, called the **reciprocal function**, is one type of rational function. Its domain is the set of real numbers excluding zero, and its range is the set of real numbers excluding zero.

One practical application of a reciprocal function is the design and layout of the cables used on a suspension bridge. The cables help keep the bridge stable, and this is achieved by engineers who design the bridge so that the opposing forces of tension and compression work together. This is a photo of the Clifton Suspension Bridge, which spans the Avon Gorge and the River Avon, in Bristol, England.

E10.2 Meet the transformers

Exploration 1

1 Complete the table of values for the function $y = \frac{1}{x}$.

x	-8	-4	-2	-1	$-\frac{1}{2}$	$-\frac{1}{4}$	$-\frac{1}{8}$	$\frac{1}{8}$	$\frac{1}{4}$	$\frac{1}{2}$	1	2	4	8
y	$-\frac{1}{8}$			-2										

2 Use the table in step **1** to help you draw the graph of $y = \frac{1}{x}$ for values of x from -8 to 8.

3 **a** Explain what happens as x gets closer to 0. Think about what happens for both positive and negative values of x that approach 0.

   b Describe the behavior of y as x gets closer and closer to 0 on the graph. Think about positive and negative values for x.

   c In a different color, draw a vertical line on your graph that the curve never touches.

   d Explain why x can never be equal to 0. Hence, state the domain of the function $y = \frac{1}{x}$.

4 **a** Find the values of x for which y gets closer to 0. Think about positive and negative values for x.

   b Explain why y can never be equal to 0. How is this represented on the graph?

   c In a different color, draw a horizontal line on your graph that the curve never touches.

   d State the range of the function $y = \frac{1}{x}$.

For a reciprocal function:

- The **horizontal asymptote** is the horizontal line that the graph of $f(x)$ approaches as x approaches positive or negative infinity.
- The **vertical asymptote** is the vertical line that the graph of $f(x)$ approaches as x approaches 0. (The denominator can never equal zero.)

Reflect and discuss 1

- Find the equation of the horizontal asymptote and the vertical asymptote for the graph of $y = \frac{1}{x}$.
- How are the domain and range of this function related to the equations of the asymptotes?

Exploration 2

1 a Copy and complete this table of values for $y = -\frac{1}{x}$.

x	-8	-4	-2	-1	$-\frac{1}{2}$	$-\frac{1}{4}$	$-\frac{1}{8}$	$\frac{1}{8}$	$\frac{1}{4}$	$\frac{1}{2}$	1	2	4	8
y														

b Draw the graph of $y = -\frac{1}{x}$ for values of x between -8 and 8.

c State the domain and range of this function.

d State the equation of the horizontal and vertical asymptotes. How do they relate to the domain and range?

e Look back at your graph of $y = \frac{1}{x}$ from Exploration 1. Explain why that graph is only in quadrants 1 and 3.

f Explain why the graph of $y = -\frac{1}{x}$ is only in quadrants 2 and 4.

g Write down a conjecture for which quadrants the graph of $y = \frac{a}{x}$ will be in for $a > 0$, and for $a < 0$. Test your conjecture with a GDC.

Reflect and discuss 2

- If $f(x) = \frac{1}{x}$ and $g(x) = -\frac{1}{x}$, is $g(x) = -f(x)$ or is $g(x) = f(-x)$? How do the effects of the two transformations $-f(x)$ and $f(-x)$ compare when $f(x) = \frac{1}{x}$?

Exploration 3

1 Graph the function $y = \frac{a}{x}$ and insert a slider for the parameter a for values from -10 to $+10$.

2 Move the slider to explore what happens to the graph for different values of a (positive, negative, fraction, zero).

3 Generalize the effect of the parameter a on the graph of $y = \frac{a}{x}, a \neq 0$. Write conclusions for when $a > 0$, and for $a < 0$.

4 Explain why the case when $a = 0$ is ignored in step **3**.

The graph of a reciprocal function $y = \frac{a}{x}$, $a \neq 0$ is a special case of a curve called a **hyperbola**. It has a horizontal asymptote at $y = 0$ and a vertical asymptote at $x = 0$. Because the asymptotes are perpendicular, it is also known as a **rectangular hyperbola**.

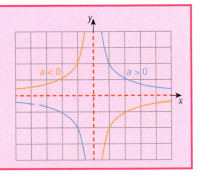

Reflect and discuss 3

- Compare the effect of parameter a on the graph of $y = \dfrac{a}{x}$ with its effect on:

    a linear function ($y = ax$)

    a quadratic function ($y = ax^2$)

- For the rational function $f(x) = \dfrac{1}{x}$, is $y = \dfrac{1}{2x}$ the same as $y = \dfrac{1}{2}f(x)$ or $y = f(2x)$? What does this mean about the effects of these transformations on the graphs of rational functions?

Example 1

For each function below:
- state the domain and range
- write down the equations of the asymptotes
- draw the graph of the function.

a $f(x) = \dfrac{3}{x}$ **b** $f(x) = -\dfrac{2}{x}$

---

a $f(x) = \dfrac{3}{x}$

Domain: $x \in \mathbb{R}, x \neq 0$; range: $y \in \mathbb{R}, y \neq 0$

Horizontal asymptote: $y = 0$

Vertical asymptote: $x = 0$

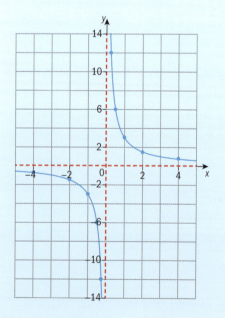

x	y
−4	−0.75
−2	−1.5
−1	−3
$-\dfrac{1}{2}$	−6
$-\dfrac{1}{4}$	−12
$\dfrac{1}{4}$	12
$\dfrac{1}{2}$	6
1	3
2	1.5
4	0.75

Make a table of values. Draw the graph by plotting several points from the table of values.

▶ Continued on next page

b $f(x) = -\dfrac{2}{x}$

Domain: $x \in \mathbb{R}, x \neq 0$; range: $y \in \mathbb{R}, y \neq 0$

Horizontal asymptote: $y = 0$

Vertical asymptote: $x = 0$

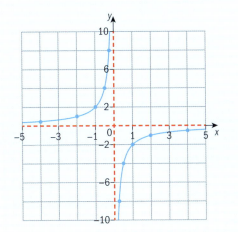

x	y
−4	0.5
−2	1
−1	2
$-\dfrac{1}{2}$	4
$-\dfrac{1}{4}$	8
$\dfrac{1}{4}$	−8
$\dfrac{1}{2}$	−4
1	−2
2	−1
4	−0.5

Tip

When $a > 0$, $y = \dfrac{a}{x}$ is always in quadrants 1 and 3. When $a < 0$, the graph is reflected in the *x*-axis, and is therefore always in quadrants 2 and 4.

For the function $y = \dfrac{1}{x}$:

$y = af(x)$ is a vertical dilation of $f(x)$, scale factor a, parallel to the *y*-axis.	$y = f(ax)$ is a horizontal dilation of $f(x)$, scale factor $\dfrac{1}{a}$, parallel to the *x*-axis.
$af(x)$	$f(ax)$
graph: $f_1(x) = \dfrac{1}{x}$ (blue), $f_2(x) = 5 \cdot \dfrac{1}{x}$ (orange)	*graph: $f_1(x) = \dfrac{1}{x}$ (blue), $f_2(x) = \dfrac{1}{5x}$ (orange)*
vertical dilation scale factor 5	horizontal dilation scale factor $\dfrac{1}{5}$

▶ Continued on next page

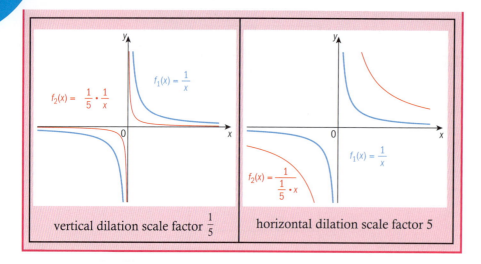

| vertical dilation scale factor $\frac{1}{5}$ | horizontal dilation scale factor 5 |

Reflect and discuss 4
- Explain why it makes sense that the vertical asymptote of $y = \frac{a}{x}$ is $x = 0$.
- Based on your answer above, what do you think the vertical asymptote will be for the function $y = \frac{2}{x-1}$?
- When does a rational function have two vertical asymptotes? Give an example of such a function.

> Rational functions with more than one vertical asymptote are beyond the level of this course.

Example 2

It takes 20 students 14 days to eat all the snacks in a vending machine.

Model this situation with a rational function and sketch its graph.

Determine how many days it would take 8 students to eat all the snacks.

Let x = number of students

Let y = number of days

Define variables. Decide which variable is dependent on the other.

$y \propto \frac{1}{x} \Rightarrow y = \frac{k}{x}$

Identify inverse proportion – as the number of students increases, the number of days decreases.

When $x = 20$, $y = 14$, so:

$14 = \frac{k}{20} \Rightarrow k = 280$

Use the initial condition to calculate the constant of proportionality k.

$\Rightarrow y = \frac{280}{x}$

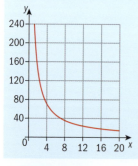

When $x = 8$ (8 students),

$y = \frac{280}{8} = 35$. Therefore, it would take 35 days for 8 students to eat the same number of snacks.

x cannot take negative values, so the graph is in the first quadrant only.

Reflect and discuss 5

- In Example 2, what kind of values can *x* take? Is it possible to have any number of students, including fractions? Does your graph represent this? If not, describe how to change your graph to reflect this limitation.

- In Example 2, do *x* and *y* have maximum or minimum values? If so, state them. Explain your reasoning. Would it make sense for either *x* or *y* to have a value of 0?

Practice 1

1 For each function:
- state the domain and range
- draw a graph of the function
- draw and label the asymptotes.

a $y = \dfrac{4}{x}$ **b** $y = -\dfrac{8}{x}$ **c** $y = \dfrac{1}{4x}$ **d** $y = -\dfrac{2}{3x}$

2 The length of time it takes to build a community shelter is inversely proportional to the number of people working on the project. It takes 600 hours for 4 people to build a shelter.

a Model this situation with a rational function and sketch its graph.

b Determine how long it would take 12 people to build a shelter.

> A 2016 Design of the Year was awarded to a flat-pack shelter kit which included all the tools required for assembling, and could be put together by four people in just four hours.

3 Below is the graph of the resistance in an electric circuit (R) versus the current (I) flowing through it for a constant voltage.

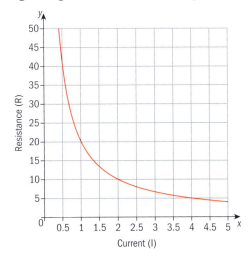

a Describe the relationship between current and resistance.

b Find the constant of proportionality using points from the graph.

c Hence write down the function relating current (I) and resistance (R).

E10.2 Meet the transformers   223

4 The graph models the number of hours (y) it takes for aid workers (x) to build a shelter for refugees.

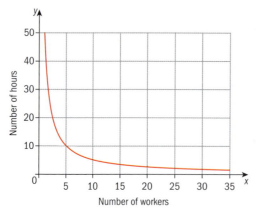

From this graph, determine the rational function and hence explain how long it would take:

a 1 person to build the shelter

b 5 people to build the shelter

c 10 people to build the shelter.

Problem solving

5 Newton's second law of motion, $F = ma$, states the relationship between force, mass and acceleration.

  a Describe the proportional relationship between:
  i force and acceleration
  ii acceleration and mass.

  b Hence sketch graphs of:
  i force versus acceleration
  ii acceleration versus mass.

Transforming rational functions

- How can understanding transformations of rational functions help you graph them?

> ### Reflect and discuss 6
> - For the function $y = f(x - h) + k$, describe the effects of the parameters h and k on the graph of $y = f(x)$. Use appropriate mathematical terms and demonstrate what you mean with an example.
> - Predict what happens to the graph of $y = \dfrac{1}{x}$ after it is transformed using $y = \dfrac{1}{x-2} + 3$. What will be the equations of its asymptotes? Explain your reasoning.

ALGEBRA

Objective B: Investigating patterns
i. select and apply mathematical problem-solving techniques to discover complex patterns

In Exploration 4 you will use graphing and exploration skills to make new findings and draw new conclusions. Explain your findings clearly with labelled sketches and state your conclusions clearly.

Exploration 4

1. Use your GDC to graph the function $y = \dfrac{a}{x-h} + k$ and insert sliders for the parameters a, h and k.

2. Set the values $h = 0$ and $k = 0$. Move the slider to explore what happens to the graph for different values of a (e.g. positive, negative, fraction).

3. Describe the effect of a on the graph of $y = \dfrac{a}{x}$. Include positive, negative and fractional values of a. State the asymptotes, domain and range.

4. Set the value $a = 1$. Move the sliders one at a time to explore what happens for different values of h and k (positive, negative, fraction).

5. Sketch several graphs to demonstrate the effect of h on the graph of $y = \dfrac{1}{x-h} + k$. Include positive, negative and fractional values of h.
Label the asymptotes and state the domain and range.

6. Sketch several graphs to demonstrate the effect of k on the graph of $y = \dfrac{1}{x-h} + k$. Include positive, negative and fractional values of k.
Label the asymptotes and state the domain and range.

7. Generalize the effects of the parameters h and k in $y = \dfrac{a}{x-h} + k$ on the graph of $y = \dfrac{1}{x}$.

8. Determine the horizontal and vertical asymptotes of a rational function, $y = \dfrac{a}{x-h} + k$, just by looking at the function.

9. Determine the domain and range of $y = \dfrac{a}{x-h} + k$ just by looking at the function.

Reflect and discuss 7

- In Reflect and discuss 6, you were asked to predict how the graph of $y = \dfrac{1}{x-2} + 3$ might look. How good was your prediction? If you weren't completely correct, what *should* you have predicted?

E10.2 Meet the transformers

The function $f(x) = \frac{a}{x-h} + k$, where $a \neq 0$, is a transformation of $f(x) = \frac{1}{x}$ (the parent function).

The domain of the rational function $f(x) = \frac{a}{x-h} + k$ is $x \in \mathbb{R}, x \neq h$.

The range of the rational function $f(x) = \frac{a}{x-h} + k$ is $y \in \mathbb{R}, y \neq k$.

The equation of the horizontal asymptote is $y = k$ and the equation of the vertical asymptote is $x = h$.

Example 3

Sketch the graph of $y = \frac{1}{x-1} + 3$.

Horizontal asymptote: $y = 3$
Vertical asymptote: $x = 1$

— Find the asymptotes.

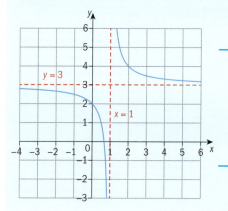

— Sketch the asymptotes on the graph using dashed lines.

— Sketch the function.

Tip

To check where a graph is located, you can substitute into its equation a value of x to the right of the vertical axis, and a value to the left.

For the graph in Example 3: when $x = 0$, $y = 2$; when $x = 2$, $y = 4$.

You can then sketch the curves through these points.

Practice 2

1 For each of the functions below:
- find the value of h and k
- write down the equations of the asymptotes
- write down the domain and range.

a $f(x) = \frac{1}{x+3} - 7$ **b** $f(x) = \frac{1}{x+2}$ **c** $f(x) = \frac{1}{x-5} + 8$

d $f(x) = \frac{1}{x} - 11$ **e** $f(x) = \frac{1}{x-1} + \pi$

2 For each function below:
- draw and label the asymptotes
- sketch the function on a graph
- state the domain and range.

a $y = \dfrac{1}{x} - 4$ **b** $y = \dfrac{1}{x-4}$ **c** $y = \dfrac{1}{x} + 3$

d $y = \dfrac{1}{x-2}$ **e** $y = \dfrac{1}{x+2} + 5$ **f** $y = \dfrac{1}{x-3} + 1$

3 Each of these functions has an equation of the form $y = \dfrac{1}{x-h} + k$. Find the equation of each graph.

a
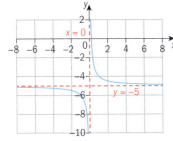

b
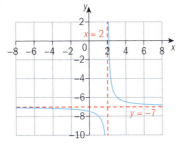

c
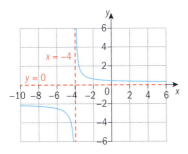

d
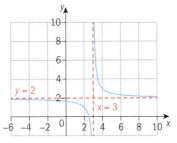

Problem solving

4 Find the equation of a curve $y = \dfrac{1}{x-h} + k$ with asymptotes:

a $y = 4$ and $x = 3$ **b** $y = 2$ and the y-axis

c $y = -1$ and $x = 2$ **d** $y = \dfrac{4}{3}$ and $x = -\dfrac{3}{4}$

5 Describe the individual transformations on $f(x) = \dfrac{1}{x}$ to obtain $f(x) = \dfrac{1}{x-h} + k$ for each set of values for h and k.

a i $h > 0$ **b i** $k > 0$
ii $h < 0$ **ii** $k < 0$

Example 4

For each function below:
- list the individual transformations on $f(x)=\frac{1}{x}$
- find the equations of the asymptotes
- hence sketch the graph of the function.

a $f(x)=\frac{10}{x-3}+2$

b $f(x)=\frac{4}{3-2x}+3$

a $f(x)=\frac{10}{x-3}+2$

Use the values a, h and k from the rational function $f(x)=\frac{a}{x-h}+k$.

$h=3 \Rightarrow$ horizontal translation of 3 units in the positive x-direction

$a=10 \Rightarrow$ vertical dilation scale factor 10

$k=2 \Rightarrow$ vertical translation of 2 units in the positive y-direction

Horizontal asymptote: $y=2$

Vertical asymptote: $x=3$

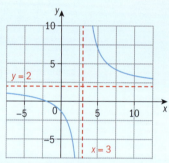

Draw the asymptotes as dashed lines before sketching the graph.

b $f(x)=\frac{4}{3-2x}+3$

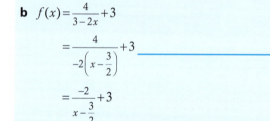

Rearrange the equation to look like $f(x)=\frac{a}{x-h}+k$ by factorizing out the coefficient of x in the denominator.

$h=\frac{3}{2} \Rightarrow$ horizontal translation 1.5 units in the positive x-direction

$a=-2 \Rightarrow$ vertical stretch of factor 2 and horizontal reflection

A negative value of a gives a reflection in the x-axis.

$k=3 \Rightarrow$ vertical translation 3 units in the positive y-direction

Horizontal asymptote: $y=3$

Vertical asymptote: $x=1.5$

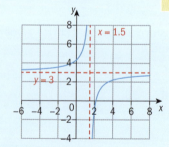

The order for performing multiple transformations on functions is:

1 Horizontal translation (h)
2 Reflection in the x-axis (if $a < 0$)
3 Vertical dilation (a)
4 Vertical translation (k)

> For a rational function $f(x) = \dfrac{a}{x-h} + k$:
> - h determines the horizontal translation; the vertical asymptote is $x = h$.
> - k determines the vertical translation; the horizontal asymptote is $y = k$.
> - a determines
>   - either the horizontal dilation of factor $\dfrac{1}{a}$ or the vertical dilation of factor a
>   - the reflection in the x-axis (if $a < 0$)

Practice 3

1 Perform these transformations on the rational function $f(x) = \dfrac{1}{x}$, then write down the function obtained as a single algebraic fraction.

   a Vertical dilation by scale factor $\dfrac{1}{2}$, vertical translation of 1 unit in the negative y-direction, horizontal translation of 4 units in the positive x-direction.

   b Horizontal dilation by scale factor 3, horizontal translation of 3 units, vertical translation of -2 units.

   c Vertical dilation by scale factor $\dfrac{1}{4}$, reflection in the x-axis, vertical translation of 8 units.

   d Vertical dilation by scale factor $\dfrac{3}{2}$, reflection in the y-axis, vertical translation of 4 units, horizontal translation of -2 units.

2 For each rational function below:
   - list the transformations applied to $f(x) = \dfrac{1}{x}$
   - find the equations of the asymptotes
   - hence sketch the graph of the rational function.

   a $f(x) = \dfrac{3}{x+6} - 5$ b $f(x) = -\dfrac{5}{2-x} + 3$ c $f(x) = \dfrac{9}{3x+1}$

   d $f(x) = \dfrac{5}{8-2x} + 2$ e $f(x) = -\dfrac{5}{2x-1} - 8$ f $f(x) = -\dfrac{13}{x} - 11$

D The inverse of a rational function

- Can a function be its own inverse?
- Does science solve problems or create them?

Exploration 5

1. Using your GDC, draw the graph of $y = \dfrac{1}{x}$ and the line $y = x$. Reflect the graph of $y = \dfrac{1}{x}$ in the line $y = x$. Describe what you notice.

2. Repeat step **1** for the graphs of:

   a $y = \dfrac{2}{x}$ b $y = \dfrac{5}{x}$ c $y = \dfrac{0.5}{x}$ d $y = \dfrac{0.1}{x}$

3. Explain what your results tell you about the inverse relation of a rational function $y = \dfrac{a}{x}$, $a \neq 0$. Justify why this inverse relation is a function.

4. a Find the inverse function $f^{-1}(x)$ of the function $f(x) = \dfrac{a}{x}$, $a \neq 0$ algebraically.

   b Determine the domain and range of $f^{-1}(x)$.

   c Summarize your findings about the inverse function of $y = \dfrac{a}{x}$, $a \neq 0$.

Reflect and discuss 8

You could say that in the reciprocal function $y = \dfrac{1}{x}$, y is the inverse of x, because y represents the multiplicative inverse (or reciprocal) of x, except when $x = 0$. This means that $yx = 1$. However, the reciprocal function $y = \dfrac{1}{x}$ is not the inverse function of $y = x$.

- How are the multiplicative inverse (reciprocals) and inverse functions different?
- What is meant by 'inverse function'?
- What is the inverse function of $y = x$?
- What is the inverse function of $y = \dfrac{1}{x}$?

Example 5

Find the inverse of $f(x) = \dfrac{-2}{x+8} - 10$.

$y = \dfrac{-2}{x+8} - 10$ ———————————————— Interchange the x and the y.

$x = \dfrac{-2}{y+8} - 10$

$x + 10 = \dfrac{-2}{y+8}$

$y + 8 = \dfrac{-2}{x+10}$ ———————————————— Isolate the variable y.

$y = \dfrac{-2}{x+10} - 8$

$f^{-1}(x) = \dfrac{-2}{x+10} - 8$

Practice 4

1 On your GDC, graph each function and its inverse on the same grid.

  a $y = \dfrac{4}{x}$
  b $y = -\dfrac{2}{x}$
  c $y = \dfrac{1}{x+4}$
  d $y = \dfrac{3}{x} - 1$
  e $y = 3 + \dfrac{3}{3-x}$
  f $y = \dfrac{1}{1-x} + 1$

2 Find the inverse of each of each function:

  a $y = \dfrac{-5}{x+1}$
  b $y = \dfrac{7}{x-9} + 6$
  c $y = -\dfrac{1}{x-12} - 8$
  d $y = \dfrac{6}{5-x} - 3$
  e $y = \dfrac{3x-1}{x+3}$
  f $y = -\dfrac{2x}{x-4}$

Summary

A **rational function** is any function which can be expressed as a quotient $\dfrac{f(x)}{g(x)}$ where $f(x)$ and $g(x)$ are functions of x, and $g(x) \neq 0$.

The function $f(x) = \dfrac{1}{x}$, $x \neq 0$, is a rational function called the reciprocal function. Its domain is the set of real numbers excluding zero and its range is the set of real numbers excluding zero.

For a reciprocal function, the **horizontal asymptote** is the horizontal line that the graph approaches as x approaches positive or negative infinity.

For a reciprocal function, the **vertical asymptote** is the vertical line that $f(x)$ approaches as the denominator approaches 0. (The denominator can never equal 0.)

For the function $f(x) = \dfrac{a}{x-h} + k$, the vertical asymptote is $x = h$ and the horizontal asymptote is $y = k$.

For a rational function $f(x) = \dfrac{a}{x-h} + k$:
- h determines the horizontal translation
- k determines the vertical translation
- a determines
  - either the horizontal dilation of factor $\dfrac{1}{a}$ or the vertical dilation of factor a
  - the horizontal reflection
  - whether the graph is reflected in the x-axis ($a < 0$) or not ($a > 0$).

The domain of the rational function $f(x) = \dfrac{a}{x-h} + k$ is $x \in \mathbb{R}, x \neq h$. The range of the rational function $f(x) = \dfrac{a}{x-h} + k$ is $y \in \mathbb{R}, y \neq k$. The horizontal asymptote is at $y = k$ and the vertical asymptote is at $x = h$.

Mixed practice

1 For each function below:
- **state** the largest possible domain and range
- **write down** the equations of the asymptotes
- **sketch** the graph.

  a $y = -\dfrac{5}{x}$
  b $y = \dfrac{1}{x} + 1$
  c $y = \dfrac{2}{x-2} - 8$
  d $y = -\dfrac{4}{x+3} + 2$

2 Perform these transformations on the rational function $f(x) = \dfrac{1}{x}$ to obtain $g(x)$. **Write down** $g(x)$ obtained as a single algebraic fraction.

  a The function $f(x) = \dfrac{1}{x}$ is vertically dilated by scale factor 3, translated vertically 5 units in the negative y-direction and horizontally by 5 units in the positive x-direction.

b The function $f(x)=\frac{1}{x}$ is vertically dilated by scale factor $\frac{1}{2}$, translated 1 unit vertically in the positive y-direction and reflected in the y-axis.

c The function $f(x)=\frac{1}{x}$ is reflected in the x-axis, translated vertically by 2 units in the negative y-direction and horizontally by 3 units in the positive x-direction.

d The function $f(x)=\frac{1}{x}$ is dilated horizontally by scale factor 12, reflected in the x-axis, translated 1 unit vertically in the negative y-direction and 4 units horizontally in the positive x-direction.

3 For each rational function:
- **list** the individual transformations applied to $f(x)=\frac{1}{x}$ to obtain the given function
- **find** the equations of the asymptotes
- **sketch** the graph of the rational function
- **find** its inverse.

a $g(x)=-\frac{1}{x-2}+9$ **b** $h(x)=\frac{3}{x+1}-3$

c $k(x)=\frac{10}{5-x}$ **d** $m(x)=-\frac{4}{3x+2}+1$

e $n(x)=\frac{5}{8-4x}+6$ **f** $p(x)=\frac{1}{4x-5}$

Problem solving

4 Young's dosage calculates drug dosages for children. For a certain pharmaceutical drug the adult dose is 50 mg.

Young's dosage says that the children's dose is $C=50\left(\frac{x}{x+12}\right)$, where x is the age of the child.

a Calculate the dosage of the medicine for a 10-year-old child.

b Rewrite the function in the form $C=\frac{a}{x-h}+k$.

c Explain the series of transformations.

Review in context

1 Catherine is travelling from New York to Tokyo and then to Sydney.

She starts with 1000 USD and converts this to Yen. She spends 40 000 Yen in Tokyo, and converts her remaining money into AUD when she reaches Sydney. Using your knowledge of reciprocal functions and the currency conversion rates below, **explain** how much money Catherine now has in AUD.

1 USD =113.4 YEN = 1.32 AUD

2 Two oil pipelines run between Abuja and Lagos in Nigeria. One pipeline can fill an oil tanker three times faster than a second pipeline. If both pipelines are operational, an oil reservoir can be filled in 14 hours.

> **Tip**
>
> $r=\frac{W}{t}$ where r is the rate, W is the work required (in this case fill 1 reservoir), t is the time taken.

a Let the hours needed to fill one reservoir from the fast pipe be p. The rate of filling is the reciprocal function $r=\frac{1}{p}$. Find the reciprocal function for the rate of filling by the slow pipe.

b Hence find an expression for the two pipes working together.

c As it takes 14 hours to fill the reservoir with both pipelines, find p. Hence find the time taken to fill the reservoir with just the slow pipe.

> **Reflect and discuss**
>
> How have you explored the statement of inquiry? Give specific examples.

> **Statement of Inquiry:**
>
> Representing change and equivalence in a variety of forms has helped humans apply our understanding of scientific principles.

E11.1 Unmistaken identities

Global context: Orientation in space and time

Objectives
- Using trigonometric identities to simplify expressions and solve equations
- Solving equations involving trigonometric functions

Inquiry questions

- How do you solve trigonometric equations?
- How many solutions does a trigonometric equation have?

- How do you prove trigonometric identities?
- How can you use trigonometric identities to solve trigonometric equations?

- Can two different solutions both be correct?
- How do we define 'where' and 'when'?

RELATIONSHIPS

ATL	Creative-thinking

Make unexpected or unusual connections between objects and/or ideas

Statement of Inquiry:

Generalizing and applying relationships between measurements in space can help define 'where' and 'when'.

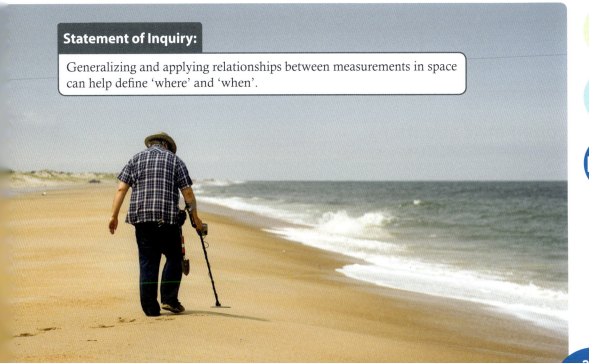

You should already know how to:

• convert angles from degrees to radians and vice versa	1 Convert: a 120° to radians b 135° to radians c $\frac{\pi}{6}$ to degrees d $\frac{5\pi}{3}$ to degrees
• find the value of the sine, cosine or tangent of any angle, including special angles	2 Find the value of: a $\sin 45°$ b $\cos\left(\frac{2\pi}{3}\right)$ c $\tan 47°$
• find a missing angle when given the sine, cosine or tangent value	3 Find x when: a $\cos x = 0.321$ b $\sin x = 0.5$
• transform trigonometric functions	4 Describe the transformations on the graph of $f(x) = \sin x$ to give the graph of $g(x) = 3\sin(2x) + 5$.
• solve quadratic equations	5 Solve $15 + 7x - 2x^2 = 0$.
• use the sine rule to find a missing side or angle in a triangle. $\frac{\sin A}{a} = \frac{\sin B}{b} = \frac{\sin C}{c}$	6 a Find $\angle B$ in this triangle. *(triangle with B at top, AC = 3 cm, BC = 6 cm, angle A = 60°)* b Find the length of side x, to 3 s.f. *(triangle with base 8, angles 30° and 40°, side x)*

F Trigonometric equations

- How do you solve trigonometric equations?
- How many solutions does a trigonometric equation have?

In problems with right-angled triangles, you solve trigonometric equations to get a single solution, such as solving $\sin x = \frac{1}{2}$ to get $x = 30°$ or $\frac{\pi}{3}$ radians. In Exploration 1 you will investigate how many solutions a trigonometric equation can have.

GEOMETRY AND TRIGONOMETRY

Exploration 1

1. Draw the graph of the sine function $f(\theta) = \sin\theta$, for $-4\pi \leq \theta \leq 4\pi$.
2. Draw the line $y = \frac{1}{2}$ on the same graph.
3. Use your graph to determine the value of each angle that satisfies the equation $\sin\theta = \frac{1}{2}$.
4. State how many angles satisfy this equation within the domain $0 \leq \theta \leq 2\pi$.
5. Generalize your findings: if $\sin\theta = b$, state which other angle within $0 \leq \theta \leq 2\pi$ also has a sine value of b.
6. State how many angles satisfy this equation
    a. within the domain $-2\pi \leq \theta \leq 2\pi$
    b. within the domain $-4\pi \leq \theta \leq 4\pi$
7. State how many angles satisfy this equation when the domain is the set of real numbers. Explain how you know.
8. Generalize your findings: if $\sin\theta = b$, state how to find all angles that satisfy this equation if the domain is the set of real numbers.
9. Repeat steps **1** to **8** for the graph of $f(\theta) = \cos\theta$, for $-4\pi \leq \theta \leq 4\pi$.

Reflect and discuss 1

- Would you find the same number of solutions to $\sin x = \frac{1}{2}$ and $\cos x = \frac{1}{2}$ in the domain $0 \leq \theta \leq 2\pi$ using the unit circle?
- In your generalization in Exploration 1, does it matter if b is positive or negative? Explain.
- Why is it important to specify a domain for the values of θ when solving trigonometric equations?

In a given domain, there may be multiple solutions to a trigonometric equation.

Trigonometric identities:
$\sin\theta \equiv \sin(\pi - \theta)$
$\cos\theta \equiv \cos(2\pi - \theta)$
$\tan\theta \equiv \tan(\pi + \theta)$

An **identity** is true for all values of the variable.

E11.1 Unmistaken identities

Example 1

Solve these equations algebraically for $0 \leq \theta \leq 2\pi$. Verify your solutions graphically using a GDC or dynamic geometry software.

a $2\sin\theta = \sqrt{3}$ **b** $9\tan^2\theta - 1 = 0$

a $\sin\theta = \dfrac{\sqrt{3}}{2}$ —— Rearrange to isolate $\sin\theta$. The domain is given in radians, so give angles in radians.

$\theta_1 = \dfrac{\pi}{3}$ —— Find one value of θ.

$\theta_2 = \pi - \dfrac{\pi}{3} = \dfrac{2\pi}{3}$ —— Use the identity $\sin\theta \equiv \sin(\pi - \theta)$.

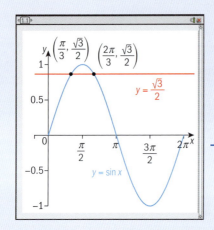

Graph the LHS and the RHS of the equation and find the x-coordinates where they intersect.

b $9\tan^2\theta - 1 = 0$

$9\tan^2\theta = 1$

$\tan^2\theta = \dfrac{1}{9} \Rightarrow \tan\theta = \pm\dfrac{1}{3}$ —— Solve each equation separately.

$\theta = \arctan\left(\dfrac{1}{3}\right) \Rightarrow \theta_1 = 0.322$

$\theta_2 = \pi + 0.322 = 3.46$ —— Use the identity $\tan\theta \equiv \tan(\pi + \theta)$.

$\theta = \arctan\left(-\dfrac{1}{3}\right) \Rightarrow \theta_3 = -0.322$ —— Answer is outside the given domain.

$\theta_4 = \pi + (-0.322) = 2.82$

$\theta_5 = \pi + 2.82 = 5.96$ —— Use the identity $\tan\theta \equiv \tan(\pi + \theta)$.

Use the same identity again, as long as the solution is within the given domain.

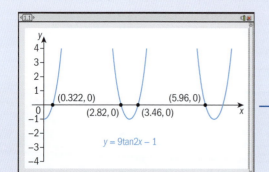

Set your window to the given domain.

Practice 1

1 Solve each equation for $0 \leq \theta \leq 360°$. Give exact answers where possible; if not possible, round to the nearest degree. Check your answers graphically.

a $\sin\theta = \frac{1}{2}$ **b** $4\cos\theta = -1$ **c** $\tan\theta = 1$

d $\sqrt{2}\sin\theta = -1$ **e** $\tan\theta = -\frac{3}{4}$ **f** $2\cos\theta = -\sqrt{3}$

2 Solve each equation for $0 \leq \theta \leq 2\pi$. Give exact answers where possible; if not possible, round to 0.01. Check your answers graphically.

a $2\cos\theta = -1$ **b** $2\cos\theta = \sqrt{2}$ **c** $2\sin\theta = -\sqrt{3}$

d $6\cos\theta = -5$ **e** $7\cos\theta = 1$ **f** $-\frac{5}{3}\sin\theta = 1$

g $\tan\theta = -1$ **h** $3\tan\theta = \sqrt{3}$ **i** $\sin^2\theta = 1$

> parts **j** to **q**, factorize the quadratic first.

j $\sin^2\theta + 2\sin\theta + 1 = 0$ **k** $4\cos^2\theta - 1 = 0$ **l** $\tan^2\theta = \frac{1}{4}$

m $2\cos^2\theta + \cos\theta = 1$ **n** $6\cos^2\theta - \cos\theta - 1 = 0$ **o** $2\sin\theta\cos\theta + \sin\theta = 0$

p $2\tan^2\theta - \tan\theta - 6 = 0$ **q** $2\sin^2\theta + (2-\sqrt{2})\sin\theta - \sqrt{2} = 0$

Problem solving

3 a Find the set of values in the domain $0 \leq \theta \leq 2\pi$ such that $\sin\theta = \cos\theta$.

b Find the set of values in the domain $0 \leq \theta \leq 2\pi$ that satisfy the equation: $\sin^2 x + \sin x \cos x = 0$.

4 a Show that $6\sin x \cos x + 3\cos x + 4\sin x + 2 \equiv (2\sin x + 1)(3\cos x + 2)$.

b Hence solve $6\sin x \cos x + 3\cos x + 4\sin x + 2 = 0$ in the domain $0 \leq x \leq 2\pi$.

Example 2

Find all solutions within the given domain.

a $\sin(2x) = \frac{\sqrt{3}}{2}, 0 \leq x \leq \pi$ **b** $\cos\left(\frac{x}{2}\right) = \frac{\sqrt{2}}{2}, 0 \leq x \leq 2\pi$

a Let $u = 2x$

$0 \leq x \leq \pi \Rightarrow 0 \leq u \leq 2\pi$

> Substitute a variable for 2x. If the domain of x is $0 \leq x \leq \pi$ then the domain of $u = 2x$ is $0 \leq u \leq \pi$.

$\sin u = \frac{\sqrt{3}}{2} \Rightarrow u = \frac{\pi}{3}$

or $u = \pi - \frac{\pi}{3} = \frac{2\pi}{3}$

> Solve for u.

$2x = \frac{\pi}{3} \Rightarrow x_1 = \frac{\pi}{6}$

$2x = \frac{2\pi}{3} \Rightarrow x_2 = \frac{2\pi}{6} = \frac{\pi}{3}$

> Substitute 2x for u and solve for x.

▶ Continued on next page

b Let $u = \frac{1}{2}x$

$0 \leq x \leq 2\pi \Rightarrow 0 \leq u \leq \pi$

> If the domain of x is $0 \leq x \leq 2\pi$ then the domain of $u = \frac{1}{2}x$ is $0 \leq x \leq \pi$.

$\cos u = \frac{\sqrt{2}}{2} \Rightarrow u = \frac{\pi}{4}$ or $u = 2\pi - \frac{\pi}{4} = \frac{7\pi}{4}$

> Reject $u = \frac{7\pi}{4}$ since it lies outside the domain $0 \leq u \leq \pi$.

$\frac{x}{2} = \frac{\pi}{4}$

$\Rightarrow x = \frac{\pi}{2}$

Practice 2

1 Solve the equations within the given domain. Check your solutions graphically.

a $\sin(2\theta) = 1$, $0 \leq \theta \leq \pi$

b $\cos\left(\frac{\theta}{2}\right) = -\frac{1}{2}$, $0 \leq \theta \leq 2\pi$

c $\sin\left(\frac{\theta}{3}\right) = \frac{\sqrt{3}}{2}$, $0 \leq \theta \leq \pi$

d $\cos(3\theta) = 1$, $0 \leq \theta \leq 2\pi$

e $3\sin(5\theta) - 1 = 0$, $0 \leq \theta \leq \pi$

f $\sqrt{2}\cos(2\theta) = 1$, $0 \leq \theta \leq 2\pi$

g $2\sin(3x) = 1$, $0 \leq x \leq \pi$

h $\sin(2x)\sqrt{2} = -\sin(2x)$, $0 \leq x \leq 2\pi$

i $\sin^2(3\theta) - \sin(3\theta) = 0$, $0 \leq \theta \leq \frac{2\pi}{3}$

j $3\cos^2\left(\frac{\theta}{2}\right) + \cos\left(\frac{\theta}{2}\right) = 0$, $0 \leq \theta \leq \pi$

k $2\sin^2(3x) - 3\sin(3x) = -1$, $0 \leq x \leq \frac{\pi}{3}$

l $3\cos^2(2x) - 5\cos(2x)$, $0 \leq x \leq 2\pi$

C Trigonometric identities

- How do you prove trigonometric identities?
- How can you use trigonometric identities to solve trigonometric equations?

Exploration 2

Identity 1

1 For this right-angled triangle, write down the trigonometric ratios:

a $\tan\theta =$

b $\sin\theta =$

c $\cos\theta =$

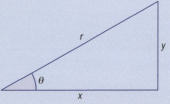

▶ Continued on next page

GEOMETRY AND TRIGONOMETRY

2 Rearrange your equation for $\sin\theta$ to make y the subject. Rearrange your equation for $\cos\theta$ to make x the subject. Hence write $\tan\theta$ in terms of $\sin\theta$ and $\cos\theta$.

3 Explain how this relates to how you calculate $\tan\theta$ using the unit circle.

Identity 2

4 For the unit circle where $r = 1$, write expressions for:

   a $\sin\theta$ **b** $\cos\theta$ **c** $\sin^2\theta + \cos^2\theta$

5 Using the right-angled triangle, write an expression for r in terms of x and y.

6 Rewrite your expression for $\sin^2\theta + \cos^2\theta$ using your result from step **4**.

7 Test that your result is true for different values of θ. Make sure you test some obtuse angles and negative angles.

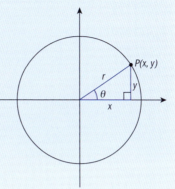

Reflect and discuss 2

- Why do you think **Identity 1** is called the Ratio identity?
- Why do you think **Identity 2** is called the Pythagorean identity?

Ratio identity: $\tan\theta \equiv \dfrac{\sin\theta}{\cos\theta}$

Pythagorean identity: $\sin^2\theta + \cos^2\theta \equiv 1$

Rearranging the Pythagorean identity gives two equivalent identities:

$\sin^2\theta \equiv 1 - \cos^2\theta$ and $\cos^2\theta \equiv 1 - \sin^2\theta$

Once you have proven an identity, you can use it to simplify and solve trigonometric equations.

Example 3

Solve $2\cos^2\theta - \sin\theta - 1 = 0$, for $0 \leq x \leq \pi$, and check graphically.

$2\cos^2\theta - \sin\theta - 1 = 0$ *Use the Pythagorean identity $\cos^2\theta \equiv 1 - \sin^2\theta$ to write the equation in $\sin\theta$.*

$2(1 - \sin^2\theta) - \sin\theta - 1 = 0$

$1 - 2\sin^2\theta - \sin\theta = 0$

$2\sin^2\theta + \sin\theta - 1 = 0$ *Multiply by -1 to avoid negative coefficients.*

$(2\sin\theta - 1)(\sin\theta + 1) = 0$ *Factorize and solve the quadratic equation.*

▶ Continued on next page

$2\sin\theta - 1 = 0$ or $\sin\theta + 1 = 0$

$\sin\theta = \frac{1}{2}$ or $\sin\theta = -1$

$\sin\theta = \frac{1}{2} \Rightarrow \theta_1 = \frac{\pi}{6}$ or $\theta_2 = \pi - \frac{\pi}{6} = \frac{5\pi}{6}$ — Find the all the values of θ in the given domain.

$\sin\theta = -1 \Rightarrow \theta_3 = \frac{3\pi}{2}$ — Reject $\theta_3 = \frac{3\pi}{2}$ as it is outside the domain.

$\Rightarrow \theta_1 = \frac{\pi}{6}(=0.524); \theta_2 = \frac{5\pi}{6}(=2.62)$

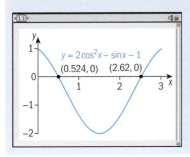

Practice 3

1 Solve these trigonometric equations for $0 \leq \theta \leq 2\pi$. Leave your answer in exact form (in radians) if possible, otherwise write your answer to 3 s.f.

a $\cos^2\theta - \sin^2\theta = 1$

b $2\sin^2\theta - \cos\theta = 1$

c $\cos^2\theta + \cos\theta = \sin^2\theta$

d $3\tan^2\theta \cos^2\theta = 2 + \cos^2\theta$

e $3\cos^2\theta - 6\cos\theta = \sin^2\theta - 3$

2 Solve these trigonometric equations for $0 \leq \theta \leq 360°$. Write your answer to 3 s.f. where appropriate.

a $2\sin^2\theta - \cos^2\theta + 1 = 0$

b $2\cos^2\theta + \sin\theta = 1$

c $4\sin\theta = \tan\theta$

d $4\sin^2\theta - \cos\theta = 1$

e $4\tan^2\theta \cos^2\theta - \sin^2\theta = \cos^2\theta$

Problem solving

3 Determine whether $1 + \frac{1}{\tan\theta} = \frac{1}{\sin^2\theta}$ is an identity. Justify your answer.

You can use the trigonometric identities you have previously proved to prove other identities. To prove an identity, work on one side until your result is the same as the other side.

In Japan, students use the memory aid shown below for remembering the sides involved in the formulas for sine, cosine and tangent. The letter superimposed on the triangle is drawn to include the two sides involved in each case.

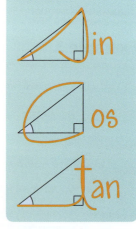

Example 4

Prove that $\tan\theta \sin\theta + \cos\theta \equiv \dfrac{1}{\cos\theta}$.

LHS:

$\tan\theta \sin\theta + \cos\theta$ — Start on one side. Tip: Start on the side that seems more complex.

$= \dfrac{\sin\theta}{\cos\theta} \sin\theta + \cos\theta$

$= \dfrac{\sin^2\theta}{\cos\theta} + \cos\theta$ — Use the Ratio identity.

$= \dfrac{\sin^2\theta + \cos^2\theta}{\cos\theta}$ — Write the expression with common denominator $\cos\theta$.

$= \dfrac{1}{\cos\theta}$ — Use the Pythagorean identity.

$= $ RHS

Example 5

Prove that $\dfrac{\sin^2 x + 4\sin x + 3}{\cos^2 x} \equiv \dfrac{3 + \sin x}{1 - \sin x}$.

LHS:

$\dfrac{\sin^2 x + 4\sin x + 3}{\cos^2 x}$ — Factorize the quadratic expression in the numerator.

$= \dfrac{(\sin x + 3)(\sin x + 1)}{1 - \sin^2 x}$ — Use the Pythagorean identity, then factorize the denominator.

$= \dfrac{(\sin x + 3)\,\cancel{(\sin x + 1)}}{(1 - \sin x)\,\cancel{(1 + \sin x)}}$ — Cancel common factors.

$= \dfrac{\sin x + 3}{1 - \sin x}$

$= $ RHS

Objective B: Investigating patterns
iii. prove, or verify and justify, general rules

To prove other identities, make sure that you only use results that have been proven to be true. If you are unsure about a statement, you can justify it by graphing both sides: if the graphs are the same, the statement is probably an identity.

Practice 4

1 Prove these identities:

  a $\dfrac{1 - \sin^2\theta}{\cos\theta} \equiv \cos\theta$

  b $\dfrac{\cos\theta \tan\theta}{\sin\theta} = 1$

  c $\cos^2\theta(\tan^2\theta + 1) \equiv 1$

  d $\sin^2\theta\left(1 + \dfrac{1}{\tan^2\theta}\right) \equiv 1$

E11.1 Unmistaken identities

e $\sin^4\theta + \cos^2\theta \sin^2\theta \equiv \sin^2\theta$

f $\sin^2\theta + \tan^2\theta + \cos^2\theta \equiv \dfrac{1}{\cos^2\theta}$

g $\dfrac{1-\sin\theta}{\cos\theta} \equiv \dfrac{\cos\theta}{1+\sin\theta}$

h $\dfrac{\dfrac{1}{\cos\theta}+\tan\theta}{\dfrac{1}{\cos\theta}-\tan\theta} \equiv \dfrac{1+2\sin\theta+\sin^2\theta}{\cos^2\theta}$

> In part **g**, multiply the RHS by $\dfrac{1-\sin\theta}{1-\sin\theta}$.

Problem solving

2 Use your GDC or graphing software to graph these expressions. Based on the graph, conjecture an identity and justify your answer.

a $\left(\dfrac{1}{\cos x}+\tan x\right)(1-\sin x)$

b $\left(\dfrac{\sin x + \tan x}{-\cos x - 1}\right)(\cos x)$

> **Tip**
>
> When both sides of an identity are complex, as in part **h**, you can work independently on both sides until you reach an equivalent expression.

Reflect and discuss 3

- Explain how verifying an identity is similar to solving an equation.
- Explain how verifying an identity is different to solving an equation.

D The ambiguous case

- Can two different solutions both be correct?
- How do we define 'where' and 'when'?

Exploration 3

1 Find two angles where $\sin B = \dfrac{\sqrt{3}}{2}$, $0 < B \leq 180°$.

2 In the diagram below, find angle B to 3 s.f.

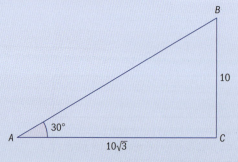

3 Hence, find the third angle of the triangle (angle C).

4 Based on your answer to step **1**, there should have been two solutions in step **2**. Write down the other possible solution for angle B.

5 Can you construct a triangle with angle A and this second value for angle B? If so, write down the third angle of the triangle.

▶ Continued on next page

E11 Equivalence

GEOMETRY AND TRIGONOMETRY

6 Based on the diagram below, find the value of angle N in this diagram (give an exact value).

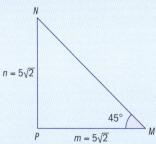

7 Using the sine rule, find the other possible value of N. Can a triangle be formed with the given angle and each of these answers? Explain.

8 Write down the only possible angles for this triangle.

> When using the sine rule to find a missing angle in a triangle, sometimes there are two possible solutions. Your calculator will give you one solution as θ; the other solution is $180° - \theta$. This is called the **ambiguous case** for the sine rule.

Reflect and discuss 4

- When solving problems with triangles, you are normally given three pieces of information: either sides (S) or angles (A) or a combination of both. For each of the following scenarios, state whether or not you would use the sine rule or cosine rule to find a missing side/angle. Explain why there is no ambiguous case for any of these:

  i SSS   ii ASA   iii SAS

- Explain why SSA produces the ambiguous case.

> The word 'sine' is the result of an error in translating a trigonometry book written by the famous Arab mathematician Al-Khwarizmi. In Arabic the word for sine is *jiba*, and it was misread as *jaib*, which means 'fold'. And the Latin word for fold is *sinus*, hence the word sine.

Example 6

In triangle ABC, $\angle B = 40°$, $BC = 14$ cm and $AC = 10$ cm. Find the other sides and angles.

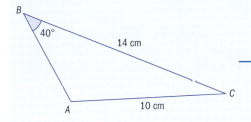

— Sketch the triangle.

▶ Continued on next page

E11.1 Unmistaken identities

$$\frac{\sin A}{14} = \frac{\sin 40°}{10}$$ — Find both possible values of $\angle A$.

$$\sin A = \frac{14 \sin 40°}{10}$$

$$\angle A = \arcsin\left(\frac{14 \sin 40°}{10}\right) = 64.1°$$

or $\angle A = 180 - 64.1 = 115.9°$

If $\angle A = 64.1°$, then $\angle C = 75.9°$ — Find $\angle C$ and side length c for the first possible value of $\angle A$.

$$\frac{14}{\sin 64.1°} = \frac{c}{\sin 75.9°}$$

$\Rightarrow c = 15.7$ cm

If $\angle A = 115.9°$, then $\angle C = 24.1°$ — Find $\angle C$ and side length c for the second possible value of $\angle A$.

$$\frac{14}{\sin 115.9°} = \frac{c}{\sin 24.1°}$$

$\Rightarrow c = 6.36$ cm

Reflect and discuss 5

- How are the lengths of the sides in a triangle related to the sizes of the angles opposite them? Why does this make sense?
- Does this also hold true for triangles produced by the ambiguous case?

Example 7

In triangle XYZ, $\angle X = 65°$, $x = 23$ cm and $y = 18$ cm. Find the other sides and angles.

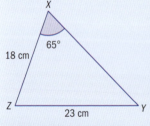

Sketch the triangle.

$$\frac{\sin 65°}{23} = \frac{\sin Y}{18}$$

$$\sin Y = \frac{18 \sin 65°}{23}$$

$$\angle Y = \arcsin\left(\frac{18 \sin 65°}{23}\right) = 45.2°$$ — Find both possible values of $\angle Y$.

or $\angle Y = 180 - 45.2° = 134.8°$

If $\angle Y = 45.2°$, then $\angle Z = 69.8°$ — Find $\angle Z$ and length of side z for the two values of $\angle Y$. Verify that a triangle can be constructed with each of these possibilities.

$$\frac{23}{\sin 65°} = \frac{z}{\sin 69.8°}$$

$z = 23.8$ cm

If $\angle Y = 134.8°$, then $\angle X + \angle Z$ is already more than $180°$. Therefore, the only solution is the one already found.

GEOMETRY AND TRIGONOMETRY

Reflect and discuss 6

Although it is sufficient for this course to check both solutions of the ambiguous case, could you find a general rule to determine whether or not there are two solutions, without checking them both?

Practice 5

1. In triangle ABC, find all possible values for the measure of angle C. Give answers to 3 s.f.

   a $A = 38°$, $a = 14$, $c = 22$
   b $A = 71°$, $a = 10$, $c = 7$
   c $A = 62°$, $a = 12$, $c = 3$
   d $A = 53°$, $a = 11$, $c = 13$

2. Find the missing sides and angles in triangle PQR. Give all values to the nearest 0.1. Make sure you find all possible solutions.

   a $r = 5$ cm, $q = 8$ cm, $\angle R = 30°$
   b $r = 26$ cm, $q = 22$ cm, $\angle R = 55°$
   c $r = 20$ m, $q = 30$ m, $\angle R = 22°$
   d $r = 7$ km, $q = 13$ km, $\angle R = 18°$

3. In triangle ABC, find all possible values for angle B to the nearest 0.1.

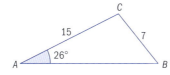

4. In $\triangle DEF$, $e = 7$ m, $f = 6$ m, and $\angle E = 19°$. Find all possible solutions for $\angle F$ to the nearest 0.1.

5. In $\triangle RST$, $r = 25$ m, $s = 30$ m, and $\angle R = 40°$. Find all possible values of t to the nearest meter.

6. Determine how many unique triangles are possible for each scenario.

   a $\triangle PQR$ with $\angle P = 50°$, $p = 10$ m, $q = 6$ m
   b $\triangle PQR$ with $\angle P = 90°$, $p = 3$ cm, $q = 2$ cm
   c $\triangle PQR$ with $\angle P = 115°$, $p = 9$ m, $q = 3$ m
   d $\triangle PQR$ with $\angle P = 62°$, $p = 2.8$ km, $q = 3.0$ km

Problem solving

7. A lighthouse at point A is 10 km from a ship at point C and 8 km from a sailboat at point B. Seen from the ship, the angle between the lighthouse and the sailboat is 48°.

   a Sketch all possible diagrams for this situation.

   b Determine the possible distances from the ship to the sailboat, to the nearest tenth of a kilometer.

8. A triangular plot of land is enclosed by a fence. Two sides of the fence are 10.8 m and 7.6 m long, respectively. The other side forms an angle of 41° with the 10.8 m side.

   a Determine how many lengths are possible for the third side.

   b Calculate all possible lengths to the nearest 0.1 m.

E11.1 Unmistaken identities

Summary

An identity is true for all values of the variable.

Trigonometric identities:

$\sin\theta \equiv \sin(\pi - \theta)$

$\cos\theta \equiv \cos(2\pi - \theta)$

$\tan\theta \equiv \tan(\pi + \theta)$

Ratio identity: $\tan\theta \equiv \dfrac{\sin\theta}{\cos\theta}$

Pythagorean identity: $\sin^2\theta + \cos^2\theta \equiv 1$

Rearranging the Pythagorean identity gives two equivalent identities:

$\sin^2\theta \equiv 1 - \cos^2\theta$ and $\cos^2\theta \equiv 1 - \sin^2\theta$

When using the sine rule to find a missing angle in a triangle, sometimes there are two possible solutions. Your calculator will give you one solution as θ; the other is $180° - \theta$. This is called the **ambiguous case** for the sine rule.

Mixed practice

1 Solve each equation for $0 \le \theta \le 360°$. Give exact answers where possible; if not possible, round to 2 d.p. Check your answers graphically.

  a $\cos\theta = 0$
  b $\sqrt{3}\tan\theta = -1$
  c $5\tan\theta = -2$
  d $\sin\theta = \dfrac{2}{7}$
  e $\sqrt{3}\cos\theta = 2$
  f $\sin\theta = -\dfrac{1}{\sqrt{2}}$

2 Without using a calculator, **find** these angles in radians, for $0 \le \theta \le 2\pi$. Give exact answers.

  a $\sin\theta = \dfrac{\sqrt{3}}{2}$
  b $\cos\theta = -\dfrac{\sqrt{2}}{2}$
  c $\tan\theta = \sqrt{3}$
  d $\sin\theta = -1$
  e $\cos\theta = \dfrac{1}{2}$
  f $\tan\theta = \dfrac{\sqrt{3}}{3}$

3 Solve each equation for $0 \le \theta \le 2\pi$. Give exact answers in radians when possible, otherwise give answers in degrees to the nearest 0.1°.

  a $2\sin\theta = -\sqrt{3}$
  b $5\cos\theta = 3$
  c $2\sin(2\theta) = 1$
  d $2\cos(3\theta) = -1$

4 a Find θ if $\sin\theta = -\dfrac{\sqrt{2}}{2}$, for $0 \le \theta \le 360°$

  b If $\cos\theta > 0$, in which quadrants is θ?

Problem solving

  c Find $\tan\theta$ if $\cos\theta < 0$.

5 Find $\tan\theta$ if $\cos\theta = -\dfrac{1}{2}$ and $\sin\theta > 0$.

6 Prove the identity: $\sin x \cos x \tan x \equiv 1 - \cos^2 x$

7 Prove the identity: $\sin x - \sin x \cos^2 x \equiv \sin^3 x$

8 Prove the identity:

$$\dfrac{1-\sin\theta}{\cos\theta} + \dfrac{1-\cos\theta}{\sin\theta} \equiv \dfrac{\sin\theta + \cos\theta - 1}{\sin\theta \cos\theta}$$

9 Prove the identity: $\dfrac{1}{\sin^2\theta} + \dfrac{1}{\cos^2\theta} \equiv \dfrac{1}{\sin^2\theta - \sin^4\theta}$

10 Solve these trigonometric equations for $0 \le \theta \le 360°$. Write your answer to 3 s.f. where appropriate.

  a $\sin^2\theta + \sin\theta = \cos^2\theta$
  b $2\sin^2\theta = \cos\theta + 2$
  c $2\sin^2\theta + 3\cos^2\theta = 3$
  d $3\sin^2\theta - 4\cos\theta - 4 = 0$

GEOMETRY AND TRIGONOMETRY

11 Solve these trigonometric equations for $0 \leq \theta \leq 2\pi$. Leave your answer in exact form in radians if possible, otherwise accurate to 3 s.f.

a $2\cos^2\theta = \sqrt{3}\cos\theta$

b $\dfrac{1}{\sin\theta} = 2\sin\theta$

c $\tan\theta \sin^2\theta = 2\tan\theta$

d $2\cos^2\theta + 3\sin\theta = 3$

e $3\cos^2\theta - 5\sin\theta - 4 = 0$

f $(\sqrt{2} - 2)\cos\theta - 2\sqrt{2}\sin^2\theta = 1 - 2\sqrt{2}$

Problem solving

12 Determine whether $\dfrac{1}{\cos^2\theta} - 1 = \tan^2\theta$ is an identity. **Justify** your answer.

13 Find both solutions for $\angle B$ for each triangle, and sketch each solution. Write your answer accurate to 1 d.p.

a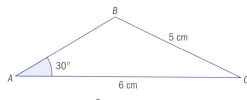

b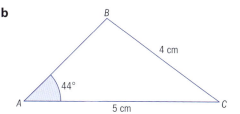

Review in context

1 Anouk and Timur are standing 30 m apart along a shoreline, looking at an island out to sea. Anouk measures a 40° angle between where Timur is standing and the island. Kelly is on the island, 20 m away from Timur.

   a **Find** the two possible angles (to the nearest 0.1°) that Kelly could measure between Timur and Anouk.

   b **Find** the two possible angles (to the nearest 0.1°) that Timur could measure between Anouk and Kelly.

2 A canoe leaves the dock and travels toward a buoy 5 km away. Upon reaching the buoy, it changes course and travels another 3 km. From the dock, the angle between the buoy and the canoe's position is 12°. **Determine** the distance of the canoe from the dock, and **explain** whether or not there is more than one possible answer.

Problem solving

3 Lucas is lost while on an orienteering course. He knows that the finish line is due east from his current position, O. He follows a path, but after 400 meters he realizes he is walking at a bearing of 125° instead of walking east. In the diagram, he is now at point A. Lucas's GPS says he is only 250 m away from the finish line. Lucas walks to point B but the finish line isn't there.

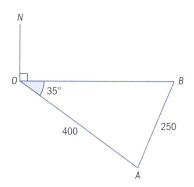

> You will have to use the sine rule more than once.

a **Find** how far east Lucas is from point O.

b **Explain** how the finish line could be 250 m away from point A, but not be located at B.

c **Determine** how far Lucas is from the actual finish line.

E11.1 Unmistaken identities 247

4 Both Earth (E) and Venus (V) orbit around the sun (S) in paths that are nearly circular. The distance from Earth to the sun is approximately 149 million kilometers, while the distance from Venus to the sun is roughly 108 million kilometers.

a **Sketch** a diagram of this part of the solar system, showing the orbits of the two planets. (Draw the planets and the sun so that they form a triangle.)

b An astronomer measures the angle SEV to be 20°. **Determine** how far Venus could be from Earth at that point.

Reflect and discuss

How have you explored the statement of inquiry? Give specific examples.

Statement of Inquiry:

Generalizing and applying relationships between measurements in space can help define 'where' and 'when'.

At the turn of the 20th century, Albert Einstein put forward his special theory of relativity which provided a new framework for the laws of physics but was at odds with the law of gravity. Some years later, Einstein unified the concepts of gravity and the special theory, resulting in the general theory of relativity.

Einstein realized that there was no difference between the forces acting on a person standing on the Earth due to gravity and a person travelling in a rocket accelerating at $9.81\ \text{ms}^{-2}$. This is a simplified description of Einstein's Equivalence Principle which was inspired by studying the effect of objects in free fall.

It is a common mistake to use the term 'zero gravity' for describing free fall. Astronauts aboard the International Space Station (ISS), for example, are in constant free fall, but the gravitational force acting upon them is as high as 90% of that on the Earth's surface. The weightlessness experienced is because the free fall towards the earth is balanced by the tangential velocity of the ISS which allows it to remain in orbital motion.

E12.1 Go ahead and log in

Global context: Orientation in space and time

Objectives
- Developing the laws of logarithms
- Using the laws of logarithms to simplify expressions and solve equations
- Proving the laws of logarithms
- Proving the change of base formula

Inquiry questions
- **F** • What are the laws of logarithms?
- **C** • How can you prove generalizations?
- **D** • How do you solve logarithmic equations?
- • Is change measurable and predictable?

RELATIONSHIPS

| ATL | Critical-thinking |

Draw reasonable conclusions and generalizations

Statement of Inquiry:

Generalizing changes in quantity helps establish relationships that can model duration, frequency and variability.

E6.1

E10.1

E12.1

You should already know how to:

• evaluate logarithms using a calculator	**1** Evaluate: **a** log 5 **b** ln 3 **c** $\log_2 6$
• solve equations involving exponents	**2** Solve the equation $3^x = 20$.
• solve equations involving logs	**3** Solve the equation $\log_7\left(\frac{1}{7}\right) = x$.
• confirm algebraically and graphically that two functions are inverses of each other	**4** Show algebraically and graphically that $f(x) = 3x - 4$ and $g(x) = \frac{x+4}{3}$ are inverses of each other.
• find the inverse of a logarithmic function	**5** Find the inverse of the function $y = 2\log_5(x+3) - 11$.

F Creating generalizations

- What are the laws of logarithms?

Exploration 1

1 Choose any positive value of *a* for the base, and use your calculator to evaluate each logarithm in this table to 3 decimal places.

$\log_a 2 + \log_a 3$	$\log_a 6$
$\log_a 8 + \log_a 4$	$\log_a 32$
$\log_a 10 + \log_a 100$	$\log_a 1000$
$\log_a\left(\frac{1}{3}\right) + \log_a 81$	$\log_a 9$

Examine the patterns and write down a generalized rule:

The **product rule**: $\log_a xy =$

2 Do the same for the expressions in this table.

$\log_a 3 - \log_a 2$	$\log_a\left(\frac{3}{2}\right)$
$\log_a 8 - \log_a 4$	$\log_a 2$
$\log_a 100 - \log_a 10$	$\log_a 10$
$\log_a 81 - \log_a\left(\frac{1}{3}\right)$	$\log_a 243$

Examine the patterns and write down a generalized rule:

The **quotient rule**: $\log_a\left(\frac{x}{y}\right) =$

▶ Continued on next page

3 Compare the values in each column to discover a pattern.

$\log_a 3^4$	$\log_a 3$
$\log_a 7^2$	$\log_a 7$
$\log_a 75^{\frac{1}{2}}$	$\log_a 75$
$\log_a 81^{-\frac{1}{3}}$	$\log_a 81$

Examine the patterns and write down a generalized rule:

The **power rule**: $\log_a(x^n) =$

4 Evaluate $\log_a 1$ for different values of a.

Write down a generalized rule:

The **zero rule**: $\log_a 1 =$

5 Evaluate $\log_a a$ for different values of a.

Write down a generalized rule:

The **unitary rule**: $\log_a a =$

6 Rewrite $b^{\log_b x} = y$ as a logarithmic statement.

Hence solve for y in terms of x.

Write this as a generalized rule.

Scottish mathematician and astronomer John Napier (1550 – 1617), most famous for discovering logarithms, is also credited for introducing the decimal notation for fractions.

Reflect and discuss 1

- How are the rules you discovered similar to the laws of exponents? Give two specific examples.
- Explain why the zero rule and the unitary rule are true.
- How can you simplify $e^{\ln x}$?

Laws of logarithms

For **all** logarithm rules: $x, y, a \in \mathbb{R}^+$

The **product rule**: $\log_a xy = \log_a x + \log_a y$

The **quotient rule**: $\log_a\left(\dfrac{x}{y}\right) = \log_a x - \log_a y$

The **power rule**: $\log_a(x^n) = n \log_a x$

The **zero rule**: $\log_a 1 = 0$

The **unitary rule**: $\log_a a = 1$

$a^{\log_a m} = m$

E12.1 Go ahead and log in

Example 1

Rewrite $6\ln x - 3\ln 3 + 2\ln x$ as a single logarithm.

$6\ln x - 3\ln 3 + 2\ln x = \ln x^6 - \ln 3^3 + \ln x^2$ — The power rule: $\log_a x^n = n\log_a x$

$\qquad = \ln(x^6 \times x^2) - \ln 3^3$ — The product rule: $\log_a xy = \log_a x + \log_a y$

$\qquad = \ln x^8 - \ln 3^3$

$\qquad = \ln\left(\dfrac{x^8}{27}\right)$ — The quotient rule: $\log_a x - \log_a y = \log_a\left(\dfrac{x}{y}\right)$

Example 2

Express $\ln\left(\dfrac{1}{\sqrt{ab}}\right)$ in terms of x and y, where $x = \ln a$ and $y = \ln b$.

$\ln\left(\dfrac{1}{\sqrt{ab}}\right) = \ln 1 - \ln\sqrt{ab}$ — The quotient rule.

$\qquad = \ln 1 - \dfrac{1}{2}\ln(ab)$ — The power rule.

$\qquad = 0 - \dfrac{1}{2}\ln(ab)$ — The zero rule: $\log_a 1 = 0$.

$\qquad = -\dfrac{1}{2}(\ln a + \ln b)$ — The product rule.

$\qquad = -\dfrac{1}{2}(x + y)$

Practice 1

1 Rewrite each expression as a single logarithm:

a $\log 4 + \log 3$

b $\log 5 - \log 2 + \log 6$

c $2\log 5 + 3\log 2 + \log 1$

d $3\log x - \log y + \dfrac{1}{2}\log z$

e $5\ln x - 3\ln 2 + 4\ln x$

f $\log y + 4\log x - (5\log 2y + \log 3x)$

2 Express in terms of $\ln a$ and $\ln b$:

a $\ln ab$

b $\ln\left(\dfrac{a}{b}\right)$

c $\ln(a^2 b)$

d $\ln\sqrt{a}$

e $\ln\left(\dfrac{1}{a^2}\right)$

f $\ln(a\sqrt{b})$

g $\ln\left(\dfrac{a^3}{b}\right)$

h $\ln\left(\dfrac{a^3}{b^2}\right)$

i $\ln\sqrt{\dfrac{a}{b}}$

Problem solving

3 Write each expression in terms of x and y, where $\log_a 3 = x$ and $\log_a 6 = y$.

a $\log_a 18$

b $\log_a 2$

c $\log_a 0.5$

d $\log_a 9$

e $\log_a 36$

f $\log_a 27$

g $\log_a 162$

4 Express in terms of x and y, where and $x = \log a$ and $y = \log b$.

a $\log(100a^2b^3)$

b $\log\left(\dfrac{a^5}{100b^4}\right)$

c $\log\sqrt[3]{1000a^6b^8}$

d $\log\left(\dfrac{1}{10a\sqrt{b}}\right)$

Example 3

Find the inverse of the function $g(x) = \ln x - \ln(x - 3)$ and confirm it algebraically using the laws of logarithms. State the domain and range of the function and its inverse.

$y = \ln x - \ln(x-3) = \ln\left(\dfrac{x}{x-3}\right)$ — Use the quotient rule.

$x = \ln\left(\dfrac{y}{y-3}\right)$ — Interchange x and y.

$e^x = \dfrac{y}{y-3}$ — Solve for y, and simplify.

$(y-3)e^x = y$

$ye^x - 3e^x = y$

$ye^x - y = 3e^x$

$y = \dfrac{3e^x}{e^x - 1}$

$g^{-1}(x) = \dfrac{3e^x}{e^x - 1}$

$g(g^{-1}(x)) = \ln\left(\dfrac{3e^x}{e^x-1}\right) - \ln\left(\dfrac{3e^x}{e^x-1} - 3\right)$ — Show that $g(g^{-1}(x)) = x = g^{-1}(g(x))$.

$= \ln\left(\dfrac{3e^x}{e^x-1}\right) - \ln\left(\dfrac{3e^x}{e^x-1} - \dfrac{3(e^x-1)}{e^x-1}\right)$

$= \ln\left(\dfrac{3e^x}{e^x-1}\right) - \ln\left(\dfrac{3}{e^x-1}\right)$

$= \ln\left(\dfrac{3e^x}{e^x-1} \div \dfrac{3}{e^x-1}\right)$

$= \ln e^x = x$

$g^{-1}(g(x)) = \dfrac{3e^{\ln\left(\frac{x}{x-3}\right)}}{e^{\ln\left(\frac{x}{x-3}\right)} - 1} = \dfrac{3\left(\frac{x}{x-3}\right)}{\frac{x}{x-3} - 1} = x$

Since $g(g^{-1}(x)) = x = g^{-1}(g(x))$, they are inverses.

Domain of g is $x > 3$; range of g is $y > 0$.

Domain of g^{-1} is $x > 0$; range of g^{-1} is $y > 3$.

Exploration 2

1. Describe the transformations on the graph of $f(x) = \log_3 x$ that give the graph of $g(x) = \log_3(9x)$.

2. Describe the transformations on the graph of $f(x) = \log_3 x$ that give the graph of $h(x) = 2 + \log_3 x$.

3. Graph the functions f, g and h on the same coordinate axes, and write down what you notice. Explain, using laws of logarithms.

4. **a** Describe the **horizontal dilations** that transform the graph of f into the graph of:

   i $y = \log_3\left(\dfrac{x}{9}\right)$ **ii** $y = \log_3(27x)$

   b By using the laws of logarithms, describe the **horizontal translations** that transform the graph of f into the graph of:

   i $y = \log_3\left(\dfrac{x}{9}\right)$ **ii** $y = \log_3(27x)$

5. Summarize what you have learned in this Exploration about the transformations of dilations and translations with logarithmic functions.

Practice 2

1. Find the inverse of each function and prove that it is an inverse, both graphically and algebraically. State the domain and range of the function and its inverse.

   a $f(x) = 2^{x-3}$ **b** $g(x) = \sqrt{\ln(2x)}$
   c $h(x) = 4\log(1 + 2x)$ **d** $j(x) = \ln(x+2) - \ln(x-1)$

2. Use the laws of logarithms to explain why the graph of $g(x) = \log x^n$ is a vertical dilation of the graph of $f(x) = \log x$ for any positive n.

3. Given $f(x) = \log_3\left(\dfrac{9}{x}\right)$, use properties of logs to find an equivalent function.

   Describe the transformations on the graph of $y = \log_3 x$ that give the graph of the equivalent function.

Problem solving

4. **a** Use properties of logs to find an equivalent expression for each function:

   i $g(x) = \log_2\left(\dfrac{8}{x}\right)$ **ii** $h(x) = \log_2(x+3)^2$

   b Then describe the sequence of transformations on the graph of $y = \log_2 x$ that give the graphs of g and h.

C Proving the generalizations

- How can you prove generalizations?

If a and b are positive real numbers, then the exponential statement $a^c = b$ is equivalent to $\log_a b = c$. Does this mean that the exponent rules are also related to the logarithm rules?

Exploration 3

1 Proving the product rule

Let $\log_a x = p$ and $\log_a y = q$.

a Write both these expressions as exponential statements.

b Multiply the two expressions together to obtain an expression for xy.

c Write this expression in logarithmic form.

d Substitute $p = \log_a x$ and $q = \log_a y$.

You should have the product rule.

2 Proving the quotient rule

Let $\log_a x = p$ and $\log_a y = q$.

a Write both these expressions as exponential statements.

b Divide to obtain an expression for $\frac{x}{y}$.

c Write this expression in logarithmic form.

d Substitute $p = \log_a x$ and $q = \log_a y$.

You should have the quotient rule.

3 Proving the power rule

Let $\log_a x = p$.

1 Write this expression as an exponential statement.

2 Raise this expression to the power n.

3 Write this expression in logarithmic form.

4 Substitute $p = \log_a x$.

You should have the power rule.

> Internet search engines rank every website to determine an approximate measure of how important the site is. Ranking is a logarithmic scale, so a site with a rank of 5 is 100 times more popular than a site with a rank of 3. The difference of $5 - 3 = 2$ signifies two orders of magnitude on the logarithmic scale, where an order of magnitude is a power of 10.

Reflect and discuss 2

- How could you use a graphing tool with a slider to demonstrate the power rule for $f(x) = \log(x^2)$?
- What transformation takes the graph of $f(x) = \log x$ to the graph of $f(x) = \log(x^2)$? Explain.

E12.1 Go ahead and log in

Example 4

Express $3\ln x + 2\ln(2x+1)$ as a single logarithm.

$3\ln x + 2\ln(2x+1) = \ln x^3 + \ln(2x+1)^2$ — The power rule.

$= \ln\left[x^3(2x+1)^2\right]$ — The product rule.

$= \ln\left[x^3(4x^2+4x+1)\right]$ — Expand.

$= \ln(4x^5 + 4x^4 + x^3)$

Practice 3

1 Express as a single logarithm:

a $4\log p + 2\log q$

b $n\log p - 3\log q$

c $3\ln(x-2) - \frac{1}{2}\ln x$

d $\ln x - 2\ln(x-2)$

e $1 - \ln x$

f $4\ln x + 2\ln(x-1)$

2 Let $a = \ln x$, $b = \ln(x-1)$ and $c = \ln 3$. Express in terms of a, b and c:

a $\ln\left(\dfrac{x}{x-1}\right)$

b $\ln(x^2 - x)$

c $\ln(3x^2)$

d $\ln\sqrt{\dfrac{x+1}{x}}$

e $\ln\sqrt{3x^2 + 6x + 3}$

Problem solving

3 Express as a sum and/or difference of linear logarithms:

a $\ln(x^2 - 9)$

b $\ln(x^2 + 5x + 6)$

c $\ln\left(\dfrac{x-3}{x+4}\right)$

d $\ln\left(\dfrac{x+1}{x^2-4}\right)$

e $\ln\left(\dfrac{x^2 - 5x - 6}{x^2 - 4}\right)$

f $\ln\left(\dfrac{x^4 - 3x^3 - 10x^2}{5x - 20}\right)$

You can use a calculator to find the logarithms of any number with any base. Until quite recently, calculators could not do this. Books of log tables gave logarithms to base 10, and mathematicians used the change of base formula to calculate logs to other bases.

> You may have used the change of base formula in E6.1.

Any positive value can be the base of a logarithm, but the two bases that are the most useful for practical applications are the 'common' logarithm (base 10) and the 'natural' logarithm (base e).

Logarithms date back to 1614, when John Napier took an interest in simplifying astronomical calculations which often involved multiplying very large numbers together.

NUMBER

ATL

Exploration 4

How to calculate $\log_2 5$ using logs to base 10

To find $\log_a b$ when you know $\log_c b$ and $\log_c a$:

Let $\log_a b = x$, $\log_c b = y$, and $\log_c a = z$.

1. Write all three expressions in exponential form.
2. Equate the two expressions for b to get an identity involving a, c, x and y.
3. Substitute for a to get an identity involving z, c, x and y.
4. Equate the exponents to find an identity linking x, y and z.
5. Rearrange to make x the subject.
6. Substitute the logarithmic expressions for x, y and z. You should have the change of base formula.
7. Use values from the log table below to find the value of $\log_2 5$.

> In step **3**, use the exponential form of statement $\log_c a = z$.

Change of base formula:

$$\log_a b = \frac{\log_c b}{\log_c a}$$

Practice 4

1. Use the log table to find each value to 3 s.f.

   a $\log_2 7$ **b** $\log_5 8$

   c $\log_3 10$ **d** $\log_9 5$

   e $\log_2 2$ **f** $\log_4 5$

   Verify your results with a calculator.

x	$\log_{10} x$
1	0
2	0.301029996
3	0.477121255
4	0.602059991
5	0.698970004
6	0.77815125
7	0.84509804
8	0.903089987
9	0.954242509
10	1

Example 5

Given that $\ln a = 4$:

a express $\log_a(x^2)$ as a simple natural logarithm

b express $\ln(x^2) + 5\log_a x$ as a single logarithm.

a $\log_a(x^2) = 2\log_a x$ — The power rule.

$= 2\dfrac{\log_e x}{\log_e a}$ — Change of base.

$= 2\dfrac{\ln x}{\ln a}$ — $\ln a = 4$ is given.

$= 2\dfrac{\ln x}{4} = \dfrac{1}{2}\ln x$

▶ Continued on next page

E12.1 Go ahead and log in

b $\ln(x^2) + 5\log_a x = \ln(x^2) + 5\dfrac{\log_e x}{\log_e a}$

$= \ln(x^2) + 5\dfrac{\ln x}{\ln a}$

$= \ln(x^2) + \dfrac{5}{4}\ln x$

$= \ln(x^2) + \ln(x^{1.25}) = \ln(x^2 \times x^{1.25}) = \ln(x^{3.25})$

Practice 5

1 Given that $\ln a = 5$, express as a simple natural logarithm:

 a $\log_a(x^3)$ **b** $\log_a(x^2)$

 c $\dfrac{1}{2}\log_a x$ **d** $\log_a\left(\dfrac{x}{3}\right)$

2 Given that $\ln a = 1$, express as a single logarithm:

 a $\ln(x^3) + 2\log_a x$ **b** $2\ln x - 3\log_a x$

 c $3\log_a x + \dfrac{1}{2}\log_a x$ **d** $2\ln\dfrac{x}{3} - \dfrac{3}{2}\log_a x$

D Solving equations

- How do you solve logarithmic equations?
- Is change measurable and predictable?

A useful technique for solving equations involving exponents is to take logarithms of both sides.

Example 6

Solve $4^x = 25$ by taking logarithms of both sides.

$\log 4^x = \log 25$

$x \log 4 = \log 25$ ——— The power rule.

$x = \dfrac{\log 25}{\log 4}$ ——— Isolate x.

$x = 2.32$ (3 s.f.) ——— Use a calculator to evaluate.

Reflect and discuss 3

Show that you can obtain the same result in Example 6 by first rewriting the original equation as a logarithmic statement. Which method do you prefer? Explain.

Example 7

A mathematician has determined that the number of people P in a city who have been exposed to a news story after t days is given by the function $P = P_0(1 - e^{-0.03t})$, where P_0 is the city population. A lawyer knows that it is very difficult to appoint an unbiased jury to determine guilt in a crime after 25% of the population has read the news.

Find the maximum number of days available to select a jury.

$P = P_0(1 - e^{-0.03t})$ ——— 25% of the population implies $\dfrac{P}{P_0} = 0.25$

$0.25 = 1 - e^{-0.03t}$

$e^{-0.03t} = 1 - 0.25$

$e^{-0.03t} = 0.75$ ——— Take natural logs of both sides.

$\ln e^{-0.03t} = \ln 0.75$

$-0.03t = \ln 0.75$

$t = \dfrac{\ln 0.75}{-0.03}$

$t = 9.56$

There is a maximum of 9 days available to select a jury.

To solve equations that contain logarithms, you often need to use the laws of logarithms first.

> Each logarithm has three parts: the base, the argument and the answer.

Example 8

Find the value of x in each equation.

a $\log_2 x + \log_2 3 = \log_2 9$

b $\log_4(x+1) + \log_4(x-2) = 1$

a $\log_2 x + \log_2 3 = \log_2 9$

$\log_2(3x) = \log_2 9$ ——— The product rule.

$3x = 9$

$x = 3$ ——— The arguments of the logarithms must be equal to each other.

b $\log_4(x+1) + \log_4(x-2) = 1$

$\log_4[(x+1)(x-2)] = 1$ ——— The product rule.

$x^2 - x - 2 = 4^1$

$x^2 - x - 6 = 0$

$(x-3)(x+2) = 0$

$x = 3$ or -2

$x \neq -2$ since the logarithm $\log_4(-2-2) = \log_4(-4)$ doesn't exist.

Therefore, $x = 3$

E12.1 Go ahead and log in

Practice 6

1. Solve these equations by taking logs of both sides:
   - **a** $3^x = 25$
   - **b** $7^x = 41$
   - **c** $12^x = 25$

2. Solve:
   - **a** $3^{5x} = 45$
   - **b** $4^{2x} = 9$
   - **c** $3^{5x+1} = 60$

3. Solve these equations for x:
   - **a** $3 \times 4^{2x} = 9$
   - **b** $7 \times 4^{2x+1} = 89$
   - **c** $5 \times 3^{3x-1} = 52$

4. Solve these exponential equations:
   - **a** $2^{2x} - 7 \times 2^x + 6 = 0$
   - **b** $4^{2x} = 10 \times 4^x + 24 = 0$

   > Look for quadratic equations.

5. Find the value of x in each equation:
   - **a** $\log_5 x + \log_5 10 = \log_5 12$
   - **b** $\log_7 (3x) + \log_7 12 = \log_7 (2x + 5)$
   - **c** $\log_3 (x + 1) - \log_3 (x + 4) = -2$
   - **d** $\log_2 (x - 1) + \log_2 (x + 6) = 3$
   - **e** $\log_{20} x = 1 - \log_{20} (x - 1)$

Objective A: Knowing and understanding
iii. solve problems correctly in a variety of contexts

You can use the rules you have found to solve these context problems involving exponential equations.

---

6. A state surveys school traffic. At the end of 2010, there were 25 000 cars per day taking children to school.

   After t years the number of cars C was modelled by the exponential equation $C = 25\,000 \times e^{0.07t}$.

   - **a** Show that by the end of 2015, there were 35 477 cars per day taking children to school.
   - **b** By taking logs of both sides, calculate the time until the number of cars taking children to school reaches 50 000.

7. The time taken for an online video to go viral can be modelled by an exponential function, $H = 15e^{0.5t}$, where H is the number of times the video is watched and t is the number of hours since the first people shared the video.

   - **a** Describe what the value 15 represents in this model.
   - **b** By taking logs of both sides, calculate the number of hours before the video is watched 500 000 times.
   - **c** By rewriting the exponential statement as a logarithmic statement, calculate the number of hours before the video is watched 1 000 000 times.
   - **d** Determine which method (**b** or **c**) is most efficient.

E12 Generalization

Problem solving

8 Solve:

a $2^{5x-2} = 5^x$

b $3^{2x-1} = 4^x$

c $2^{x+1} = 4^{2x}$

d $3^{x+2} = 9^{2x-2}$

> Rearrange to get all the x terms on one side, then factorize to isolate x.

9 If $\log_a x = p$, show that rewriting it as a logarithmic statement and solving the resulting equation proves the power rule for logarithms.

Summary

Laws of logarithms

For **all** logarithm rules: $x, y, a \in \mathbb{R}^+$

The **product rule**: $\log_a xy = \log_a x + \log_a y$

The **quotient rule**: $\log_a \left(\dfrac{x}{y}\right) = \log_a x - \log_a y$

The **power rule**: $\log_a (x^n) = n \log_a x$

The **zero rule**: $\log_a 1 = 0$

The **unitary rule**: $\log_a a = 1$

$a^{\log_a m} = m$

Change of base formula:

$\log_a b = \dfrac{\log_c b}{\log_c a}$

Mixed practice

1 Rewrite in terms of $\log a$, $\log b$ and $\log c$:

a $\log(abc)$

b $\log\left(\dfrac{ac}{b}\right)$

c $\log(a^3 b^2 c^4)$

d $\log(a\sqrt{b})$

e $\log\left(a\sqrt{\dfrac{b}{c^2}}\right)$

2 Express as a single logarithm:

a $2 \log 5 + \log 4 - \log 8$

b $3 \ln 5 - \ln 4 + \ln 8$

c $3 \log_4 5 - 2 \log_4 5 + \log_4 8$

d $2 \log_a \left(\dfrac{1}{2}\right) + \log_a 4 - \log_a \left(\dfrac{1}{4}\right)$

e $6 \ln x - 2 \ln(x+1) - 3 \ln(2x+6)$

f $\log_4 2 + 3 \log_4 3$

g $\dfrac{1}{2} \log_7 16 + \dfrac{1}{3} \log_7 x - 2 \log_7 y$

3 Given that $\log_a 3 = x$ and $\log_a 4 = y$, **write down** each expression in terms of x and y.

a $\log_a 0.25$

b $\log_a 48$

c $\log_a \left(\dfrac{1}{12}\right)$

d $\log_a 144$

e $\log_a \left(\dfrac{27}{16}\right)$

4 Express in terms of x and y, given that $x = \log a$ and $y = \log b$.

a $\log(a^5 b)$

b $\log(10\sqrt[4]{ab^3})$

c $\log(0.01 a^3 b^4)^2$

d $\log\left(\dfrac{50 a^7}{5\sqrt{b}}\right)$

5 **Find** the inverse of each function and **state** the domain and range of both the function and its inverse. **Prove** they are inverses both graphically and algebraically.

a $f(x) = 6(5^{2x+1})$

b $g(x) = \sqrt[3]{\log x} - 2$

c $h(x) = -\dfrac{2}{3} \ln(3x - 1)$

d $k(x) = \log_4(7x) - \log_4(x+5)$

6 **Use** properties of logarithms to **find** an equivalent expression for each function. Then **describe** the sequence of transformations on the graph of $y = \log_3 x$ that give the graphs of the equivalent expressions for f and g.

a $f(x) = \log_3(27x)$

b $g(x) = \log_3(x-9)^4$

7 Let $a = \ln x$, $b = \ln(x+2)$ and $c = \ln 4$. Express in terms of a, b and c:

a $\ln\left(\dfrac{4}{x+2}\right)$ **b** $\ln(4x^2 + 8x)$

c $\ln\left(\dfrac{x^3}{16}\right)$ **d** $\ln\sqrt{\dfrac{x^2 + 4x + 4}{4x}}$

e $\ln\sqrt{x^4 + 2x^3}$

Problem solving

8 Express as a sum and/or difference of logarithms:

a $\ln(x^2 - 25)$ **b** $\ln(x^2 - 3x - 28)$

c $\ln\left(\dfrac{x+1}{x-8}\right)$ **d** $\ln\left(\dfrac{x^2 - 4x - 5}{x^2 + 6x + 8}\right)$

e $\ln\left(\dfrac{x^2 - 10x - 24}{x^3 - 4x}\right)$

9 **Solve** these exponential equations:

a $3^{2x+1} + 3 = 10 \times 3^x$

b $2^{2x} - 3 \times 2^x - 4 = 0$

10 **Solve** these logarithmic equations:

a $\log_a x + \log_a 4 - \log_a 5 = \log_a 12$

b $\log_4 x - \log_4 7 = 2$

c $\log_9 x + \log_9(x^2) + \log_9(x^3) + \log_9(x^4) = 5$

d $1 - \log(x+3) = \log x$

e $\log(x-5) - \log(x+4) = -1$

11 $\log_a 2 = x$ and $\log_a 5 = y$. **Find**, in terms of x and y, expressions for:

a $\log_5 2$ **b** $\log_a 40$

c $\log_a 400$ **d** $\log_a 2.5$

e $\log_a 0.4$

12 **Solve** by taking logs of both sides:

a $19^x = 2$ **b** $\left(\dfrac{1}{2}\right)^x = 25$

c $2^{5x-2} = 80$ **d** $3^{5x-6} = 111$

13 Let $\log_a 3 = x$ and $\log_a 7 = y$. **Find** expressions, in terms of x and y, for:

a $\log_3 7$ **b** $\log_7 3$ **c** $\log_3 49$

14 Given that $\ln a = 8$:

a express $\log_a(x^3)$ as a simple natural logarithm

b express $3\log_a x - \ln(a^2)$ as a single logarithm.

15 Reproduction in a colony of sea urchins can be modelled by the equation $N = 100e^{0.03x}$ where N is the number of sea urchins and x is the number of days since the original 100 were introduced to the reef.

a **Find** the number of days it takes for the population to double.

b **Find** how many sea urchins were born on day 19.

Problem solving

16 A mathematician calculated that the number of people infected after the first case of a disease was identified followed the model $P = P_0(1 - e^{-0.001t})$ where P is the number of infected people, P_0 is the population of the city, and t is the number of days after the first case was identified.

a **Calculate** how long it would take for 5% of the population to be infected.

b In a city, 58 187 people are infected after 21 days. **Calculate** the population of the city.

17 A game park in Swaziland recorded 15 rhinoceros in 2000, and 26 rhinoceros in 2011. Assuming the number of rhinoceros in the game park follows an exponential function, **find** the function.

Review in context

In E6.1 and E10.1 you used this formula for modelling the magnitude M of an earthquake using the Richter scale:

$$M = \log_{10} \frac{I_c}{I_n}$$

where I_c is the intensity of the 'movement' of the earth from the earthquake, and I_n is the intensity of the 'movement' on a normal day-to-day basis.

You can also write this as:

$$M = \log_{10} I$$

where I is the intensity ratio $\frac{I_c}{I_n}$.

1 a The San Francisco earthquake of 1906 measured magnitude 8.3 on the Richter scale. Shortly afterward, an earthquake in South America had an intensity ratio four times greater than the San Francisco earthquake. **Find** the magnitude on the Richter scale of the earthquake in South America.

b A recent earthquake in Afghanistan measured 7.5 on the Richter scale. **Find** how many times more intense the San Francisco earthquake was.

c Find how much larger an earthquake's magnitude is, if it is 20 times as intense on the Richter scale as another earthquake.

d Determine how much more intense is an earthquake whose magnitude is 7.8 on the Richter scale than an earthquake whose magnitude is 6.2.

e Compare the intensities of two earthquakes: Nevada in 2008 with magnitude 6.0 and Eastern Sichuan in 2008 with magnitude 7.6.

2 The pH value of a solution is used to determine whether a solution is basic (alkaline) or acidic. A pH value of 7 is neutral, less than 7 is acidic, and more than 7 is basic. To calculate the pH of a liquid you need to know the concentration of hydrogen ions (H^+) in moles per liter (mol/l) of the liquid.

The pH is then calculated using the logarithmic formula:

$$pH = -\log_{10}(H^+)$$

a Milk has $H^+ = 1.58 \times 10^{-7}$ mol/l. **Determine** whether milk is acidic or basic.

b Find the pH of vinegar with concentration: $H^+ = 1.58 \times 10^{-3}$ mol/l.

c A sample of pool water has concentration: $H^+ = 6.3 \times 10^{-7}$ mol/l. **Determine** the pH of the sample of pool water.

d Some chemicals are added to the water because the answer in **c** indicates that the pH level of the water is not optimal. A sample of water is taken after adding the chemicals, and now the concentration is $H^+ = 7.94 \times 10^{-8}$ mol/l.

Determine if the water is now within optimum levels: $7.0 < pH < 7.4$.

e Find the values for H^+ that give a pH in the range $7.0 < pH < 7.4$.

3 The volume of sound D is measured in decibels (db) and I is the intensity of the sound measured in Watts/m², using the formula $D = 10\log_{10} I$.

a An anti-theft car alarm has an intensity of 5.8×10^{13} W/m². **Find** the volume of the alarm in decibels.

b Jill's scream measured 56 db, and Jack's was 48 db. **Find** how much more intense Jill's scream was than Jack's.

Reflect and discuss

How have you explored the statement of inquiry? Give specific examples.

Statement of Inquiry:

Generalizing changes in quantity helps establish relationships that can model duration, frequency and variability.

E13.1 Are we very similar?

Global context: Personal and cultural expression

Objectives
- Identifying congruent triangles based on standard criteria
- Using congruent and similar triangles to prove other geometric results

Inquiry questions

F
- What conditions are necessary to establish congruence?
- Why is 'Side-Side-Angle' *not* a condition for congruence?

C
- How can congruence be used to prove other results?

D
- Which is more useful: establishing similarity or establishing congruence?

LOGIC

ATL | **Communication**

Understand and use mathematical notation

Statement of Inquiry:

Logic can justify generalizations that increase our appreciation of the aesthetic.

13.1

13.2

E13.1

GEOMETRY AND TRIGONOMETRY

You should already know how to:

• use Dynamic Geometry Software (DGS) to draw basic shapes	**1 a** Use DGS to construct a triangle of side lengths 4, 6 and 7 cm. **b** Use DGS to construct triangle *ABC* with $\angle ABC = 40°$, $\angle ACB = 80°$ and $BC = 5$ cm.
• recognize reflections, rotations and translations	**2** Describe the single transformation that takes: **a** *A* to *B* **b** *A* to *C* **c** *C* to *D* **d** *D* to *E* **e** *B* to *D* **f** *E* to *B*
• find angles in parallel lines	**3** Find the size of the angles marked *a* and *b*.
• find lengths in similar triangles	**4** Triangles *ABC* and *DEF* are similar. Find lengths *AC* and *ED*.

F Establishing congruent triangles

• What conditions are necessary to establish congruence?
• Why is 'Side-Side-Angle' *not* a condition for congruence?

Exploration 1

1 Use dynamic geometry software to construct:

   a triangle *A* with sides 5 cm, 6 cm and 7 cm

   b triangle *B* with a side of 8 cm and all angles 60°

   c triangle *C* with angles 45°, 60° and 75°

   d triangle *D* with angles 40°, 60° and 80° and a side of 5 cm

> If you do not have access to dynamic geometry software, construct these triangles accurately using pencil, compasses and protractor.

▶ Continued on next page

E13.1 Are we very similar?

e triangle *E* with sides 6 cm and 8 cm, and an angle of 50°

f triangle *F* with one side 4 cm, and with angles of 30° and 45° at each end of the 4 cm side

g *G*, a right-angled triangle with one side 4 cm and hypotenuse 5 cm.

2 Compare your triangles with others. Identify which triangles everyone drew exactly the same. Discuss which of the descriptions **a** to **g** lead to more than one possible triangle.

Two shapes are **congruent** if they have the same side lengths and the same size angles.

Reflect and discuss 1

- Are the triangles in each pair congruent?

a 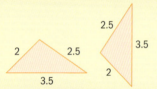 **b**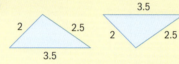

c 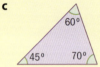 **d**

- Look at the diagrams you drew in Exploration 1. Which descriptions create triangles which are always congruent, and which descriptions create triangles which might not be congruent?

- What do you call triangles that have the same angles, but are not the same size?

Two shapes are **congruent** if one can be transformed into the other by a reflection, rotation or translation, or a combination of these.

> The word *congruence* comes from the Latin *Congruō*, meaning 'I agree' or 'I come together'.

When working with congruent triangles you need to identify the **corresponding** sides and angles:

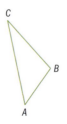 

GEOMETRY AND TRIGONOMETRY

△DEF is the reflection of △ABC in the line, so the two triangles are congruent. D is the image of A, so D and A are corresponding vertices.

DE is the reflection of AB, so DE and AB are corresponding sides.

∠DEF is the reflection of ∠ABC, so ∠DEF and ∠ABC are corresponding angles.

ATL The ≅ symbol represents congruence. In the convention for congruent triangle notation the order of the letters is important.

The statement △ABC ≅ △DEF means 'triangle ABC is congruent to triangle DEF, so A corresponds to D, B corresponds to E, and C corresponds to F'. Using this notation helps you show your working clearly.

To identify corresponding vertices, you can identify the transformation that takes the first shape into the second, using rotation, reflection or translation, and then find the images of the vertices. You can also match pairs of shortest/longest sides, or largest/smallest angles.

Example 1

△PQR and △LMN are congruent. Identify the corresponding sides.

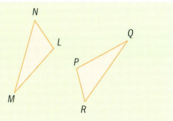

△PQR is a rotation of △LMN.

LN corresponds to PR. ——— The shortest side of each triangle.

∠L corresponds to ∠P. ——— The obtuse angle in each triangle.

MN corresponds to QR. ——— The side opposite the obtuse angle in each triangle.

LM corresponds to PQ. ——— The last pair of sides.

Practice 1

In questions **1** to **4**, identify the corresponding sides in each pair of triangles.

1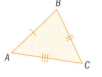

2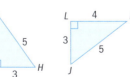

3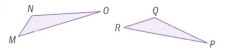

4

E13.1 Are we very similar?

5 Identify the pairs of congruent triangles in this diagram. Use congruence symbol notation, with the corresponding vertices in the correct order.

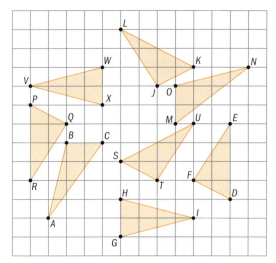

To show that two triangles are congruent, you need to show that they satisfy one of the conditions for congruence.

Exploration 2

The aim of this exploration is to discover what information guarantees that two triangles will be congruent.

1 Choose any three side lengths that could make a triangle.
   Construct a triangle with these side lengths.
   Now try to construct another triangle that is not congruent to the original. Discuss whether or not it is possible.

2 Record the results from step **1** in a table like this:

3 elements	Sketch of the two triangles (indicate the 3 elements)	Always congruent? YES or NO
3 sides		
3 angles		
2 sides and the angle between them		
2 sides and an angle *not* between them		
2 angles and the side between them		
2 angles and a side *not* between them		

3 Now look at the other rows of the table. In the same way as in step **1**, decide on measurements for the elements and try to construct different triangles. Record whether or not it is possible in the right-hand column.

▶ Continued on next page

GEOMETRY AND TRIGONOMETRY

4 Based on your results, discuss the necessary conditions for two triangles to be congruent.

5 Compare your results here to your results from Exploration 1.

Conditions for congruence

- Side-Side-Side congruence (SSS):
  If the three sides of triangle A are the same lengths as the three sides of triangle B, the two triangles are congruent.

- Side-Angle-Side congruence (SAS):
  If two sides of triangle A are the same lengths as two corresponding sides of triangle B, and the angle between those pairs of sides is the same in both triangles, the two triangles are congruent.

- Angle-Side-Angle congruence (ASA or AAS):
  If two angles in triangle A are the same size as two corresponding angles in triangle B, and there is a corresponding side whose length is the same in both triangles, the two triangles are congruent.

- Right Angle-Hypotenuse-Side congruence (RHS):
  If two right-angled triangles have equal hypotenuses, and one of the remaining two sides is the same length in both triangles, the two triangles are congruent.

Reflect and discuss 2

Explain how the Right Angle-Hypotenuse-Side congruence condition can be derived using each of the other congruence conditions.

Exploration 3

1 Draw:
   a a line segment AB of length 6 cm
   b a circle of radius 7 cm centered on B
   c a line from A that makes an angle of 40° with AB.

2 In $\triangle ABC$, $AB = 6$ cm, $BC = 7$ cm and $\angle CAB = 40°$.
   Determine the position of point C on your diagram.

3 Draw:
   a a line segment DE of length 6 cm
   b a circle of radius 4 cm centered on E
   c a line from D that makes an angle of 40° with DE.

4 In $\triangle DEF$, $DE = 6$ cm, $EF = 4$ cm and $\angle FDE = 40°$.
   Explain why it is not possible to determine the position of point F on your diagram.

5 Explain why Side-Side-Angle (SSA) is not a condition for congruence.

E13.1 Are we very similar?

Example 2

Prove that in this diagram, $\triangle ABC \cong \triangle DEC$.

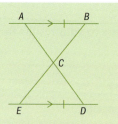

Given: AB and ED are parallel
AB = ED

Prove: $\triangle ABC \cong \triangle DEC$

Separate the question into the part you are given, and the part you need to prove.

Proof

$\angle BAC = \angle EDC$ (alternate angles)
$\angle ABC = \angle DEC$ (alternate angles)
$AB = DE$ (given)
Therefore $\triangle ABC \cong \triangle DEC$ (ASA)

Give a reason for each statement.

Both the equals sign (=) and the congruence sign (≅) are used to indicate that lengths or angles are equal. Both are acceptable in general use, though some examinations or schools might expect you to use one instead of the other.

Example 3

Prove that in a rectangle ABCD, $\triangle ABC \cong \triangle CDA$.

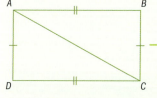

Draw a diagram.

Given: ABCD is a rectangle.

Prove: $\triangle ABC \cong \triangle CDA$

Separate the question into the part you are given, and the part you need to prove.

Proof

$\angle ABC = \angle CDA = 90°$ because all angles in a rectangle are right angles.

$\triangle ABC$ and $\triangle CDA$ are right-angled triangles with equal hypotenuses because both have hypotenuse AC.

$AB \cong CD$ because the opposite sides of rectangle are equal.

Give a reason for each statement.

$\triangle ABC \cong \triangle CDA$ (RHS)

You could also prove this result by showing all the sides are equal (SSS) or by using the two congruent pairs of opposite sides and the right angle in between them (SAS).

GEOMETRY AND TRIGONOMETRY

> **Objective C:** Communicating
> **v.** Organize information using a logical structure
>
> *In Practice 2, question 4, organize your answers systematically to help make sure that you include all possible triangles.*

Practice 2

1 Determine whether the triangles in each pair are congruent. If they are, state the conditions for congruence.

a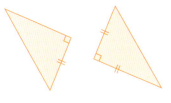

b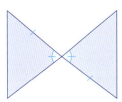

c

d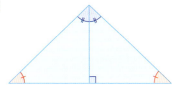

2 a Prove that $\triangle PQR$ is congruent to $\triangle STR$.

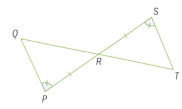

b Prove that $\triangle GHK \cong \triangle JHK$.

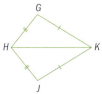

c In triangle *ABC*, *CD* is the perpendicular bisector of *AB*. Prove that $\triangle ADC$ is congruent to $\triangle BDC$.

3 A regular pentagon has vertices *ABCDE*. Explain why $\triangle ABC$ is congruent to $\triangle CDE$.

> In **3**, draw and label a diagram to help you.

Problem solving

4 A regular hexagon has vertices *ABCDEF*. List all the triangles that are congruent to $\triangle ADE$. State the condition used to establish congruence.

Example 4

Circles C_1 and C_2 have centers O_1 and O_2 respectively, and intersect at points X and Y.

Prove that $\triangle O_1 X O_2$ is congruent to $\triangle O_1 Y O_2$.

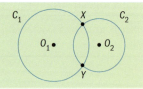

Given: circles C_1 and C_2 with centers O_1 and O_2 respectively intersect at points X and Y. ——— Separate into 'given' and 'prove'.

Prove: $\triangle O_1 X O_2 \cong \triangle O_1 Y O_2$.

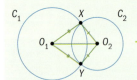

——— Draw the triangles $\triangle O_1 X O_2 \cong \triangle O_1 Y O_2$ on the diagram.

Proof ——————————————————— Give a reason for each statement.

$O_1 X = O_1 Y$ (both radii of C_1)

$O_2 X = O_2 Y$ (both radii of C_2)

$O_1 O_2 = O_1 O_2$ (same line segment)

$\triangle O_1 X O_2 \cong \triangle O_1 Y O_2$ (SSS)

ATL Practice 3

1 Triangle ABC is isosceles with $AB = AC$.
M is the midpoint of BC.
Prove that $\triangle ABM$ and $\triangle ACM$ are congruent.

2 Points A, B, C and D lie on a circle with center O.
The chords AB and CD are equal in length.
Prove that $\triangle ABO$ and $\triangle CDO$ are congruent.

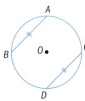

3 PQR is an isosceles triangle with $PQ = QR$. A line from Q meets PR at X, at an angle of $90°$. Prove that $\triangle QPX \cong \triangle QRX$.

4 Points A, B and C lie on a circle with center O. $\angle ABC = \angle ACB$.
Prove that $\triangle AOB$ and $\triangle AOC$ are congruent.

5 Circle C_1 has center A, and circle C_2 has center B which lies on the circumference of C_1. Points D and E are the intersections of C_1 and C_2.

Prove that $\triangle ABD$ and $\triangle ABE$ are congruent.

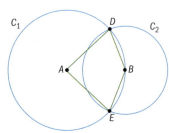

6 *HIJK* is a square. *H* and *J* lie on a circle with center *F*. Prove that △*FKH* and △*FKJ* are congruent.

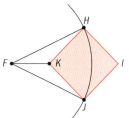

7 Prove that if lines *PQ* and *RS* are parallel, and *X* is the midpoint of *PS*, then △*PXQ* and △*SXR* are congruent.

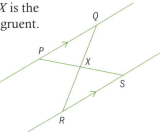

C Using congruence to prove other results

• How can congruence be used to prove other results?

All mathematical systems start by establishing some agreed facts. These are known as axioms. There are many different systems of axioms which can be used to describe geometry; the most famous is Euclidean geometry, which was formalized by the Greek mathematician Euclid around 300 BCE.

The rest of the geometry in this topic takes the congruence conditions to be axioms. In other words, it assumes them to be true, and then uses them to prove other results.

Exploration 4

1 In the diagram, *PS* and *TQ* bisect each other at *U*. Prove that △*STU* ≅ △*PQU*.

2 Find the length of *ST*. Justify your reasoning.

3 Identify the congruent angles in △*STU* and △*PQU*. Justify your reasoning.

4 If you can prove two triangles are congruent using SSS, SAS, ASA/AAS or RHS, explain what you can deduce about the other corresponding angles and sides in the triangles.

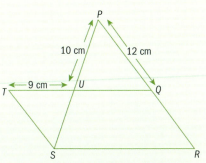

> Euclid's most famous achievement is his publication of *The Elements* – 13 books containing mathematical definitions, theorems and proofs. Many of the results were already known, but Euclid's gathering of them in a single place earned him the modern nickname "the father of geometry".

Corresponding parts of congruent triangles are congruent. When using this as part of a proof, you can write CPCTC for short.

Reflect and discuss 3

Read and discuss this proof to make sure you understand it.

Isosceles triangle theorem

If $\triangle ABC$ is isosceles with $AB = AC$, then $\angle ABC \cong \angle ACB$.

Proof

$AB = AC$	(given)
$\angle BAC = \angle CAB$	(the same angle)
$AC = AB$	(given)
$\triangle ABC \cong \triangle ACB$	(SAS)
$\angle ABC \cong \angle ACB$	(CPCTC)

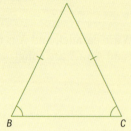

The isosceles triangle theorem was known as the *pons asinorum*, or 'bridge of asses', because students in medieval times who could not understand the proof, or why a proof was needed, were thought to be ignorant. This proof is by Pappus of Alexandria.

Practice 4

1 Copy this proof and add the reasons.

Theorem
If $ABCD$ is a rhombus then $\angle DAC \cong \angle BCA$.

Proof

$AB = CD$, $BC = DA$ because _____

$AC = CA$ because _____

$\triangle ABC \cong \triangle CDA$ because _____

$\angle DAC \cong \angle BCA$ because _____

2 $\triangle ACE$ is an isosceles right-angled triangle with the right angle at C. B lies on AC and D lies on CE such that $BE = DA$.
Copy and complete the proof that $\angle CEB \cong \angle CAD$.

Theorem _____

Proof

$BE = DA$ because _____

$\angle BCE = \angle DCA = 90°$ because _____

$CE = CA$ because _____

$\triangle BCE \cong \triangle DCA$ because _____

$\angle CEB \cong \angle CAD$ because _____

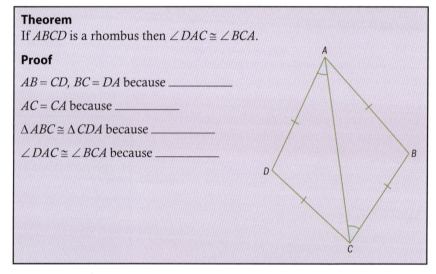

E13 Justification

GEOMETRY AND TRIGONOMETRY

Problem solving

3 Copy and complete this skeleton proof.

> **Theorem**
>
> If △ABC is isosceles with AB = AC then the angle bisector of ∠BAC is perpendicular to BC.
>
> **Proof**
>
> Let X be the point where the angle bisector of ∠BAC meets BC. — *Draw a diagram, showing X.*
>
> Show that △ABX ≅ △ACX: — *Give reasons to prove congruence.*
>
> AB = AC because _____.
>
> ∠BAX ≅ ∠____ because _____.
>
> ____ is common.
>
> Therefore △ABX ≅ △ACX (____). — *Give the condition for congruence.*
>
> Hence ∠____ ≅ ∠AXC. (____). — *Explain the reason for this.*
>
> ∠____ + ∠____ = 180° because _____.
>
> ⇒ 2 × ∠AXC = 180°
>
> ⇒ _____.
>
> Therefore the angle bisector is perpendicular to BC.

4 Copy and complete this skeleton proof. Include a suitable diagram.

> **Theorem**
>
> If AB is a chord to a circle C with center O, and M is the midpoint of AB, OM is perpendicular to AB.
>
> **Proof**
>
> OA = OB because _____.
>
> AM = BM because _____.
>
> ____ is common.
>
> Therefore △OAM ≅ △OBM (____).
>
> Hence ∠____ ≅ ∠____.
>
> ∠____ + ∠____ = ____ because _____.
>
> ⇒ 2 × ____ = ____
>
> ⇒ _____.
>
> Therefore OM is perpendicular to AB.

E13.1 Are we very similar?

5 Circle C has center O. L_1 is a tangent to C at X. L_2 is a tangent to C at Y. L_1 and L_2 meet at Z.

   a Sketch L_1, L_2 and C. Mark points O, X, Y and Z on your diagram.

   b Prove that triangles $\triangle OXZ$ and $\triangle OYZ$ are congruent.

   c Hence show that $XZ = YZ$.

6 A parallelogram is defined as having two pairs of parallel sides. Prove that opposite sides are equal in length.

7 Prove that the diagonals of a kite are perpendicular.

8 Prove that the diagonals of a parallelogram bisect each other.

9 Square $TUVW$ is intersected by the straight line $ABXCD$. X is the midpoint of AD and also the midpoint of UW.

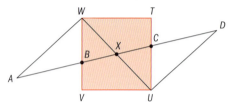

Prove that $\angle AWB \cong \angle DUC$.

Tip

In **6**, divide the parallelogram into two congruent triangles.

Problem solving

10 The Pentagon, the headquarters of the US military, is among the world's largest buildings. It takes the shape of a regular pentagon, contains over 28 km of corridors, and has an internal courtyard about the size of three American football fields.

A statue (K) is placed directly in front of one face of the building, so that it is equidistant from the two nearest corners of the building (D and E). Prove that it is also equidistant from the next two corners of the building (A and C).

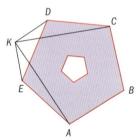

Tip

Remember that 'P is equidistant from A and B' means that $AP = BP$.

 Using similarity to prove other results

- Which is more useful: establishing similarity or establishing congruence?

You have seen how establishing congruence can lead to the proof of other geometric results. Sometimes, simply establishing similarity is enough to lead to important mathematical results.

GEOMETRY AND TRIGONOMETRY

Exploration 5

1 Copy and complete this proof to discover the relationship between segments created by intersecting chords.

Theorem: If A, B, C and D are points on the circumference of a circle such that the chord AB meets the chord CD at point X interior to the circle, then $AX \times BX = CX \times DX$.

$\angle AXC = \angle$ _____ because _____ . ——— Reproduce the diagram showing $\triangle AXC$ and $\triangle DXB$.

$\angle CAB = \angle CDB$ because _____ . ——— Show that corresponding angles are equal.

$\angle ACD = \angle$ _____ because _____ .

Since all three angles are equal, we know that $\triangle AXC$ is similar to $\triangle DXB$.

$\dfrac{AX}{BX} = \dfrac{}{}$. ——— Corresponding sides of similar triangles are proportional.

Rearranging gives: $AX \times BX = CX \times DX$. ——— This the 'intersecting chords theorem'.

Reflect and discuss 4

- Are all similar triangles congruent? Are all congruent triangles similar? Justify your reasoning.
- Which is more useful: establishing similarity or establishing congruence?

Summary

Two shapes are **congruent** if they have the same side lengths and same size angles.

Two shapes are **congruent** if one can be transformed into the other by a reflection, rotation or translation, or a combination of these.

Conditions for congruence

- Side-Side-Side congruence (SSS): If the three sides of triangle A are the same lengths as the three sides of triangle B, the two triangles are congruent.

- Side-Angle-Side congruence (SAS): If two sides of triangle A are the same lengths as two corresponding sides of triangle B, and the angle between those pairs of sides is the same in both triangles, the two triangles are congruent.

- Angle-Side-Angle congruence (ASA or AAS): If two angles in triangle A are the same size as two corresponding angles in triangle B, and there is a corresponding side whose length is the same in both triangles, the two triangles are congruent.

- Right Angle-Hypotenuse-Side congruence (RHS): If two right-angled triangles have equal hypotenuses, and one of the remaining two sides is the same length in both triangles, the two triangles are congruent.

Corresponding parts of congruent triangles are congruent. You can write CPCTC for short.

E13.1 Are we very similar?

Mixed practice

1 Are the two triangles in each diagram congruent? If so, give the conditions for congruence used to prove it.

   a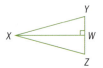

   b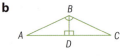

   c

   d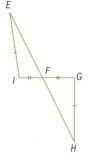

2 a In the diagram, given that $AD = BC$ and $\angle DAC = \angle ACB$, **prove** that $\angle ADC \cong \angle ACB$.

   b In the diagram, given that EF bisects IG, and EI is parallel to IGH, **prove** that $\angle FEI \cong \angle GFH$.

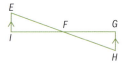

   c In the diagram, given that $\angle J = \angle L$ and that MK bisects $\angle JML$, **prove** that $\triangle JMK \cong \triangle LMK$.

3 $ABCDEFGH$ is a regular octagon. By considering $\triangle AEF$ and $\triangle ADE$, **prove** that $\triangle ADF$ is an isosceles triangle.

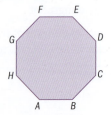

4 $ABCDE$ is a regular pentagon. X is the midpoint of CD. **Prove** that AX is the perpendicular bisector of BE.

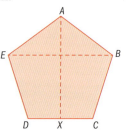

> To prove AX is the perpendicular bisector of BE, show that AX is perpendicular to BE **and** that X is the midpoint of BE.

5 Part of an origami construction consists of a parallelogram and three fold lines: one diagonal joining the two most distant corners of the parallelogram, and two short parallel lines joining the other corners to the diagonal, as illustrated.

Prove that the short parallel lines are equal in length.

6 The diagram shows points A, B, C and D. **Prove** that triangles ABD and ACD are congruent.

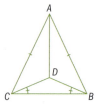

7 In the diagram, AB is parallel to DE and $AB = AC$, and $CAB = 60°$. **Prove** that CDE is an isosceles triangle.

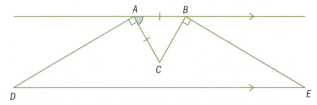

8 A small radio mast is formed by joining three large triangular frames together. One of the three frames, illustrated below, is made from four lengths of metal: *AC*, *AD*, *BD* and *CE*. It stands on level ground.

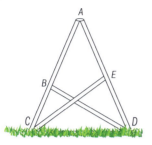

Given that *AD* and *AC* are equal in length, and *BD* and *CE* are equal in length, **prove** that lengths *AB* and *AE* are equal.

9 A King Post Truss Bridge has a simple structure and is suitable for short spans.

The king post *PQ* is a single piece of wood or metal that helps prevent the tie beam *AB* from sagging.

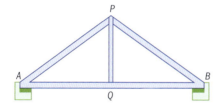

Prove that if *Q* is the midpoint of *AB*, the king post *PQ* is vertical, and *AB* is horizontal then *AP* and *BP* must be equal in length and ∠*PAQ* = ∠*PBQ*.

10

The three blades of a wind turbine are all of equal length and have the same angle between each pair. **Prove** that the tips of the blades form an equilateral triangle.

11 Railway tracks are set in parallel lines at a fixed distance, called the gauge. Different countries have different national gauges. The Republic of Ireland uses Irish Gauge, where the two rails are 1600 mm apart. The photograph shows a railway junction in Limerick, Ireland. The diagram below it represents an overhead view of the layout of the lines.

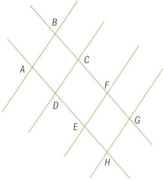

Assuming the tracks to be straight and parallel, and all of Irish Gauge:

a prove that △ *ABD* is congruent to △ *CDB*

b prove that △ *BAC* ≅ △ *BCA*

c hence show that *ABCD* is a rhombus

d prove that △ *ABC* ≅ △ *EFG* given that the two sets of tracks are parallel.

Problem solving

12 The horizontal bars on a stepladder make it more stable. **Use** the ideas of congruence to **explain** clearly how positioning the horizontal bars at the same height on both sides of the ladder ensures that both sides of the ladder open out to the same angle.

13 Similar triangles can be used to develop the Angle Bisector Theorem. Copy and complete this proof.

Theorem: An angle bisector of an angle of a triangle divides the opposite side in two segments that are proportional to the other two sides of the triangle.

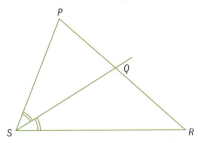

Given: SQ is the bisector of angle S.

Prove: $\dfrac{SR}{SP} = \dfrac{QR}{QP}$.

Proof:

Draw PT parallel to SR.

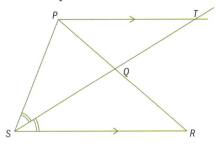

$\angle SQR = \angle PQT$ because _____

$\angle RSQ = \angle$ _____ because _____

$\angle$ _____ $= \angle TPQ$ because _____

Since all three angles are equal, we know that _____.

$\dfrac{SR}{} = \dfrac{}{QP}$

$\angle PST = \angle QSR$ because _____

Hence $\angle PST = \angle PTS$ because _____

Therefore, $PS = PT$ because _____

Hence, $\dfrac{SR}{SP} = \dfrac{QR}{QP}$. □

Review in context

The ancient art of origami has roots in Japanese culture and involves folding flat sheets of paper to create models. Some models are designed to resemble animals, flowers or buildings; others just celebrate the beauty of geometric design.

1 An origami artist takes a square sheet of paper, and folds it in half, as shown.

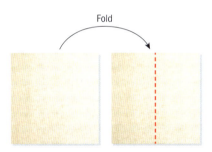

a Explain why folding the sheet in half produces two congruent rectangles.

The artist then unfolds the sheet, takes one corner and folds it up to meet the center line, not quite to the top edge, as illustrated. The fold created runs straight from one corner to the opposite side. Finally, he folds a crease parallel to the top edge of the sheet at the height of the folded corner.

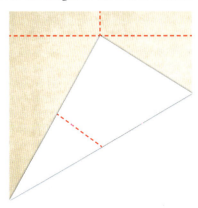

The resulting creases are illustrated in this diagram.

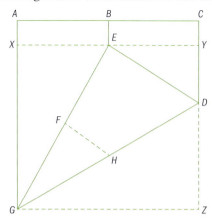

The artist thinks lengths FH and DY are equal.

b **Explain** why $EX = \frac{1}{2} EG$.

c **Hence use** trigonometry to find $\angle EGX$.

d **Explain** why $\triangle GED \cong \triangle GZD$. Hence find $\angle EGD$ and $\angle DGZ$.

e **Find** $\angle YED$.

f **Show** that $GF = EY$.

g **Show** that $\angle GFH = \angle EYD$.

h **Hence show** that the artist's claim – that lengths FH and DY are equal – is correct.

Reflect and discuss

How have you explored the statement of inquiry? Give specific examples.

Statement of Inquiry:

Logic can justify generalizations that increase our appreciation of the aesthetic.

E13.1 Are we very similar?

E14.1 A world of difference

Global context: Identities and relationships

Objectives
- Solving quadratic and rational inequalities both algebraically and graphically
- Solving other non-linear inequalities graphically
- Using mathematical models containing non-linear inequalities to solve real-life problems

Inquiry questions

F
- What is a non-linear inequality?
- How do you solve non-linear inequalities graphically?

C
- How does a model lead to a solution?

D
- Which is more efficient: a graphical solution or an algebraic one?
- Can good decisions be calculated?

ATL Critical-thinking

Use models and simulations to explore complex systems and issues

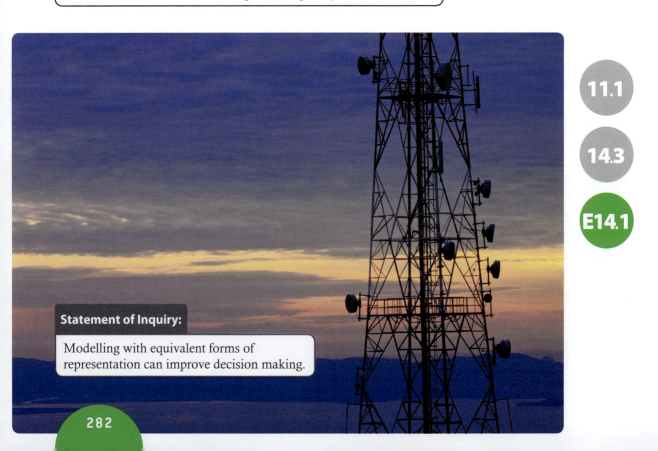

11.1

14.3

E14.1

Statement of Inquiry:
Modelling with equivalent forms of representation can improve decision making.

You should already know how to:

• solve linear inequalities	**1** Solve $-3x - 2 < 5 - 4x$.
• sketch graphs of linear, quadratic, rational, radical, exponential and logarithmic functions	**2** Sketch the graphs of: **a** $y = x^2 + 3x - 4$ **b** $y = 2e^x - 3$ **c** $y = 4\log 2x$ **d** $y = \dfrac{-2}{x-4}$ **e** $y = \dfrac{2x-3}{3x+6}$ **f** $y = \sqrt{x-5}$
• solve quadratic equations by completing the square	**3** Solve $x^2 + 6x + 3 = 0$.

F   Non-linear inequalities in one variable

- What is a non-linear inequality?
- How do you solve non-linear inequalities graphically?

Objective D: Applying mathematics in real-life contexts
iii. apply the selected mathematical strategies successfully to reach a solution

In Exploration 1, when you have selected your solution to the quadratic inequality, determine graphically if the range of the hip angle guarantees a lift area of at least 0.5 m².

Exploration 1

In ski jumping, athletes descend a take-off ramp, jump off the end of it and 'fly' as far as possible. The angle of the skier's hip α determines the 'lift area' A (in m²) which determines how far the skier jumps. The relationship between hip angle and lift area can be modelled by the function:

$$A(\alpha) = -\dfrac{0.3}{400}(\alpha - 140)^2 + 0.8$$

1. Use this function to write a mathematical model (inequality) of the range of the hip angle that would ensure a lift area of at least 0.5 m².
2. Use a GDC to graph both sides of this inequality and find:
   a the point where both graphs intersect
   b the region of the graph that models the set of points where $A(\alpha) > 0.5$
   c the region of the graph that models the set of points where $A(\alpha) < 0.5$.
3. From your graph, write down:
   a the range of the skier's hip angle to ensure a lift area of 0.5 m² or more
   b the range of the skier's hip angle that gives a lift area less than 0.5 m².
4. Explain how you can test whether or not you have selected the correct ranges in step **2**.
5. Rearrange the inequality in step **1** to the form $A(\alpha) \geq 0$.
6. Graph $A(\alpha)$, and explain how you would use the inequality to answer step **2**.
7. Explain how you could use this graph to check your answer.

E14.1 A world of difference

Reflect and discuss 1

- How do you know whether or not to include *x*-values where the two graphs intersect (or the *x*-intercepts, as in step **6**) in your solution set?
- Which method did you prefer: the one where you graph each side of the original inequality and compare, or the one where you rewrite the inequality? Explain.

To solve a quadratic inequality graphically:

- Graph both sides of the inequality, or rearrange into the form $f(x) > 0$ or $f(x) < 0$ and graph.
- Determine the points of intersection of both functions, or find the zeros of the function.

To check your solution algebraically:

- Choose a point in the region you think satisfies the inequality, and test this point in the inequality. If it satisfies the inequality, then the *x*-values in this region satisfy the inequality. If your test point does not satisfy the inequality, then test the *x*-values in the other region(s).

Example 1

Solve the inequality $28 - x \leq 16 + 12x - x^2$. Check your solution algebraically.

Method 1

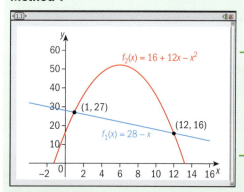

Graph both sides of the inequality and determine the points of intersection.

The *x*-values of points on the line (LHS) are lower than those on the quadratic (RHS) for $1 < x < 12$.

Test a point with *x*-value between 1 and 12:

When $x = 2$: $28 - 2 = 26$

$16 + 12 \times 2 - 2^2 = 36$

$26 \leq 36$ ✓

The solution is $1 \leq x \leq 12$.

Select an *x*-value in the solution interval, and check that it satisfies the inequality. Because the original inequality was "less than or equal to", the endpoints ($x = 1$ and $x = 12$) are included.

▶ Continued on next page

Method 2

$$28 - x \leq 16 + 12x - x^2$$
$$\Rightarrow 0 \leq -12 + 13x - x^2$$

Rearrange to $f(x) \geq 0$.

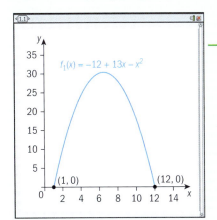

Graph the quadratic function.

The solution is $1 \leq x \leq 12$.

The zeros of the quadratic are $x = 1$ and $x = 12$. $f(x) \geq 0$ for the x-values between these zeros.

Check: When $x = 2$: $-12 + 13 \times 2 - 2^2 = 10 \geq 0$.

Select an x-value in the solution interval, and check that the inequality is satisfied.

Example 2

Solve the inequality $-2x^2 + 4x < x^2 - x - 6$. Check your solution algebraically.

Method 1

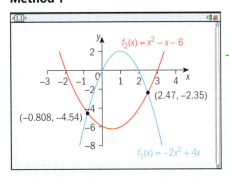

Graph both sides of the inequality and determine the points of intersection of the graphs.

$x < -0.808$ or $x > 2.47$

The blue curve, for $-2x^2 + 4x$ lies below the red curve in the regions $x < -0.808$, $x > 2.47$.

Check:

When $x = -1$: LHS $= -6$, RHS $= -4$

$-6 < -4$

When $x = 3$: LHS $= -6$, RHS $= 0$

$6 < 0$

▶ Continued on next page

Method 2

$-2x^2 + 4x < x^2 - x - 6$ ———— Rearrange to $f(x) > 0$.

$\Rightarrow 3x^2 - 5x - 6 > 0$

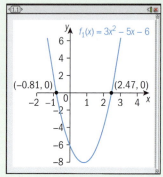

———— Graph the quadratic function.

$x < -0.808$ or $x > 2.47$ ———— The quadratic is greater than 0 above the x-axis.

ATL Practice 1

1 Solve each inequality graphically, and check your answers algebraically.

   a $x^2 - 2x - 9 > 2x + 3$ **b** $2x^2 - 5 \leq x + 1$ **c** $5x^2 - 10 \geq 23x$

   d $x^2 + 3x - 8 < 4 - 2x - x^2$ **e** $2x^2 - 7x + 7 > 8x^2 - 2x + 1$

> Use whichever method you prefer.

2 For astronaut training, weightlessness is simulated in a reduced gravity aircraft that flies in a parabolic arc. The function $h(t) = -3t^2 + 191t + 6950$ models the relationship between the flight time t in seconds, and the height h of the aircraft in meters. The astronauts experience weightlessness when the aircraft's height is above 9600 m.

   a Write an inequality that describes this situation.

   b Determine how long the astronauts experience weightlessness when the aircraft flies in a parabolic arc.

3 A rectangle is 6 cm longer than it is wide. Its area is greater than 216 cm². Write and solve an inequality to find its possible dimensions.

4 A gardener has 24 meters of fencing to build a rectangular enclosure. The enclosure's area should be less than 24 m². Write and solve an inequality to find possible lengths of the enclosure.

5 For drivers aged between 16 and 70, the reaction time y (in milliseconds) to a visual stimulus (such as a traffic light) can be modelled by the function $y = 0.005x^2 - 0.23x + 22$, where x is the driver's age in years. Write an inequality to determine the ages for which a driver's reaction time is more than 25 milliseconds, and solve.

Problem solving

6 A baseball is thrown from a height of 1.5 m. The relationship between the time t (in seconds) and the height h (in meters) of the baseball while in the air is modelled by the function $h(t) = -4.9t^2 + 17t + 1.5$.

   a Write an inequality that describes how long the baseball is in the air.

   b Determine how long the baseball is in the air.

7 In a right-angled triangle, one of the perpendicular sides is 2 cm longer than the other. The hypotenuse is more than twice the length of the shortest side. Determine the possible lengths for the shortest side.

8 A 2 cm square is cut from each corner of a square piece of card, and the sides are then folded up to form an open box. Pia needs a box with maximum volume 40 cm³. Find the smallest possible dimensions of the box.

Example 3

Find the set of values such that $\frac{1}{x} > \sqrt{x}$ in the domain $\mathbb{R}^+$ and check your answer.

The solutions to Examples 3 and 4 can also be found using the same **Method 2** approach as described in Examples 1 and 2.

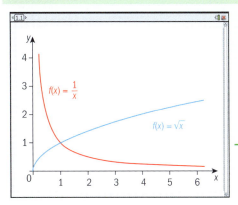

Graph both sides of the inequality, and find the point of intersection.

$\frac{1}{x} > \sqrt{x}$ in the interval $0 < x < 1$.

Select an x-value in the solution interval and test it in the inequality.

Check: When $x = 0.5$, $\frac{1}{0.5} = 2$ and $\sqrt{0.5} \approx 0.707 < 2$

Example 4

Solve $\frac{2x-7}{x-5} \leq 3$, $x \neq 5$, and check your answer.

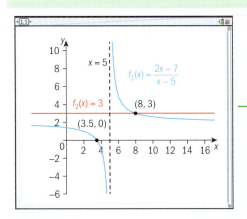

Locate where the LHS function (blue) $\leq$ RHS function (red).

$x \geq 8$ or $x < 5$

▶ Continued on next page

Check:

When $x = 10$, $\frac{2x-7}{x-5} = \frac{13}{5}$ and $\frac{13}{5} \leq 3$.

When $x = 0$, $\frac{2x-7}{x-5} = \frac{7}{5}$, and $\frac{7}{5} \leq 3$.

Select an *x*-value in each region, and check in the original inequality.

Practice 2

1 Find the set of values that satisfy each non-linear inequality:

a $\frac{1}{x} - x < 0$

b $\frac{1}{x} - 4 \geq 0$

c $\frac{x+6}{x+1} > 2$

d $\ln x < \log_2(x-2)$

e $x^2 > \sqrt{2x+3}$

f $\frac{4}{x+5} < \frac{1}{2x+3}$

g $\frac{1}{x} < \frac{1}{x+3}$

h $\frac{3x}{x-1} \geq 3 + \frac{x}{x+4}$

Problem solving

2 When two resistors, R_1 and R_2 are connected as shown, the total resistance R is modelled by the equation $\frac{1}{R} = \frac{1}{R_1} + \frac{1}{R_2}$.

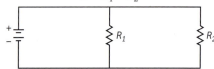

a $R_2 = 2$ ohms and R must be at least 1 ohm. Find R_1.

b $R_1 = 2.2$ ohms and R must be at least 1.7 ohms. Find R_2.

C Non-linear inequalities in two variables

- How does a model lead to a solution?

When you solve an inequality in one variable, the solution is a set of *x*-values. When you solve an inequality in two variables, the solution is a set of points.

For this graph of $y = x^2 - 2x - 3$, the blue shaded area represents the set of points that satisfy $y < x^2 - 2x - 3$. You can test that you have the correct region by substituting the coordinates of a point in this region into the original inequality. For example, for the point $(0, -4)$: $x^2 - 2x - 3 = -3 > -4$, so the correct region is shaded.

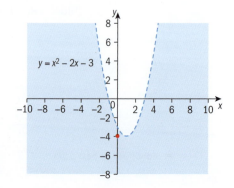

288   E14 Models

The unshaded region of the graph on the previous page represents the set of points that satisfy the inequality $y > x^2 - 2x - 3$. The quadratic graph is drawn with a dashed line, to show that values on the line are not included in the solution set.

> To solve non-linear inequalities in two variables:
> - Graph the region defined by the equality.
> - Select a point that is not on the curve and substitute the values into the original inequality. If the inequality is true, shade the region where the point is (e.g. inside the curve). If the inequality is not true, shade the region where the point is not (e.g. outside the curve).

Example 5

By sketching a suitable graph, solve the inequality $y < -x^2 + 7x - 10$. Shade the region containing the possible solutions on your graph.

$-x^2 + 7x - 10 = (-x + 2)(x - 5)$ — Factorize to find the roots of $f(x) = 0$.

$(-x + 2) = 0 \Rightarrow x = 2$

$(x - 5) = 0 \Rightarrow x = 5$

x-coordinate of vertex $= \dfrac{-7}{-2} = 3.5$ — Use the formula $x = \dfrac{-b}{2a}$ to find the coordinates of the vertex.

Vertex $= (3.5, 2.25)$

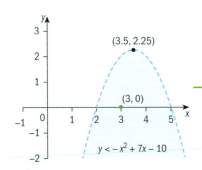

Draw the parabola with a dashed line, as it represents a true inequality.

Check:

For $(3, 0)$: $-x^2 + 7x - 10 = 2 > 0$, hence the region shaded is correct.

Test a point in the original inequality, then shade the correct region.

Practice 3

1 Sketch a suitable graph and shade the region on the graph that shows the solution of the inequality.

a $y > x^2 + 3x - 1$ **b** $y \leq -x^2 + 3x - 2$ **c** $y \geq 4 - 3x - 2x^2$

d $y > \ln(x - 2) + 3$ **e** $y \leq e^{(x+1)} - 3$ **f** $y < \dfrac{x+3}{x-3}$

> In the case of a strict inequality, the line is dotted as in the graph above. In the case of $\geq$ or $\leq$, the graph of the quadratic would be a solid line.

E14.1 A world of difference

2 Write an inequality to describe the shaded region in each graph.

a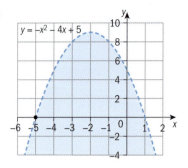

b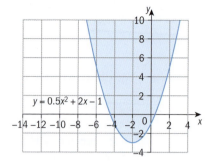

c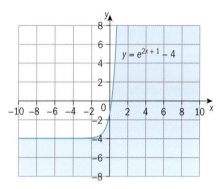

d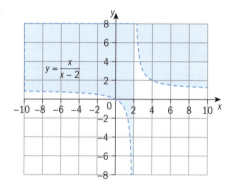

Problem solving

3 For each graph:

  i find the quadratic function for the parabola
  ii write an inequality that describes the shaded region.

a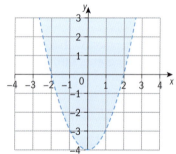

b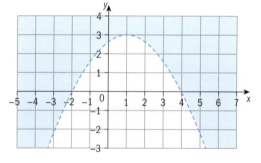

c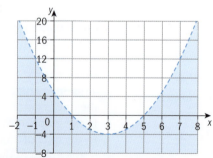

4 A parabolic dam is built across a river, as shown in the graph. The maximum height of the dam above the water level is 3 m, and the length of the dam across the river is 90 m.

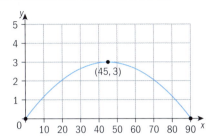

a Write a quadratic function that models the parabolic arch of the dam.

b State the domain and range of the quadratic function.

c Write an inequality that defines the region between the dam and the river.

D Algebraic solutions to non-linear inequalities

- Which is more efficient: a graphical solution or an algebraic one?
- Can good decisions be calculated?

Exploration 2

1 Factorize the quadratic function in the inequality $x^2 - 3x - 4 < 0$ into two linear factors such that $(x+a)(x+b) < 0$.

2 For the product of two factors to be less than zero, one factor must be positive and one negative. You need to solve the two cases separately.

   Case 1: $(x+a) > 0$ and $(x+b) < 0$

   Case 2: $(x+a) < 0$ and $(x+b) > 0$

3 Create the two cases for the inequality in step **1** and solve each one.

4 Determine which case is the solution to the original inequality. Check your answer graphically.

You can also solve a quadratic inequality algebraically by first finding the zeros of the quadratic function.

The quadratic function in Exploration 2 factorizes into $(x-4)(x+1)$, hence the zeros of the function are $x = 4$ and $x = -1$. The x-axis is divided into intervals by the roots of the equation. Choose an x-value in each interval, test it in the inequality to determine if it satisfies the inequality.

Interval	$x < -1$	$-1 < x < 4$	$x > 4$
Test point	-2	0	7
Substitution in $x^2 - 3x - 4$	$(-2)^2 - 3 \times -2 - 4 = 6$ $6 > 0$	$0^2 - 3 \times 0 - 4 = -4$ $-4 < 0$	$7^2 - 3 \times 7 - 4 = 24$ $24 > 0$
Is $x^2 - 3x - 4 < 0$?	No	Yes	No

The values satisfying the quadratic inequality $x^2 - 3x - 4 < 0$ are $-1 < x < 4$. Written in set notation, the solution set is $\{x : -1 < x < 4\}$.

Reflect and discuss 2

- How would you solve the inequality in Exploration 2 graphically?
- Do you prefer the algebraic or the graphical approach for this question?
- Which two cases would you need to consider to solve $x^2 - 3x - 4 > 0$ algebraically? What is the solution set of this inequality?

To solve a quadratic inequality algebraically:

- Rearrange the inequality so that either $f(x) < 0$ or $f(x) > 0$.
- Factorize the quadratic and find its zeros.
- Place these two x-values on a number line, dividing it into intervals.
- Select an x-value in one of the intervals and substitute it into the original inequality. If the inequality is true, then that interval is part of the solution. If the inequality is false, then it is not.
- Repeat for all intervals.

To check your solution algebraically:

- Choose a point in the region you think satisfies the inequality, and test this point in the inequality. If it satisfies the inequality, then the x-values in this region satisfy the inequality.

Example 6

Solve the quadratic inequality $x^2 - x > 12$ and check your answer graphically.

$$x^2 - x > 12$$
$$x^2 - x - 12 > 0$$
$$(x-4)(x+3) > 0$$

▶ Continued on next page

Method 1

Case 1: $x-4<0$ and $x+3<0$
$\Rightarrow x<4$ and $x<-3$,
$\Rightarrow x<-3$

Case 2: $x-4>0$ and $x+3>0$
$\Rightarrow x>4$ and $x>-3$,
$\Rightarrow x>4$

For a product to be positive, the factors must have the same signs.

Because the value of x needs to be both less than 4 and less than -3, using $x<-3$ covers all possibilities.

As before, using $x>4$ covers both of the inequalities, $x>4$ and $x>-3$.

Check your solutions by sketching a graph.

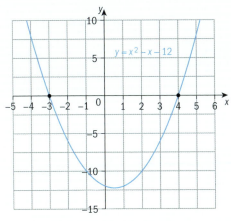

From the graph, you can see that both cases solve the original inequality since $x^2-x-12>0$ when $x<-3$ or when $x>4$.

The solution is $x<-3$ or $x>4$.

Method 2

The zeros of x^2-x-12 are $x=-3$ and $x=4$.

Interval	$x<-3$	$-3<x<4$	$x>4$
Test point	-4	0	5
Substitution in x^2-x-12	$(-4)^2-(-4)-12=8$ $8>0$	$0^2-0-12=-12$ $-12<0$	$5^2-5-12=8$ $8>0$
Is $x^2-x-12>0$?	Yes	No	Yes

From the table, the solution is $x<-3$ or $x>4$.

Hence, $f(x)>0$ for $x<-3$ or $x>4$.

Reflect and discuss 3

Which algebraic method for solving quadratic inequalities is more efficient? Explain your reasons.

Practice 4

1. Solve these quadratic inequalities algebraically and confirm graphically.

   a $x^2 - 2x - 3 \leq 0$
   b $x^2 - 10x + 16 > 0$
   c $12 + x - x^2 \geq 0$
   d $x^2 + 6x \leq 40$
   e $2x^2 + 5x < 3$
   f $3x^2 - 1 > 2x$
   g $6x^2 + 4x + 1 < 5x + 3$

2. One of the perpendicular sides of a right-angled triangle is 3 cm longer than the other. Determine the possible lengths of the shorter side that give a triangle with area less than 4cm².

Problem solving

3. Each inequality is the solution set of a quadratic inequality. Find a possible quadratic inequality.

   a $-1 < x < 1$
   b $x < -1$ or $x > 2$
   c $x \leq 2$ or $x \geq 10$
   d $x < \frac{1}{2}$ or $x > -\frac{2}{3}$
   e $\frac{1}{3} \leq x \leq 6$

Exploration 3

1. By using the quadratic formula or the method of 'completing the square', solve the equation $x^2 + 3x - 7 = 0$.

2. From the roots of the equation in step **1**, explain how you would obtain the solution to the inequality **a** $x^2 + 3x - 7 > 0$ and **b** $x^2 + 3x - 7 < 0$.

3. Confirm your solutions graphically.

4. Write out the steps to solve a non-factorizable quadratic inequality.

5. Check that your steps work when solving:

   a $x^2 + x - 4 < 0$
   b $2x^2 + 4x - 1 > 0$

Practice 5

Solve each inequality algebraically, accurate to 3 s.f.

1. a $x^2 - 5x + 1 > 0$
   b $x^2 + 6x - 2 < 0$
   c $2x^2 + 3x - 3 \leq 0$
   d $3x^2 - 7x + 2 > 0$
   e $-x^2 - x + 5 \geq 0$
   f $-2x^2 + 4x + 2 \leq 0$
   g $5x^2 - 2x - 1 < 0$
   h $x^2 + 10x \geq -8$
   i $-3x^2 - 2x + 1 < x^2 + x - 1$
   j $x^2 - 5x \leq -1$

ALGEBRA

Problem solving

2 Solve each inequality. Explain your result by considering the discriminant Δ of the quadratic function.

a $x^2 + 2x + 1 \leq 0$
b $x^2 - 4x + 4 \geq 0$
c $x^2 + 2x + 4 > 0$
d $-(x-1)^2 - 3 < 0$

$\Delta = b^2 - 4ac$

Reflect and discuss 4

How can the discriminant help you solve quadratic inequalities?

Example 7

Solve $\dfrac{2-x}{x+1} < 1$, $x \neq -1$, and confirm your answer graphically.

$\dfrac{2-x}{x+1} < 1$ ———————————————— Rearrange and simplify.

$\dfrac{2-x}{x+1} - 1 < 0$

$\dfrac{2-x-(x+1)}{x+1} < 0$

$\dfrac{1-2x}{x+1} < 0$ ———————————————— A fraction is negative when the numerator and denominator have opposite signs.

Method 1

Case 1: $1 - 2x < 0$ and $x + 1 > 0$

Then $x > \dfrac{1}{2}$ and $x > -1$, hence $x > \dfrac{1}{2}$.

Case 2: $1 - 2x > 0$ and $x + 1 < 0$

Then $x < \dfrac{1}{2}$ and $x < -1$, hence $x < -1$.

Solution $x > \dfrac{1}{2}$ or $x < -1$.

Method 2

$1 - 2x = 0$; $x = \dfrac{1}{2}$ ———————————————— Find the x-intercept by setting the numerator equal to 0, and the asymptote by setting the denominator equal to 0.

$x + 1 = 0$; $x = -1$

Interval	$x < -1$	$-1 < x < \dfrac{1}{2}$	$x > \dfrac{1}{2}$
Test point	-2	0	1
Substitution into $\dfrac{2-x}{x+1}$	$\dfrac{2-(-2)}{-2+1} = -4$	$\dfrac{2-0}{0+1} = 2$	$\dfrac{2-1}{1+1} = \dfrac{1}{2}$
Is $\dfrac{2-x}{x+1} < 1$?	Yes	No	Yes

Consider the intervals to the left, right, and between these values.

Solution: $x < -1$ or $x > \dfrac{1}{2}$

▶ Continued on next page

E14.1 A world of difference

Check:

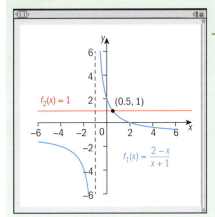

From the graph, $f_1(x)$ is below $f_2(x)$ for x-values greater than 0.5 or less than -1.

Reflect and discuss 5

- Decide which algebraic method for solving rational inequalities you feel is more efficient, and explain your reasons why.
- Does the type of inequality (quadratic, rational, etc.) affect your choice of which algebraic method to use?

Practice 6

1 Solve each inequality algebraically and check your solution graphically.

   a $\dfrac{1+x}{1-x} < 0,\ x \neq 1$ b $\dfrac{x+6}{x+1} \geq 2,\ x \neq -1$

   c $\dfrac{x+1}{x-4} > 0,\ x \neq 4$ d $\dfrac{8-x}{x-2} \leq 5,\ x \neq 2$

   e $\dfrac{3x-1}{x} + 1 < 0,\ x \neq 0$ f $\dfrac{5x-8}{x-5} \geq 2,\ x \neq 5$

Problem solving

2 Determine the numbers that can be added to the numerator and denominator of the fraction $\dfrac{2}{3}$ so that the resulting algebraic fraction is always less than $\dfrac{1}{2}$.

3 Gretchen would like to improve her free throw average in basketball. She currently scores about 9 free throws in 20 trials. Determine how many more consecutive free throws she would need to successfully make in order to bring her average up to 70%.

ALGEBRA

Summary

To solve a quadratic inequality graphically:

- Graph both sides of the inequality, or rearrange into the form $f(x) > 0$ or $f(x) < 0$ and graph.
- Determine the points of intersection of both functions, or find the zeros of the function.

Remember to check your solution algebraically:

- Choose a point in the region you think satisfies the inequality, and test this point in the inequality. If it satisfies the inequality, then the x-values in this region satisfy the inequality.

To solve non-linear inequalities in two variables:

- Graph the region defined by the equality.
- Select a point that is not on the curve and substitute the values into the original inequality. If the inequality is true, shade the region where the point is (e.g. inside the curve). If the inequality is not true, shade the region where the point is not (e.g. outside the curve).

To solve a quadratic inequality algebraically:

- Rearrange the inequality so that either $f(x) < 0$ or $f(x) > 0$.
- Factor the quadratic and find its zeros.
- Place these two x-values on a number line, dividing it into several intervals.
- Select an x-value in one of the intervals and substitute it into the original inequality. If the inequality is true, then that interval is part of the solution. If the inequality is false, then it is not.
- Repeat for all intervals.

To check your solution algebraically:

- Choose a point in the region you think satisfies the inequality, and test this point in the inequality. If it satisfies the inequality, then the x-values in this region satisfy the inequality.

Mixed practice

1 **Solve** each inequality graphically, giving exact answers or accurate to 3 s.f.

 a $(x+2)^2 > 1$
 b $x^2 - 3x - 1 \geq -11x + 4$
 c $-2x^2 + 15 < 3x$
 d $2x^2 - 3x + 1 < 3x^2 + 1$
 e $\frac{2}{x} - 3 > 4$, $x \neq 0$
 f $\frac{5 - 2x}{5 - x} < -1$, $x \neq 5$

2 Shade the region representing the graph of each inequality. Check you have shaded the correct region using a test point.

 a $y > x^2 - 4x + 2$
 b $y \geq -2x^2 + 6x$
 c $y < 2x^2 - 8x - 3$
 d $y < \log(x + 1)$
 e $y \geq e^{(x-2)} + 5$
 f $y > \frac{x-2}{x+4}$

3 **Solve** algebraically and confirm graphically.

 a $x^2 + 5x - 6 < 0$
 b $x^2 - 10x - 24 > 0$
 c $2x + x^2 \leq 35$
 d $(x+3)^2 < 25$
 e $2x^2 - 4x < 7x$
 f $5x^2 - 13x \geq -6$
 g $\frac{x-3}{x+5} > 0$, $x \neq -5$
 h $\frac{3x-5}{x-1} > 4$, $x \neq 1$

4 A rectangular playing field with a perimeter of 100 m is to have an area of no more than 500 m². **Determine** the maximum possible length of the playing field.

5 A tourist agency's profits P (in dollars) can be modelled by the function $P(x) = -25x^2 + 1000x - 3000$, where x is the number of tourists. **Determine** how many tourists are needed for the agency's profit to be at least $5000.

Problem solving

6 Orange juice is to be sold in 2-liter cylindrical cans. To keep costs down, the surface area of a can must be less than 1000 cm². **Determine** possible values for the radius and height of the cans, accurate to 1 d.p.

7 A packing company designs boxes with a volume of at least 600 cubic inches. Squares are cut from the corners of a 20 inch by 25 inch rectangle of cardboard, and the flaps are folded up to make an open box. **Determine** the size of the squares that should be cut from the cardboard, accurate to 3 s.f.

E14.1 A world of difference

Review in context

1 A small satellite dish has a parabolic cross-section. Its diameter is 30 cm and it is 15 cm deep.

  a **Sketch** the parabola on the coordinate axes.

  b When designing the satellite dish, engineers decide where to place a receiver so that it receives a maximum number of signals. **Determine** the inequality that represents the region inside the cross section, which is where the signals are likely to be the strongest. **State** its domain.

2 A parabolic microphone uses a parabolic reflector to collect and then focus sound waves onto a receiver. Its cross-section has a maximum width of 50 cm and a maximum depth of 20 cm.

  a **Draw** a graph to represent this information.

  b The microphone needs to be placed in the interior of the parabola's cross-section. **Determine** the mathematical model (inequality) that describes this region and **state** its domain.

  c On your graph from **a**, shade the region in **b**, and check your solution.

Reflect and discuss

How have you explored the statement of inquiry? Give specific examples.

Statement of Inquiry:

Modelling with equivalent forms of representation can improve decision making.

E15.1 Branching out

Global context: Identities and relationships

Objectives
- Drawing tree diagrams to represent conditional probabilities
- Calculating conditional probabilities from tree diagrams, Venn diagrams and two-way tables
- Determining whether two events are independent

Inquiry questions
- **F**
  - What is conditional probability?
  - How can you represent conditional probability on a tree diagram?
- **C**
  - In which ways can conditional probability be represented?
- **D**
  - What affects the decisions we make?

LOGIC

ATL	Communication

Understand and use mathematical notation

Statement of Inquiry:
Understanding health and making healthier choices result from using logical representations and systems.

4.4

15.1

E15.1

You should already know how to:

• draw tree diagrams	**1** Caesar's new playlist contains ten songs from the 1980s and eight songs from the 1990s. He selects two songs at random, without choosing the same song twice. **a** Draw a tree diagram to represent this situation. **b** Find the probability that the two songs are both from the 1980s.
• use the addition and multiplication rules	**2** You have a standard deck of 52 playing cards. **a** You take one card. What is the probability that you take either a red card or a face card? **b** You take one card, replace it, and then take another. What is the probability that you take a 5 followed by a face card? **c** You take one card and then take another, without replacement. What is the probability that you take a 5 followed by a face card?
• calculate probabilities of mutually exclusive events	**3** A box contains 12 soft-center chocolates and 15 hard-center chocolates. One chocolate is picked at random from the box. Let A be the event 'pick a hard center' and let B be the event 'pick a soft center'. **a** Find $P(A)$. **b** Find $P(A \cup B)$.
• use Venn diagrams and set notation	**4** There are 33 students taking Art and 28 students taking Biology. There are 7 students taking both subjects. Represent this with a Venn diagram. **5** Make a copy of the Venn diagram below for each part of this question. Shade the part of the Venn diagram that represents: **a** B **b** B' **c** $A \cap B$ **d** $A \cup B'$ **e** $(A \cap B)'$

STATISTICS AND PROBABILITY

F Conditional probability

- What is conditional probability?
- How can you represent conditional probability on a tree diagram?

Exploration 1

The probability that it will rain tomorrow is 0.25. If it rains tomorrow, the probability that Amanda plays tennis is 0.1. If it doesn't rain tomorrow, the probability that she plays tennis is 0.9.

Let A be the event 'rains tomorrow' and B be the event 'plays tennis'.

1 Copy and complete this tree diagram for the events A and B.

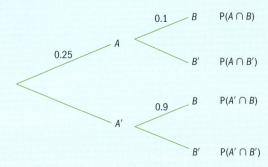

2 State whether A and B are independent events.

3 Find the probability that:

  a Amanda plays tennis tomorrow, given that it is raining

  b Amanda does not play tennis tomorrow, given that it is raining

  c Amanda plays tennis tomorrow, given that it is not raining

  d Amanda does not play tennis tomorrow, given that it is not raining.

> If the outcome of an event in one experiment does *not* affect the probability of the outcomes of the event in the second experiment, then the events are **independent**.

In step **3** of Exploration 1 you should have found that the probability of Amanda playing tennis is conditional on whether or not it is raining.

> In mathematical notation, $P(B|A)$ is 'the probability of B occurring **given that** the condition A has occurred', or 'the probability of B given A'.

Labelling the conditional probabilities on the tree diagram looks like this:

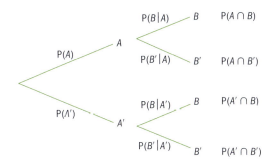

E15.1 Branching out

Multiplying along the branches for A and B gives: $P(A \cap B) = P(A) \times P(B|A)$.

This rearranges to: $P(B|A) = \frac{P(A \cap B)}{P(A)}$

> The conditional probability axiom:
> $$P(B|A) = \frac{P(A \cap B)}{P(A)}$$

You can use this formula to calculate conditional probabilities.

From MYP Mathematics Standard, recall:

Theorem 1 The addition rule is:
For any events A and B, $P(A \cup B) = P(A) + P(B) - P(A \cap B)$.

Theorem 2 The multiplication rule is:
For independent events A and B, $P(A \cap B) = P(A) \times P(B)$.

> You can use Theorem 2 as a test for independent events.
> If A and B are independent events, then $P(A \cap B) = P(A) \times P(B)$.

A and B are independent events if the outcome of one does not affect the outcome of the other. Writing this using probability notation gives another theorem in the probability axiomatic system.

> **Theorem 3**
> If A and B are independent events then $P(B|A) = P(B|A') = P(B)$.

Example 1

> The probability that a particular train is late is $\frac{1}{4}$. If the train is late, the probability that Minnie misses her dentist appointment is $\frac{3}{5}$. If the train is not late, the probability that she misses her dentist appointment is $\frac{1}{5}$.
>
> Let A be the probability that the train is late and let B be the probability that Minnie misses her dentist appointment.
>
> **a** Draw a tree diagram to represent this situation.
> **b** Find $P(A \cap B)$.
> **c** Find $P(B)$.
> **d** Show that the train being late and Minnie missing her dentist appointment are not independent events using **i** Theorem 2 and **ii** Theorem 3.
>
> ▶ Continued on next page

a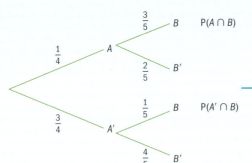

B (missing the appointment) is conditional on A (train being late).

b $P(A \cap B) = \frac{1}{4} \times \frac{3}{5} = \frac{3}{20}$ — $P(A \text{ and } B) = P(A \cap B)$.

c $P(A \cap B) = \frac{3}{20}$, $P(A' \cap B) = \frac{3}{4} \times \frac{1}{5} = \frac{3}{20}$

$P(B) = \frac{3}{20} + \frac{3}{20} = \frac{6}{20}$

B occurs either by $P(A \cap B)$ or $P(A' \cap B)$. These two events are mutually exclusive, so add the probabilities.

d i $P(A) \times P(B) = \frac{1}{4} \times \frac{6}{20}$ — Use Theorem 2.

$= \frac{6}{80}$

$\frac{6}{80} \neq \frac{3}{20}$

So $P(A) \times P(B) \neq P(A \cap B)$ and events A and B are not independent.

d ii $P(B|A) = \frac{3}{5}$ — Use Theorem 3.

$P(B|A') = \frac{1}{5}$

$P(B) = \frac{6}{20}$

$P(B|A) \neq P(B|A') \neq P(B)$ and therefore events A and B are not independent.

In Example 1 the probability of B differs depending on whether or not A has happened. So the outcome of A affects the outcome of B and thus the events A and B are not independent.

Reflect and discuss 1

- Describe another type of problem in probability where events are not independent. Explain how you know they are not independent.
- How is the tree diagram for the probability problem you thought of different from the one in Example 1?

Practice 1

For questions **1–4**, draw a tree diagram and determine whether or not events A and B are independent by using either Theorem 2 or Theorem 3.

1. One coin is flipped and two dice are thrown. Let A be the event 'coin lands tails side up' and let B be the event 'throwing two sixes'.

2. Jar X contains 3 brown shoelaces and 2 black shoelaces. Jar Y contains 5 brown shoelaces and 4 black shoelaces. Norina takes a shoelace from Jar X and a shoelace from Jar Y. Let A be the event 'taking a brown shoelace from Jar X' and let B be the event 'taking a brown shoelace from jar Y'.

3. There are 10 doughnuts in a bag: 5 toasted coconut and 5 Boston creme. Floris takes one doughnut and eats it, then he takes a second doughnut.

   Let A be the event 'the first doughnut is toasted coconut' and let B be the event 'the second doughnut is toasted coconut'.

4. The probability that Mickey plays an online room escape game given that it is Saturday is 60%. On any other day of the week, the probability is 50%.

   Let B be the event 'playing a room escape game' and let A be the event 'the day is Saturday'.

> As you ramble on through life, my friend,
> Whatever be your goal,
> Keep your eye upon the doughnut,
> And not upon the hole.
> – *The modified Optimist's Creed*

In questions **5–8**, determine whether or not the two events are independent. Justify your answer.

5. There are 10 students in your class: 6 girls and 4 boys. Your teacher selects two students at random after putting everyone's name in a hat. When the first person's name is selected, it is not put back in the hat. Let A be the event that the first person selected is a girl. Let B be the event that the second person selected is a girl.

6. Tokens numbered 1 to 12 are placed in a box and two are drawn at random. After the first token is drawn, it is not put back in the box, and a second token is selected. Let A be the event that the first token drawn is number 4. Let B be the event that the second number selected is an even number.

7. According to statistics, 60% of boys eat a healthy breakfast. If a boy eats a healthy breakfast, the likelihood that he will exercise that day is 90%. If he doesn't eat a healthy breakfast, the likelihood that he exercises falls to 65%. Let A be the event 'a boy eats a healthy breakfast'. Let B be the event 'exercises'.

Problem solving

8. Blood is classified in a variety of ways. It may or may not contain B antibodies.

   Its rhesus status can be positive (Rh+) or negative (Rh−).

   The likelihood of someone having B antibodies is 85%.

   The likelihood of being Rh+ is 80%.

   The likelihood of being B+ (having B antibodies and being Rh+) is 68%.

   Let X be the event 'having B antibodies'. Let Y be the event 'being Rh+'.

STATISTICS AND PROBABILITY

C Representing conditional probabilities

• In which ways can conditional probability be represented?

The Venn diagram represents two events, A and B. Find the probability $P(B|A)$.

In this scenario, it is taken that A has already occurred. The sample space is reduced to the elements of A. This is represented in the Venn diagram at the right by the shading on event A.

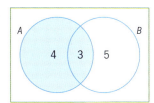

The region of B that is also in A, or $A \cap B$, represents 'B given A'. Here, the darker shading represents this region.

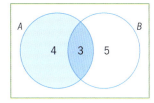

$$P(B|A) = \frac{\text{number of elements in } A \cap B}{\text{number of elements in } A} = \frac{n(A \cap B)}{n(A)} = \frac{3}{7}$$

In a Venn diagram you can calculate conditional probability by reducing the sample space to the **given** condition.

Example 2

> In a group of 30 girls, 12 girls like milk chocolate, 15 like dark chocolate, 6 do not like chocolate. One girl is selected at random.
>
> **a** Draw a Venn diagram to find the probability the girl likes chocolate.
>
> **b** Given that she likes chocolate, find the probability she likes milk chocolate.
>
> **c** Find the probability that she likes dark chocolate, given that she likes milk chocolate.
>
> **d** Test whether liking milk and dark chocolate are independent events by using **i** Theorem 2, and **ii** Theorem 3.

Let x be girls who like both milk and dark chocolate.

$P(A \cup B) = P(A) + P(B) - P(A \cap B)$ ——— Draw a Venn diagram.
Use Theorem 1 to find $n(A \cap B)$.

$24 = 12 + 15 - x$

$24 = 27 - x$

$x = 3$

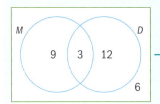

D is the event 'likes dark chocolate'.
M is the event 'likes milk chocolate'.

▶ Continued on next page

E15.1 Branching out

a P(likes chocolate) = $\frac{24}{30}$

b P(M | likes chocolate) = $\frac{12}{24}$ — Reduce the sample space to only the **given** condition (likes chocolate). Ignore the 6 people who do not like chocolate.

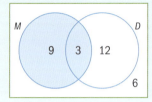

c P(D | M) = $\frac{3}{12}$ — Reduce the sample space to M.

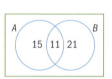

d i M and D are independent events if and only if: — Using Theorem 2.

P(M) × P(D) = P(M ∩ D)

P(M) = $\frac{12}{30}$, P(D) = $\frac{15}{30}$, P(M ∩ D) = $\frac{3}{30}$

P(M) × P(D) ≠ P(M ∩ D), therefore they are not independent events.

ii M and D are independent events if and only if: — Using Theorem 3.

P(D | M) = P(D | M′) = P(D)

P(D | M) = $\frac{3}{12}$, P(D | M′) = $\frac{12}{18}$, P(D) = $\frac{15}{30}$

P(D | M) ≠ P(D | M′) ≠ P(D), therefore they are not independent events.

ATL Practice 2

1 From the Venn diagram at the right, find:

   a P(B | A)   **b** P(A | B)

2 On a field trip, 50 students choose to do kayaking, caving, neither activity, or both.

32 students chose kayaking, 26 students chose caving, and 4 students chose neither activity.

   a Represent this information on a Venn diagram.

   b Find the probability that a student picked at random:

      i chose kayaking

      ii chose caving, given that they chose kayaking

      iii chose caving

      iv chose kayaking, given that they chose caving.

STATISTICS AND PROBABILITY

3 In a group of 30 students, 12 students study Art, 15 study Drama, and 8 study neither subject.

   a Represent this information on a Venn diagram.

   b Find the probability that a student picked at random:

      i studies Art

      ii studies Drama, given that the student studies Art

      iii studies either Drama or Art

      iv studies **both** Drama and Art, given the student studies **either** Drama or Art.

4 A group of 25 students are in a show. Of these, 10 of the students sing, 6 sing and dance, 5 students do not sing or dance. A student is selected at random. Determine the probability that, in the show, the student:

   a does not sing

   b does not dance

   c sings, given that the student does not dance.

Problem solving

5 The Venn diagram shows 35 students' homework subjects.

   a Find the probability that a student selected at random had:

      i Art homework

      ii Biology homework

      iii Chemistry homework.

   b Find:

      i $P(A|B)$ **ii** $P(B|C)$ **iii** $P(C|A')$

   c Determine the probability that a student picked at random had homework in all three subjects, given that the student had homework.

Exploration 2

In the first semi-final of a European Cup tournament, Italy play Spain. 42 students were asked to predict the winner.

- 10 males predicted Italy
- 6 males predicted Spain
- 15 females predicted Italy
- 11 females predicted Spain

1 Represent this information in a two-way table.

2 A student was selected at random. Find the probability that this student was male.

▶ Continued on next page

E15.1 Branching out

3 By altering the sample space as you did in the Venn diagrams, find the probability the student predicted Italy, given that the student was male.

4 Find the probability that the student was female given that the student predicted Spain.

5 Find:

  a $P(M)$ b $P(F)$ c $P(S)$
  d $P(I)$ e $P(F|S)$ f $P(S|F)$
  g $P(M|S)$ h $P(S|M)$ i $P(F|I)$
  j $P(I|F)$ k $P(M|I)$ l $P(I|M)$

M = Male
F = Female
I = Italy
S = Spain

6 In the other semi-final, England play Germany. The two-way table shows how 42 students predicted the winner.

	England (E)	Germany (G)	Total
Male (M)	6	8	14
Female (F)	12	16	28
Total	18	24	42

  a Find the probability that a student selected at random is male.
  b Find: i $P(M|E)$ ii $P(M|G)$

Reflect and discuss 2

Compare the probabilities for the two semi-finals in Exploration 2. Hence make a statement about which events, if any, are dependent and which are independent.

Practice 3

1 There are 28 participants in a golf competition. Of these, 5 are professional male players and 12 are professional female players. There are a total of 18 females playing.

  a Use this information to complete the two-way table.

	Male (M)	Female (F)	Total
Professional (P)			
Amateur (A)			
Total			

  b Use the table to calculate the following probabilities:
    i $P(M|P)$ ii $P(M|A)$ iii $P(M)$

  c Hence, using a theorem, determine whether being male and being a professional are independent events.

STATISTICS AND PROBABILITY

2 100 students were asked whether they were left- or right-handed, and whether or not they were on the school soccer team.

Use the results below to complete a two-way table and decide whether or not the events are independent.

- P(left-handed) = $\frac{4}{10}$
- P(left-handed and not on the soccer team) = $\frac{3}{10}$
- P(right-handed and not on the soccer team) = $\frac{9}{20}$

Problem solving

3 From the table, determine if being a science teacher is independent of gender (male or female) in the school where this data came from. Justify your answer.

	Science teacher	Other teacher
Male	8	25
Female	16	50

D Making decisions with probability

- What affects the decisions we make?

Objective D: Communication
iii. move between different forms of mathematical representation

You will need use the table and also a tree diagram in order to solve the problem in Exploration 3.

Exploration 3

A new test can quickly detect kidney disease. Dr Julia Statham performs the test on 140 patients. She knows that 65 of them have kidney disease and 75 of them do not.

A positive test result indicates kidney disease.

A negative test result indicates no kidney disease.

Here are Dr Statham's results:

	Positive test result	Negative test result	Total
Has kidney disease	30	35	65
Does not have kidney disease	15	60	75
Total	45	95	140

▶ Continued on next page

E15.1 Branching out   309

Let T be the event 'patient test result is positive'.

Let D be the event 'patient has kidney disease'.

1 A patient is selected at random. Calculate:

   a $P(T)$ **b** $P(T')$ **c** $P(D)$ **d** $P(D')$

The test is successful if it gives a positive result for people with kidney disease and a negative result for people who do not have kidney disease. Dr Statham will use the test if it is 90% accurate.

2 Use the information in the table and your knowledge of conditional probability to decide whether or not the doctor should use the test.

3 Construct a tree diagram that represents the same information as the table.

4 Use the information in your tree diagram to confirm whether or not the doctor should use the test.

5 For this problem, which representation do you prefer: the table or the tree diagram? Explain.

Practice 4

1 After having an operation for an eye condition, a patient lost his sight and claimed that this was caused by the operation. The table shows the hospital's data on treatment for this eye condition.

	Lost sight	Did not lose sight
Patient had the eye operation	25	225
Patient did not have the eye operation	75	675

   a Using the information in the table, determine whether the patient's claim is valid. Justify your answer.

   b Represent this information in a tree diagram and use it to determine if the patient's claim is justified.

   c Determine which representation (the table or the tree diagram) best illustrates the validity of the patient's claim. Give evidence for your decision.

ATL 2 The Venn diagram represents 65 members of a sports club.

T is the event 'plays tennis' and G is the event 'plays golf'.

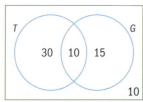

> If you are using a tree diagram, put T and T' on the first pair of branches.

   a Represent this information in a two-way table.

   b Use all three representations to determine if playing tennis and playing golf are independent events.

   c Explain, with reasons, which representation was the easiest to work with.

Problem solving

3 The tree diagram shows information about 100 students at a university.

L is the event 'studies Law' and E is the event 'studies Economics'.

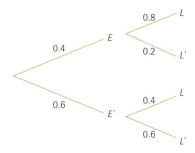

a Use the tree diagram to help you find the following probabilities.

   i $P(E)$ ii $P(L|E)$ iii $P(L|E')$ iv $P(E \cap L)$

   v $P(E \cap L')$ vi $P(E' \cap L)$ vii $P(E' \cap L')$ viii $P(L)$

b Determine whether studying Law or studying Economics are independent events.

Reflect and discuss 3

- Can you use any probability diagram to represent any situation?
- How do you decide which diagram is most useful for a given situation?

Practice 5

1 The cafeteria sells apples and waffles at morning break. There were 51 students in the cafeteria one morning. Of these, 33 students had apples (A), 18 had waffles (W), and 10 students had neither.

  a From the information, and using Theorem 1, draw a Venn diagram.

  b Hence, calculate:

     i $P(A)$ ii $P(W)$ iii $P(A|W)$ iv $P(W|A)$ v $P(W|A')$

  c Make a two-way table to represent the information.

  d Verify that the results are the same for each form of representation.

  e Using Theorem 2 and Theorem 3, test whether having waffles and having apples are independent events.

In questions **2–4**, use the most appropriate representation to solve the problem.

2 70 students who worked in the holidays, 30 of whom are boys, were surveyed about their work. 12 boys and 28 girls worked full-time. The rest worked part-time.

  a Given that a student worked part-time, what is the probability she is female?

  b What is the probability that a person chosen at random is male and worked full-time?

  c Using a theorem as a justification, determine whether 'being male' and 'working part-time' are independent events.

3 The probability of Gregoire arriving at school on time is 60% if there is fog, and 95% if it is clear. The probability of fog is 20%.

   a Calculate the probability of Gregoire arriving at school on time.

   b Calculate the probability that it is clear, given that Gregoire arrives late.

   c Determine whether the events 'getting to school on time' and 'a clear morning' are independent. Justify your answer.

4 Out of 150 students, 30 are left-handed (L) and 80 are male (M). There are 14 left-handed students who are female (F).

   a Find $P(F \mid L)$.

   b Find $P(L \mid F)$.

   c Determine whether the events 'being male' and 'being left-handed' are independent. Justify your answer.

Summary

In mathematical notation, $P(B \mid A)$ is 'the probability of B occurring **given that** the condition A has occurred', or 'the probability of B given A'.

The conditional probability axiom:

$$P(B \mid A) = \frac{P(A \cap B)}{P(A)}$$

You can use this formula to calculate conditional probabilities.

You can use Theorem 2 (the multiplication rule) as a test for independent events.

If A and B are independent events, then $P(A \cap B) = P(A) \times P(B)$.

A and B are independent events if the outcome of one does not affect the outcome of the other.

Theorem 3

If A and B are independent events then:

$$P(B \mid A) = P(B \mid A') = P(B)$$

Mixed practice

1 A lab has 100 blood samples that need to be tested. Of these, 45 are type O, 40 are type A and 15 are type B. Two samples are to be drawn at random to check for contamination.

   a **Draw** a tree diagram to represent this.

   b **Find** the probability that the first two samples have the same blood group.

   c **Find** the probability that the second sample is type A, given that the first one was type B.

   d **Find** the probability that the second sample is not type O, given that the first one was.

2 From the Venn diagram below, **find**:

   a $P(A)$
   b $P(A \cap B)$
   c $P(A \mid B)$
   d $P(B \mid A)$
   e $P(B)$
   f $P(A \mid B')$
   g $P(B \mid A')$

Venn diagram: A contains 20, intersection 10, B contains 15, outside 20.

3 The probability that it will be sunny tomorrow is $\frac{4}{5}$. If it is sunny, the probability that the cafeteria will sell ice creams is $\frac{2}{3}$. If it is not sunny, the probability that the cafeteria will sell ice creams is $\frac{1}{3}$.

   a Represent this information in a tree diagram.

   b Find the probability that it will be sunny and the cafeteria will sell ice creams.

   c Determine if the sun shining and the selling of ice creams are independent events.

4 S is the event 'Tim is wearing shorts' and R is the event 'it is raining'.

   It is known that $P(R) = 0.5$, $P(S|R) = 0.2$, and $P(S|R') = 0.8$.

   a Represent this situation in a tree diagram.

   b Use both Theorem 2 and Theorem 3 to **verify** that events S and T are not independent.

5 Sahar is trying to determine if being tall (here defined as being 180 cm or more) is somehow related to liking basketball. She surveyed 70 students, 30 of which were over 180 cm. 48 students surveyed like basketball, 29 of which were under 180 cm.

   a Draw a two-way table to represent this.

   b Find P(tall).

   c Find P(not tall | likes basketball).

   d Are the events 'being tall' and 'liking basketball' independent? **Justify** your answer.

6 In a netball match there are 14 players on the court: 8 have brown hair and 6 have blue eyes. There are 4 students who have neither brown hair nor blue eyes. Let H be the event 'has brown hair' and B be the event 'has blue eyes'.

   a Draw a Venn diagram to represent this.

   b Determine these probabilities using correct mathematical notation:

   i $P(H)$ **ii** $P(B)$
   iii $P(H \cup B)$ **iv** $P(H \cap B)$
   v $P(H')$ **vi** $P(B')$
   vii $P(H \cup B')$ **viii** $P(H \cap B')$
   ix $P(H \cup B)'$

   c Hence **state** all the conditional probabilities, using the correct mathematical notation.

7 A survey was carried out regarding the school dress code. The two-way table shows the results:

	Should have dress code	Should not have dress code	Total
Middle school		30	45
High school	40		
Total	55	110	

   a Copy and complete the table.

   b Use Theorems 2 and 3 to **determine** whether type of school and wanting a dress code are independent events.

Review in context

1 The table shows the occurrence of diabetes in 200 people.

	Diabetes	No diabetes
Not overweight	10	90
Overweight	35	65

Let D be the event 'has diabetes'.

Let N be the event 'not overweight'.

   a From the table **find**:

   i $P(D)$ **ii** $P(N)$
   iii $P(D|N)$ **iv** $P(D|N')$

   b Determine whether having diabetes and not being overweight are independent events.

2 The probability that a randomly selected person has a bone disorder is 0.01. The probability that a test for this condition is positive is 0.98 if the condition is there, and 0.05 if the condition is not there (a false positive).

Let B be the event 'has the bone disorder' and T be the event 'test is positive'.

a **Draw** a tree diagram to represent these probabilities.

b **Calculate** the probability of success for the test. Note the test is successful if it tests positive for people with the disorder and negative for people without the disorder.

3 Michele is making drinks to sell at the school play. She makes a 'Super green smoothie' with baby spinach, cucumber, apple, kiwi and grapes with a calorific content of 140 kcals, and a 'Triple chocolate milkshake' with chocolate milk, chocolate syrup and chocolate ice-cream with a calorific content of 370 kcal per cup.

She records the number of MYP and DP students buying each product. MYP students purchase 26 super green smoothies and 48 triple chocolate milk shakes. DP students purchase 13 super green smoothies and 24 triple chocolate milk shakes.

Determine, using both theorem 2 and theorem 3, whether the age of the student (MYP vs DP) is independent to their choice of drink. Use a two-way table to answer this question.

Reflect and discuss

How have you explored the statement of inquiry? Give specific examples.

Statement of Inquiry:

Understanding health and making healthier choices result from using logical representations and systems.

4 A survey of 200 000 people looked at the relationship between cigarette smoking and cancer. Of the 125 000 non-smokers in the survey, 981 had cancer at some point in their lifetime. Of the smokers in the survey, there were 1763 people who battled cancer.

a **Find** the probability that an individual selected from the study who battled cancer was a smoker.

b **Find** the probability that someone who had cancer never smoked.

c Let B be the event 'having cancer' and A be the event 'being a smoker' (includes being a former smoker). **Demonstrate** whether or not these two events are independent. **Justify** your answer mathematically.

d Does your answer in **c** encourage or discourage people from choosing to smoke? **Explain**.

5 The following information was extracted from a skin cancer website.

Statement 1: About 69% of skin cancers are associated with exposure to indoor tanning machines.

Statement 2: In the United States, 3.3 million people a year are treated for skin cancer out of a total population of 330 million.

Statement 3: 10 million US adults use indoor tanning machines.

Let A be the event 'developing skin cancer'.

Let B be the event 'exposure to indoor tanning'.

a **Write down** statements 1, 2 and 3 in probability notation.

b **Determine** whether developing skin cancer and exposure to indoor tanning are independent events.

Answers

E1.1

You should already know how to:

1. **a** 11 204 **b** 2996 **c** 1161 **d** 999 538
2. **a** Fifty **b** Five hundred thousand **c** Five **d** Five tenths

Practice 1

1. **a** 23 **b** 25 **c** 109
2. **a** 191 **b** 226 **c** 107 **d** 92 **e** 52 **f** 70
3. $1011_3 = 31_{10}$, $100011_2 = 35_{10}$, $1010_4 = 68_{10}$, $1111_5 = 156_{10}$
4. **a** Every digit has moved one place to the right, and the final zero has been lost. So the number has halved.
   b 107_{10}

Practice 2

1. **a** 1111100111_2 **b** 1101000_3 **c** 33213_4 **d** 12444_5
2. **a** $472_8 = 314_{10}$ **b** $472_8 = 2224_5$
3. **a** $223_5 = 63_{10} = 120_7$ **b** $431_8 = 281_{10} = 100011001_2$
   c $214_6 = 82_{10} = 1010010_2$ **d** $1011_2 = 11_{10} = 15_6$
   e $110111_2 = 55_{10} = 67_8$ **f** $110213_4 = 1319_{10} = 1725_9$
   g $8868_9 = 6542_{10} = 22222022_3$ **h** $101101_2 = 45_{10} = 231_4$
   i $2468_9 = 1844_{10} = 3464_8$
4. **a** Donatello used the largest base, because his number was the smallest when written out.
   b c, b, a, d
   c b is at least 7, because the number contains a 6.
   d $a = 8$, $b = 7$, $c = 6$ and $d = 9$
   e 729_{10}

Practice 3

1. **a** $210_{12} = 300_{10}$ **b** $301_{16} = 769_{10}$
   c $BAA_{12} = 1714_{10}$ **d** $G0_{20} = 320_{10}$
2. **a** $190_{10} = BE_{16}$ **b** $2766_{10} = ACE_{16}$
   c $47806_{10} = BABE_{16}$ **d** $48879_{10} = BEEF_{16}$
   e $51966_{10} = CAFE_{16}$ **f** $64206_{10} = FACE_{16}$
3. **a** The possible codes run from 00_{16} to FF_{16}, or 0_{10} to 255_{10}, so there are 256 codes.
   b $16^{10} = 1\,099\,511\,627\,766$ (Roughly 1.1 trillion)

Practice 4

1. **a** 445_8 **b** 360_8 **c** 2003_8
2. **a** 222_6 **b** 144_6 **c** 425_6
3. **a** 137_8 **b** 1554_7 **c** 8670_9
4. **a** 11717_{12} **b** $BA99_{12}$
5. Since $216_8 + 165_8 = 403_8$, $216_8 - 165_8 = 31_8$

Practice 5

1. **a** 4112_5 **b** 114333_5 **c** 340030_5
2.

×	1	2	3	4	5	10
1	1	2	3	4	5	10
2	2	4	10	12	14	20
3	3	10	13	20	23	30
4	4	12	20	24	32	40
5	5	14	23	32	41	50
10	10	20	30	40	50	100

a 4533_6 **b** 112024_6

3. $1215_6 = 299_{10} = 13_{10} \times 23_{10} = 21_6 \times 35_6$

Mixed practice

1. **a** 13 **b** 54 **c** 69
   d 52 **e** 135 **f** 2387
   g 14 559 **h** 2418 **i** 20 136
2. **a** 1100100_2 **b** 10201_3 **c** 1210_4 **d** 11111111_2
   e 1101100_3 **f** 230122_5 **g** $1A7_{12}$ **h** 320_{12}
   i $B89_{12}$ **j** 1741_{12} **k** 419_{16} **l** $3BEE_{16}$
3. **a** $42°\,48'$ **b** $90°\,9'$ **c** $2°\,31'\,12''$
   d $25.25°$ **e** $62.508\dot{3}°$ **f** $58.20\dot{5}°$
4. **a** 10110_2 **b** 1001010_2 **c** 10001111_2 **d** 10000_3
   e 22211_3 **f** 20010_4 **g** 1011323_4 **h** 1081_{12}
   i 2193_{12} **j** $1BA8_{12}$ **k** $11A_{16}$
   l $1EADB_{16}$
5. **a** 10110_2 **b** 11110_2 **c** 1111_2
   d 2101_3 **e** 2120_4 **f** 34535_6
   g 34535_6 **h** $29B1_{12}$ **i** $3FFE_{16}$
6. **a** 11:20 **b** 13:15 **c** 14:25
   d 09:26 **e** 03:55
7. **a** 1100_2 **b** 100001_2 **c** 1000001_2
   d 121121_3 **e** 432_5 **f** 4244_5
8. **a** 30 g
   b 16 g, 4 g, 4 g, 1 g, 1 g.
   c $2313_4 = 183_{10}$
   Because the weight system is equivalent to counting in base 4.
   d $331213_4 + 231232_4 = 1223111_4$
   1 g, 4 g, 16 g, 3 × 64 g, 2 × 256 g, 2 × 1024 g, 1 × 4096 g
   e $1330_4 \times 12_4 = 23220_4$
   2 × 4 g, 2 × 16 g, 3 × 64 g, 2 × 256 g
9. **a** It might be thought a good idea so that the dollar was worth five times the largest coin, but redefining the relationship between the cent and the dollar could be highly confusing.
   b $540_{10} = 4130_5$
   Hence eight notes.
   c $732_{10} = 10412_5$
   $246_{10} = 1441_5$
   $10412_5 - 1441_5 = 3421_5$
   So ten notes.
   d It would require a lot of notes to pay amounts that were close to the next bill, such as $990.

Review in context

1 a 2, 3, 5, 7, 11
  b 13, 17, 19, 23, 29, 31, 37, 41, 43, 47
  c 10, 11, 101, 111, 1011, 1101, 10001, 10011, 10111, 11101, 11111, 100101, 101001, 101011, 101111

2 "|" represents 1; no number starts with a leading zero.
  a 1, 4, 9, 16, 25; the square numbers;
  36 = |———|——, 49 = ||————|, 64 = |————————
  b 1, 1, 2, 3, 4, 8, 13, 21; the Fibonacci numbers;
  34 = |————|—, 55 = ||—|||, 89 = |—||—|
  c 1, 3, 9, 27, 81; the powers of 3; 243 = ||||——|,
  729 = |—||—||——|, 2187 = |————|————|—||

3 a 16 487.5… hours
  b 659.5… days
  c 1 010 010 100 days
  d Jupiter: 10 100 010 011 111
  Saturn: 101 101 110 110 010
  Uranus: 1 010 100 101 011 000
  Neptune: 10 110 000 010 111 010
  e A day on Venus is longer than its year, so the number of days per year is less than 1.

E2.1

You should already know how to:

1

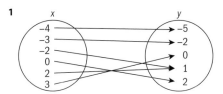

2 a Domain and Range are both $\mathbb{R}$.
  b Domain is $x \geq 0$, Range is $y \geq 0$.
  c Domain is $x \neq 0$; Range is $y \neq 0$.
  d Domain is $x \neq 2$; Range is $y \neq 0$.
  e Domain is $-2 \leq x \leq 2$; Range is $y \geq 0$.

3 a $f(4) = 5$ **b** $f(-1) = -5$ **c** $f(3x) = 6x - 3$

Practice 1

1 Domain of h is $\mathbb{R}$, domain of k is $\mathbb{R}$;
  $(h + k)(a) = 2a^3 - a - 2$; domain of $h + k$ is $\mathbb{R}$.

2 Domain of h is $\mathbb{R}$, domain of k is $\mathbb{R}$;
  $(g - h)(x) = -x^3 + 5x^2 - x + 5$; domain of $(g - h)$ is $\mathbb{R}$.

3 Domain of h is $\mathbb{R}$; domain of k is $x \geq -\frac{1}{3}$;
  $(h + k)(x) = 2x^2 - x + \sqrt{3x + 1}$; domain of $(h + k)$ is $x \geq -\frac{1}{3}$.

4 $(k - h)(c) = \sqrt{2 - 3c} - 2c + 5$; domain of $(k - c)$ is $c \leq \frac{2}{3}$.

5 $(g + h - k)(x) = 6x - 8 + \frac{1}{2x} - \sqrt{x + 4}$; domain is $x \geq -4, x \neq 0$;

  $(k - h - g)(x) = \sqrt{x + 4} - \frac{1}{2x} - 6x + 8$; domain is $x \geq -4, x \neq 0$.

6 a Domain of f is $x \neq 0$; domain of g is $x \neq 2$.
  b Domain of $f + g$ is $x \neq 0, x \neq 2$.

7 Students' own answers

Practice 2

1 a

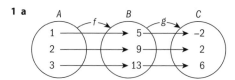

  b $A = \{1, 2, 3\}, B = \{5, 9, 13\}, C = \{-2, 2, 6\}$
  c $-2, 2, 6$
  d Domain of $g \circ f$ is A. Range of $g \circ f$ is C.

2 a

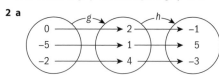

Domain of $h \circ g$ is $\{1, 2, 4\}$; range of $h \circ g$ is $\{-1, -3, 5\}$.

  b

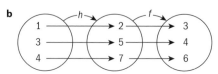

Domain of $f \circ h$ is $\{2, 5, 7\}$; range of $f \circ h$ is $\{3, 4, 6\}$.

  c

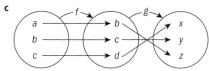

Domain of $g \circ f$ is $\{b, c, d\}$; range of $g \circ f$ is $\{x, y, z\}$.

3 a 4.5, 8, 12.5
  b Domain is $\{6, 8, 10\}$; range is $\{4.5, 8, 12.5\}$.

4 a

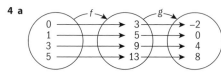

  b Range of $g \circ f$ is $\{-2, 0, 4, 8\}$.

5 9

Practice 3

1 a 12 **b** 7 **c** 18 **d** 2.25
  e 1 **f** −36 **g** 6 **h** 11
  i $3x^2 + 6$ **j** $3x + 1$ **k** $x + 2$ **l** $x^2 + 2x + 3$

2 a $x^2 - 5$
  b $(2x + 1)^2 - 5 + 1 = (4x^2 + 4x + 1) - 5 = 4x^2 + 4x - 4$

3 a $f(x) = x^2$, Domain is $\mathbb{R}$, range is $x \geq 0$
  b $f \circ g(x) = \left(\frac{1}{x - 1}\right)^2, x \neq 1$

4 5

5 a 3.5 **b** 2 **c** 0 **d** 0
  e 2 **f** 0 **g** 3.5

6 a x represents the amount of sales in the week; $f(x)$ is the commission paid, and g represents the commission she earns 2% on.
  b $f \circ g(x)$

7 $x = 1, -3$

Practice 4

1. **a** 15 cents; 45 cents
   b 210 g; 265 g
   c 33 cents. An orange of radius 5.1 cm weights 250 g, and costs 33 cents.
   d $f(g(x))$

2. **a**

Time t (seconds)	0	1	2	3	4
Radius r (cm)	0	2.5	5	7.5	10

   b $f(t) = 2.5t$
   c $g(r) = \pi r^2$
   d $A = g(f(t)) = \pi r^2 = \pi(2.5t)^2 = 6.25\pi t^2$

3. **a** $v(t) = 10t$
   b $v = \pi r^2 h, h = \dfrac{v}{\pi r^2}$
   c $h \circ v(t) = \dfrac{10t}{\pi r^2}$
   The height of the water in the cylinder as a function of time t.
   d $t = 5\pi \approx 15.8$ seconds

4. **a** $r = 2t$
   b Area as a function of the radius as a function of time.
   c 45 289 cm^2

5. **a**
   i Fuel cost is a function of vehicle velocity.
   ii Volume of a tank is a function of time.
   iii Volume of a circle is a function of time.
   iv Size of each angle of a regular polygon is a function of the number of sides of a polygon.
   v Interest earned is a function of time.
   b Students' own answers

Practice 5

1. $f(x) = x^3; g(x) = 2x - 1$
2. $f(x) = 2x - 1; g(x) = x^3$
3. $f(x) = \sqrt{x}; g(x) = 6x - 5$
4. $f(x) = 6x - 5; g(x) = \sqrt{x}$
5. $f(x) = \sqrt[3]{x}; g(x) = 2x - 1$
6. $f(x) = 2^x; g(x) = 3x - 2$

Mixed practice

1. **a** $(f+g)(x) = x^2 - 9x + 2$ **b** $(g-f)(x) = x^2 + x - 2$
   c $(f+h)(x) = \dfrac{-19x + 6}{4}$ **d** $(g-h)(x) = \dfrac{4x^2 - 17x - 2}{4}$
   e $(h+k)(x) = \dfrac{x^2 - 2x - 8}{4x}; x \neq 0$
   f $(k-h)(x) = \dfrac{-x^2 + 2x - 8}{4x}; x \neq 0$
   g $(f+k)(x) = \dfrac{-5x^2 + 2x - 2}{x}; x \neq 0$

2. **a** Domain of f is $\mathbb{R}$; domain of g is $x \geq \dfrac{1}{2}$;
   $f \circ g(x) = \sqrt{2x-1} + 1$, domain of $f \circ g$ is $x \geq \dfrac{1}{2}$;
   $g \circ f(x) = \sqrt{2x+1}$, domain of $g \circ f$ is $x \geq \dfrac{1}{2}$.

   b Domain of f is $\mathbb{R}$; domain of g is $x \geq 1$; $f \circ g(x) = x - 1$; domain of $f \circ g$ is $\mathbb{R}$; $g \circ f(x) = \sqrt{x^2 - 1}$; domain of $g \circ f$ is $x \geq 1$ or $x \leq -1$.

   c Domain of f is $\mathbb{R}$; domain of g is $\mathbb{R}$; $f \circ g(x) = 9x^2 - 15x + 6$; $g \circ f(x) = -3x^2 - 3x + 2$; Both domains are $\mathbb{R}$.

3. **a** $p(q(t)) = 4t^2 - 15$; domain is $\mathbb{R}$. **b** $q(r(t)) = \dfrac{1}{t^2} - 4; t \neq 0$
   c $p(r(t)) = \dfrac{4}{t} + 1; t \neq 0$ **d** $r(q(t)) = \dfrac{1}{t^2 - 4}; t \neq \pm 2$

4. **a** $g(f(x))$ **b** $h(g(x))$

5. **a** $g(f(a)) = 2\left(\dfrac{a-2}{3}\right) + 1$ **b** $f(g(-b)) = \dfrac{-2b-1}{3}$

6. $k = 4$

7. **a** $f(g(x-1)) = 4(x-1)^2 - 3 = 4x^2 - 8x + 1$
   b $g(f(-3x)) = 2(-6x - 3)^2$

8. $g(f(3x^2 - 1)) = 4(3x^2 - 2) + 3$

9. **a** 7 **b** -2 **c** $\dfrac{8}{3}$

10. **a** $B(T(t)) = 20(4t+2)^2 - 80(4t+2) + 500$; Bacteria as a function of the amount of time food is outside the refrigerator.
    b 3.8 hrs.

11. **a** $s = 1.5w$ **b** $d = 0.8s$
    c $d = 0.8(1.5w) = 1.2w$ **d** $86.40

12. **a** **i** $f(x) = x - 5$ **ii** $g(x) = 0.75x$
    iii $g \circ f(x) = 0.75(x - 5)$ **iv** $f \circ g(x) = 0.75x - 5$
    b From **iii**, the price is $33.75. From **iv** the price is $32.50. The difference is $1.25.

13. For example:
    a $f(x) = -x^4 + 7; g(x) = x + 1$ **b** $f(x) = \sqrt[5]{x}; g(x) = 4 - 3x$
    c $f(x) = -5x - 8; g(x) = x^2$ **d** $f(x) = \dfrac{3}{x}; g(x) = x - 4$
    e $f(x) = 10^x; g(x) = \sqrt{x}$

    Note: Other answers to Q13 are possible.

Review in context

1. **a** $r(t) = 0.8t + 70$
   b $A \circ r(t) = \pi(0.8t + 70)^2$
   c 135.5 minutes or approximately $2\dfrac{1}{4}$ hours

2. **a** $K(C(F)) = \dfrac{5}{9}(F - 32) + 273$
   b 373, 273

3. **a** $r(v(t)) = \left(\dfrac{40 + 3t + t^2}{500} - 0.1\right)^2 + 0.2$
   b 0.202 liters per km

4. **a** $f(x) = 4x + 6; g(x) = \dfrac{x}{2} + 3$
   $g(f(x)) = \dfrac{4x+6}{2} + 3 = 2x + 3 + 3 = 2x + 6$
   b Set the above expression equal to the friend's result and mentally solve the equation.

5. Students' own answers

E2.2

You should already know how to:

1 a Not a function, since $b \in A$ does not have an image in B.
b Not a function, since $c \in A$ has two images in B, p and r.
c It is a function.

2 a Not a function, since 2 is mapped to 1 and 3.
b It is a function.

3 a It is a function.
b It is not a function.

4 a $f \circ g(x) = 2x^2 + 3$
b $g \circ f(x) = 4x^2 + 4x + 2$

Practice 1

1 a Neither **b** Neither
c Onto, not one-to-one **d** Neither
e Onto, not one-to-one **f** Onto, not one-to-on
g Both **h** One-to-one, not onto

2 Students' own functions

Practice 2

1 a $f^{-1}(x) = \{(1, 3), (1, -1), (1, -3), (1, 1)\}$; not a function
b $g^{-1}(x) = \{(4, 2), (3, -5), (-3, -2), (1, 1), (-4, -4), (-2, -1), (-1, 3)\}$; is a function.
c $h^{-1}(x) = \{(0, 0), (4, -2), (-4, -1), (2, 2), (-3, -3), (4, -4), (-1, 1)\}$; not a function.

2 a The inverse is also a function.
b The inverse is also a function.
c The inverse is also a function.
d The inverse is also a function.
e The inverse is also a function.
f The inverse is not a function.

3 a $y = \dfrac{1-x}{4}$; function **b** $y = -\dfrac{2(x-4)}{3}$; function
c $y = -5 - 3x$; function **d** $y = -\dfrac{1}{x+1}$; $x \neq -1$, function
e $y = \dfrac{1-5x}{x}$; $x \neq 0$, function **f** $y = \pm\sqrt{x-1}$; not a function
g $y = \sqrt[3]{x}$; function **h** $y = x^3 - 2$; function
i $y = \dfrac{3+2x}{x-1}$; $x \neq 1$, function

Practice 3

1 a $f^{-1}(x) = \dfrac{x+2}{5}$. The inverse is a function.
f and its inverse are symmetrical about the line $y = x$.

2 a $f^{-1}(x) = 2(7-x)$. The inverse is a function.
f and its inverse are symmetrical about the line $y = x$.

3 a $f^{-1}(x) = \pm\sqrt{x+3}$
f and its inverse are symmetrical about the line $y = x$.
b Domain of f is $x \geq 0$, $f^{-1}(x) = \sqrt{x+3}$, $x \geq -3$ or domain of f is $x \leq 0$, $f^{-1}(x) = -\sqrt{x+3}$; $x \geq -3$

4 a $y = \pm\sqrt{1-x}$
f and its inverse are symmetrical about the line $y = x$.
b Domain of f is $x \geq 0$, $f^{-1}(x) = \sqrt{1-x}$; $x \geq 1$, or domain of f is $x \leq 0$, $f^{-1}(x) = -\sqrt{1-x}$; $x \geq 1$.

5 a $f^{-1}(x) = \dfrac{1}{x-1}$; the inverse is a function with domain $x \neq 1$.
f and its inverse are symmetrical about the line $y = x$.
b Domain of f is $x \neq 1$.

6 a $y = \pm\sqrt{x+2}$
f and its inverse are symmetrical about the line $y = x$.
b Domain of f is $x \geq 2$, $f^{-1}(x) = 2 + \sqrt{x}$, $x \geq 0$, or domain of f is $x \leq 2$, $f^{-1}(x) = 2 - \sqrt{x}$, $x \geq 0$.

7 a $f^{-1}(x) = \sqrt{x+5} - 1$
f and its inverse are symmetrical about the line $y = x$.
b Domain of f is $x \geq -1$ for $f^{-1}(x) = \sqrt{x+5} - 1$; or domain of f is $x \leq -1$ for $f^{-1}(x) = -\sqrt{x+5} - 1$.

8 a $y = \dfrac{\pm\sqrt{2(x-3)} + 2}{2}$
f and its inverse are symmetrical about the line $y = x$.
b Domain of f is $x \geq 1$ for $f^{-1}(x) = \dfrac{\sqrt{2(x-3)} + 2}{2}$, $x \geq 3$, or
domain of f is $x \leq 1$, $f^{-1}(x) = \dfrac{-\sqrt{2(x-3)} + 2}{2}$, $x \geq 3$

9 Either $f(x) = x$ for $x \geq 0$, or $f(x) = x$ for $x \leq 0$.
For $f(x) = x$, $x \geq 0$, $f^{-1}(x) = x$; for $f(x) = x$, $x \leq 0$, $f^{-1}(x) = x$.

10 a $s \geq 0$, since $h(s)$ can never be less than 0.
b $h^{-1}(s) = \dfrac{198 - s}{0.9}$
The inverse is also a function.
c 203

Practice 4

1 Yes **2** No
3 Yes **4** No
5 Yes **6** Yes
7 Yes **8** No, $f(g(x)) = x$ and $g(f(x)) = |x|$

Mixed practice

1 a i Yes **ii** $R^{-1} = \{(2, 1), (3, 2), (4, 3), (5, 4)\}$ **iii** Yes
b i Yes **ii** $S^{-1} = \{(0, -2), (1, 3), (1, -5), (7, 6), (0, 0)\}$
iii No
c i No **ii** $T^{-1} = \{(-1, 0), (1, 3), (-1, -4), (6, -3), (-1, 3)\}$
iii No

2 a Neither **b** Both
c Both

3 a No. f is not into. **b** Yes. f is into and onto.
c No. f is not into (nor onto) **d** No. f is not into.

4 i a $f^{-1}(x) = \dfrac{8-x}{3}$
f and its inverse are symmetrical about the line $y = x$.
b function

ii a $f^{-1}(x) = \dfrac{4(x+1)}{3}$
f and its inverse are symmetrical about the line $y = x$.
b function

iii a $f^{-1}(x) = \pm\sqrt{4-x}$
f and its inverse are symmetrical about the line $y = x$.
b For domain $f(x) \geq 0$ $f^{-1}(x) = \sqrt{4-x}$, $x \leq 4$

iv a $f^{-1}(x) = \pm\sqrt{\dfrac{x+2}{3}}$

f and its inverse are symmetrical about the line $y = x$.

b For domain $f(x) \geq 0$ $f^{-1}(x) = \sqrt{\dfrac{x+2}{3}}$, $x \geq -2$

v a $f^{-1}(x) = \dfrac{x}{2x-1}$

f and its inverse are symmetrical about the line $y = x$.

b $x \neq \dfrac{1}{2}$

5 a $f^{-1}(x) = \dfrac{x}{5}$; function

b $f^{-1}(x) = \dfrac{x-2}{6}$; function

c $f^{-1}(x) = \dfrac{3}{2}\left(x + \dfrac{1}{2}\right)$; function

d $g^{-1}(x) = 5x - 3$; function

e $h^{-1}(x) = \sqrt{x+3}$; $x \geq -3$ if the domain of f is $x \geq 0$.
$h^{-1}(x) = -\sqrt{x+3}$, if the domain of f is $x \leq 0$.

f $p^{-1}(x) = \dfrac{1}{x-2}$; $x \neq 2$

g $q^{-1}(x) = \dfrac{3x+2}{x-1}$; $x \neq 1$

h $s^{-1}(x) = x^2 - 3$; $x \geq 0$

6 For each pair of functions f and g, $f(g(x)) = g(f(x)) = x$.

7 Length = 39.5 cm

8 a 412; After 24 months, a computer has depreciated from $700 to $412.

b 54.2 months; it takes 54.2 months for a computer's value to depreciate from $700 to $50.

c $d^{-1}(t) = \dfrac{700 - d(t)}{12}$; $d^{-1}(t)$ is the number of months for a computer to depreciate to $d(t)$.

d 16.7

e $t = -500$, which means that after 100 months the computer is worthless.

Review in context

1 Add 8 to 34, multiply this result by 2, divide then by 6, subtract 5, to get the answer 9.

2 a The result is the original number, since $\dfrac{2(x+6) - 12}{2} = x$.

b Choose a number, multiply by 2, add 12, divide the result by 2, and then subtract 6. The result is the number.

3 $f(x) = 2x^2 + 3$; $f^{-1}(x) = \sqrt{\dfrac{x-3}{2}}$, $x \geq 3$; The puzzle would not work for real numbers less than 3.

4 Students' own answers

E3.1

You should already know how to:

1 a $\sqrt{53} = 7.28$ (3 s.f.)   b 5

2 $x = 0$, $y = 1$

3 94.1°

Practice 1

1 a i $\begin{pmatrix} 2 \\ 4 \end{pmatrix}$ ii $\begin{pmatrix} 4 \\ -3 \end{pmatrix}$ iii $\begin{pmatrix} -2 \\ -2 \end{pmatrix}$

iv $\begin{pmatrix} 2 \\ -4 \end{pmatrix}$ v $\begin{pmatrix} 2 \\ -4 \end{pmatrix}$ vi $\begin{pmatrix} 5 \\ -2 \end{pmatrix}$

b i $\vec{AD} = \begin{pmatrix} 1 \\ -5 \end{pmatrix}$ $\vec{DA} = \begin{pmatrix} -1 \\ 5 \end{pmatrix}$

ii $\vec{EG} = \begin{pmatrix} 1 \\ 6 \end{pmatrix}$ $\vec{GE} = \begin{pmatrix} -1 \\ -6 \end{pmatrix}$

c The components have the same size, but opposite sign.

d i $\begin{pmatrix} 1 \\ 1 \end{pmatrix}$ ii $\begin{pmatrix} 3 \\ 5 \end{pmatrix}$ iii $\begin{pmatrix} 5 \\ -2 \end{pmatrix}$ iv $\begin{pmatrix} 0 \\ 0 \end{pmatrix}$

e i F   ii G   iii D

2 a C to A b E to C c A to D d H to B

3 a (5, 9)   b (1, 8)   c (0, 3)   d (14, −5)

4 a $\vec{MN} = \begin{pmatrix} 3 \\ 2 \end{pmatrix}$ b $\vec{PQ} = \begin{pmatrix} 7 \\ -3 \end{pmatrix}$

c $\vec{RS} = \begin{pmatrix} -4 \\ 7 \end{pmatrix}$ d $\vec{TU} = \begin{pmatrix} -1 \\ -4 \end{pmatrix}$

5 $\vec{DC} = \begin{pmatrix} 2 \\ -4 \end{pmatrix}$

6 b $\begin{pmatrix} 3 \\ 1 \end{pmatrix}$

7 E.g. (0, 0) and (−5, 3); (3, 1) and (−2, 4); (6, 11) and (1, 14)

Practice 2

1 a 25   b 5   c 10

2 a 5   b $\sqrt{10}$ c $2\sqrt{5}$ d $5\sqrt{2}$ e 13   f $\sqrt{281}$

3 a 6.40   b 8.25   c 4.47   d 13.2   e 8.94   f 12.5

4 a 6   b 26   c 25   d 1   e 10   f 5

5 a 9.06   b 3.16   c 13.3   d 6.80   e 20.8   f 21.0

7 Any four from (8, 1), (8, −5), (9, 0), (9, −4), (4, 1), (4, −5), (3, 0), (3, −4)

8 $x = 2$, $y = 1$ 9 $x = 2$

Practice 3

1 a $\begin{pmatrix} 1 \\ 5 \end{pmatrix}$ b $\begin{pmatrix} 8 \\ -1 \end{pmatrix}$ c $\begin{pmatrix} -5 \\ 17 \end{pmatrix}$ d $\begin{pmatrix} 0 \\ 0 \end{pmatrix}$ e $\begin{pmatrix} -6 \\ -4 \end{pmatrix}$

2 a $\vec{AB} = \begin{pmatrix} -1 \\ -7 \end{pmatrix}$ $\vec{BA} = \begin{pmatrix} 1 \\ 7 \end{pmatrix}$

b $\vec{AB} = \begin{pmatrix} 5 \\ -1 \end{pmatrix}$ $\vec{BA} = \begin{pmatrix} -5 \\ 1 \end{pmatrix}$

c $\vec{AB} = \begin{pmatrix} -11 \\ 10 \end{pmatrix}$ $\vec{BA} = \begin{pmatrix} 11 \\ -10 \end{pmatrix}$

d $\vec{AB} = \begin{pmatrix} 9 \\ 0 \end{pmatrix}$ $\vec{BA} = \begin{pmatrix} -9 \\ 0 \end{pmatrix}$

e $\vec{AB} = \begin{pmatrix} -9 \\ -7 \end{pmatrix}$ $\vec{BA} = \begin{pmatrix} 9 \\ 7 \end{pmatrix}$

3 $(-15, -4)$
4 a $a = 0, b = -1$ **b** $a = 1, b = -2$ **c** $a = 3, b = 3$

Practice 4

1 $a = -1$ **2** $b = 8$ **3** $\overrightarrow{OM} = \begin{pmatrix} 10.5 \\ 5.5 \end{pmatrix}$

4 b $|\overrightarrow{AB}| = \sqrt{17}$ and $|\overrightarrow{CD}| = 2\sqrt{17}$. They are different lengths.

6 a $\overrightarrow{AB} = \overrightarrow{DC} = 10u + 5v$

7 ABCD or ABDE

Practice 5

1 a $127°$ **b** $67°$ **c** $25°$ **d** $35°$
3 b $111°$
4 $37°$
5 $m = 4$
6 a $ab + (a+2)(2-b) = 0$, so $a - b = -2$
  b It is an equation in two unknowns.
  c $a = -\dfrac{1}{2}, b = \dfrac{3}{2}$

7 Any non-zero multiple of $\begin{pmatrix} 1 \\ 4 \end{pmatrix}$

8 a $OC = \begin{pmatrix} 2 \\ -3 \end{pmatrix}$ **b** $14.5°$

9 -8
10 LHS = RHS = 16

Practice 6

1 a $-u + v$ **b** $-v - w$ **c** $-u + v + w$
2 a $b - a$ **b** $b - a$ **c** $b - a$ **d** $b - 2a$ **e** $a + b$
3 a $r = 5q - 4p$
4 $s = 3t - 2u$
5 a $b + \dfrac{3}{2}a$
  c A rectangle with sides in the ratio 2 : 3

Mixed practice

1 a $\begin{pmatrix} 5 \\ 4 \end{pmatrix}$ **b** $\begin{pmatrix} -2 \\ 3 \end{pmatrix}$ **c** $\begin{pmatrix} 3 \\ 4 \end{pmatrix}$ **d** $\begin{pmatrix} -11 \\ -19 \end{pmatrix}$

2 a $\overrightarrow{AB} = \begin{pmatrix} -3 \\ -3 \end{pmatrix}$ $\overrightarrow{BA} = \begin{pmatrix} 3 \\ 3 \end{pmatrix}$

  b $\overrightarrow{AB} = \begin{pmatrix} -7 \\ 3 \end{pmatrix}$ $\overrightarrow{BA} = \begin{pmatrix} 7 \\ -3 \end{pmatrix}$

  c $\overrightarrow{AB} = \begin{pmatrix} 3 \\ -5 \end{pmatrix}$ $\overrightarrow{BA} = \begin{pmatrix} -3 \\ 5 \end{pmatrix}$

  d $\overrightarrow{AB} = \begin{pmatrix} 13 \\ -14 \end{pmatrix}$ $\overrightarrow{BA} = \begin{pmatrix} -13 \\ 14 \end{pmatrix}$

3 a $\begin{pmatrix} -3 \\ -3 \end{pmatrix}$ **b** $\begin{pmatrix} 9 \\ -6 \end{pmatrix}$ **c** $\begin{pmatrix} 3 \\ -5 \end{pmatrix}$ **d** $\begin{pmatrix} -2 \\ 15 \end{pmatrix}$

4 a $a = 0, b = 1$ **b** $a = 2, b = -3$
  c $a = 5, b = -4$ **d** $a = -2, b = 4$

5 a $\sqrt{29}$ **b** 5 **c** $3\sqrt{41}$ **d** $5\sqrt{17}$

6 a $\begin{pmatrix} 2 \\ 5 \end{pmatrix}$ **b** $\begin{pmatrix} 5 \\ 3 \end{pmatrix}$ **c** $\begin{pmatrix} -5 \\ -3 \end{pmatrix}$

7 a $87°$ **b** $84°$ **c** $97°$

8 $\begin{pmatrix} -6 \\ 20 \end{pmatrix}$ and $\begin{pmatrix} 9 \\ -30 \end{pmatrix}$ $\begin{pmatrix} 16 \\ 20 \end{pmatrix}$ and $\begin{pmatrix} 36 \\ 45 \end{pmatrix}$

$\begin{pmatrix} 12 \\ -18 \end{pmatrix}$ and $\begin{pmatrix} -4 \\ 6 \end{pmatrix}$ $\begin{pmatrix} 15 \\ 25 \end{pmatrix}$ and $\begin{pmatrix} -9 \\ -15 \end{pmatrix}$

$\begin{pmatrix} 5 \\ 8 \end{pmatrix}$ is the odd one out.

10 $a = -1$ or $a = 4$
11 $b = -3$
13 a $\overrightarrow{AB}$ is parallel to $\overrightarrow{OC}$ because $\overrightarrow{OC} = 2\overrightarrow{AB}$.
  b $\overrightarrow{OB} = \overrightarrow{OA} + \overrightarrow{AB} = b + a = a + b$
  c $\overrightarrow{AC} = \overrightarrow{AO} + \overrightarrow{OC} = 2a - b$
  d Since X lies on OB, $\overrightarrow{OX}$ is parallel to $\overrightarrow{OB}$, hence $\overrightarrow{OX} = k\overrightarrow{OB} = k(a + b)$.
  e Since X lies on AC, $\overrightarrow{AX}$ is parallel to $\overrightarrow{AC}$, hence $\overrightarrow{AX} = m\overrightarrow{AC}$.
  So $\overrightarrow{OX} = \overrightarrow{OB} + \overrightarrow{AX} = b + m(2a - b)$
  f $k(a + b) = b + m(2a - b)$
  $m = \dfrac{1}{3}, k = \dfrac{2}{3}$, so ratio is 1 : 2

Review in context

2 a 76 cm   **b** 45 cm   **c** 67 cm
  d 95 cm   **e** 292 cm   **f** 447 cm

3 a 15 m^2

  b $\begin{pmatrix} -25 \\ -60 \end{pmatrix}$

  c $250 + 600 + 650 = 1500 = 15$ m

4 a $\overrightarrow{AE} = \begin{pmatrix} 36 \\ 55 \end{pmatrix}$, 657 cm

  b $\overrightarrow{OF} = \begin{pmatrix} 22 \\ -30 \end{pmatrix}$, $\overrightarrow{CF} = \begin{pmatrix} 24 \\ -45 \end{pmatrix}$, $CF = 510$ cm

  c $\overrightarrow{AC} = \begin{pmatrix} 12 \\ 45 \end{pmatrix}$, $|\overrightarrow{AC}| = \sqrt{2169} = 3\sqrt{241}$

  $\overrightarrow{CE} = \begin{pmatrix} 24 \\ 20 \end{pmatrix}$, $|\overrightarrow{CE}| = \sqrt{676}$

  AC is longer
  d Perimeter 18.2 m, area 18.6 m^2

5 a $\overrightarrow{PQ} = \begin{pmatrix} -24 \\ 32 \end{pmatrix}$

  $\overrightarrow{PS} = \begin{pmatrix} 60 \\ 45 \end{pmatrix}$

  c $\overrightarrow{SR} = \begin{pmatrix} -25 \\ 30 \end{pmatrix}$, hence not parallel to IQ. Not rectangular, opposite sides not parallel.

6 a $\overrightarrow{OU} = \begin{pmatrix} 18 \\ 22 \end{pmatrix}$, so $\overrightarrow{UV} = \begin{pmatrix} 22 \\ -22 \end{pmatrix}$, $\overrightarrow{WX} = \begin{pmatrix} 47 \\ -47 \end{pmatrix}$

b $\overrightarrow{UV}$ is parallel to $\overrightarrow{WX}$ so $UVWX$ is a trapezium.

c $\overrightarrow{VW} = \begin{pmatrix} -25 \\ -25 \end{pmatrix}$, $\overrightarrow{VW} \cdot \overrightarrow{UV} = 0$, $\overrightarrow{VW} \cdot \overrightarrow{XW} = 0$

d $|\overrightarrow{VW}| = 25\sqrt{2}$, $|\overrightarrow{UV}| = 22\sqrt{2}$, $|\overrightarrow{XW}| = 47\sqrt{2}$
Area = 17.25 m²

e 45°

E4.1

You should already know how to:

1 a 3.9 **b** 80.5

2 a A population is the whole of all outcomes in question.
b A sample is any subset of the population.

Practice 1

1 a $\mu = 5, \sigma = 2.19$. Mean is 5, but on average the spread of the data is ±2.19 on each side.
b $\mu = 16.54, \sigma = 3.07$.
This implies that the mean is 16.54, but on average the spread of the data is ±3.07 on each side.
c $\mu = 15$ cm, $\sigma = 2.05$ cm.
This implies that the mean is 15 cm, but on average the spread of the data is ±2.05 cm on each side.
d $\mu = 40.8$ kg, $\sigma = 4.75$ kg.
This implies that the mean is 40.8 kg, but on average the spread of the data is ±4.75 kg on each side.

2 a $\mu = 25.4$ min, $\sigma = 2.82$ min.
b This implies that the mean length of journey is 25.4 min, but on average the spread length of journey ±2.82 mins.

3 a motorway: $\mu = 19.8$ min, $\sigma = 4.57$ min
country lanes: $\mu = 20$ min, $\sigma = 1.41$ min
b i the country lanes, because the two means are equal but in the country lanes there is less deviation
ii the motorway, because the mean time is faster
iii the motorway, as it has a lower minimum value
iv the country roads, as the maximum is lower

4 $\mu = 8, \sigma = 4.32$

5 a $\mu = 1.72, \sigma = 0.0618$ **b** $\mu = 1.74, \sigma = 0.0906$

6 a $\mu = 7, \sigma = 2.45$
b Student's own answer
c $\mu = 11, \sigma = 2.45$
d The mean is affected but the standard deviation is not.
e $x = -7.5$ or 20
$y = 3.5$ or 9

Practice 2

1 a $\mu = 2.94, \sigma = 1.27$
b $\mu = 15.03, \sigma = 6.67$
c $\mu = 10.46, \sigma = 5.34$

2 a i 63 kg **ii** 9.45 kg
b 53.55 kg

3

Σf	Σfx	$\Sigma f(x-\mu)^2$	μ	σ
10	50	20	5	$\sqrt{2}$
20	123	41.2	6.15	$\sqrt{2.06}$
39	197	80.34	5.05	$\sqrt{2.06}$
43	223	102.34	5.186	$\sqrt{2.38}$
23	115	39.56	5	$\sqrt{1.72}$

4

Length (cm)	mid-value (x)	f	xf	$x - \mu$	$f(x-\mu)^2$
$0 < x \leq 5$	2.5	3	7.5	−11.4	389.88
$5 < x \leq 10$	7.5	4	30	−6.4	163.84
$10 < x \leq 15$	12.5	6	75	−1.4	11.76
$15 < x \leq 20$	17.5	7	122.5	3.6	90.72
$20 < x \leq 25$	22.5	5	112.5	8.6	369.8
		25	$\Sigma xf =$ 347.5		$\Sigma xf(x-\mu)^2$

The mean is 13.9 and the standard deviation is 6.41.

Practice 3

1 $\mu = 11$ kg, $\sigma = 1.5$, kg. Fully grown mute swans will on average weigh between 9.5 and 12.5 kg.

2 a $\Sigma x = 27$ m, $\Sigma x^2 = 48.628$
b $\mu = 1.811$ m, $\sigma = 0.067$ m
c There is not a dramatic change in the mean, however the standard deviation has changed.
68% of the team should lie within 6.7 cm of the mean.

3 You would expect the rainfall to be within the range of 1109–1420 mm, so:
a 1100 mm is exceptional
b 1400 mm is not exceptional

4 a 49.55 min **b** 48.48 min
c Either the raw actual data or Σx^2

5 $\mu = 90.63, \sigma = 33.02$

Practice 4

1 $s_{n-1} = 4.36$ **2** $\bar{x} = 460.16$ $s_{n-1} = 18.72$

3 $\bar{x} = 1550$ hours, $s_{n-1} = 39.53$ hours

4 b The data is symmetrically distributed about the mean and appears to follow a normal distribution.
c $\bar{x} = 7.84$ people, $s_{n-1} = 2.85$
d Katie could assume that 68% of carriages would have between 5 and 11 people in them.

Mixed practice

1 a $\mu = 4.4, \sigma = 1.36$ **b** $\mu = 24.5, \sigma = 2.93$
c $\mu = 4, \sigma = 0.76$ **d** $\mu = 13.3, \sigma = 6.06$

2 With water: $\mu = 13.6, \sigma = 2.2$
With food: $\mu = 15.7, \sigma = 2.15$
On average the strawberries given food gave a better yield, the mean was greater and the standard deviation was lower.

3 a $\mu = 55.2$, $\sigma = 5.96$
  b On an average day the hamster will eat about 55.2 g of food. About 68% of the time, the hamster will eat between 49.2 and 61.2 g of food.
  c This is 5 standard deviations less than normal, so it is an abnormal result, and he should be worried.

4 b $4.25 - x(1.25) = 2$, $x = 1.5$
  $4.25 + x(1.25) = 6.5$, $x = 1.5$

5 $\mu = 28.15$, $\sigma = 3.84$

6 a $x = 200.5$, $s_x = 2.655$ **b** $x = 200.5$, $s_{x-1} = 2.80$
7 $x = 9.5$, $s_x = 2.49$
  $x = 9.5$, $s_{x-1} = 2.62$
  Yes, because mean is close to 10 and assuming a normal distribution, 34% of the values will be between 9.5 and 12.12, so many will be over 10.

Review in context

1 Patient 1:
  $x = 91.3$, $s_{x-1} = 5.12$ high blood pressure treatment
  Patient 2:
  $x = 78.3$, $s_{x-1} = 2.62$ no treatment
  Patient 3:
  $x = 78.7$, $s_{x-1} = 15$ more tests needed

2 $x = 75$, $s_{x-1} = 5.53$, $s_{x-1} = 5.6$
  Provided the data is normally distributed you can assume 68% lie between 70 and 80.

3 a $x = 3.2$, $s_x = 0.49$, $s_{x-1} = 0.5$
  b $x = 1.95$ kg

4 a $x = 31.79$, $s_x = 6.48$, $s_{x-1} = 7.82$
  b Provided the data is normally distributed you can assume that now 68% of the ages of first time mothers lie between 25 and 39, however 30 years ago 68% of the ages of first time mothers were between 22 and 28 years old.

E5.1

You should already know how to:

1 a x^6 **b** 2^8 **c** $8x^6$
2 a $3\sqrt{5}$ **b** 4   **c** $6\sqrt{5}$
3 a 2   **b** 2   **c** 2
  d 8   **e** 4   **f** 0.1
4 a $\frac{1}{4}$ **b** $\frac{1}{36}$ **c** $\frac{1}{8}$
  d 2   **e** $\frac{3}{2}$ **f** $\frac{25}{4}$

Practice 1

1 a $\frac{1}{2}$ **b** 2   **c** $\frac{1}{2}$ **d** 5
  e $\frac{2}{3}$ **f** $\frac{1}{3}$ **g** $\frac{5}{2}$ **h** $\frac{10}{3}$
2 a $2\sqrt{x}$ **b** $3x$ **c** $\frac{1}{2x^2}$ **d** $\frac{10}{x^2}$
  e 1   **f** $x^{-\frac{1}{6}}$ **g** $x^{-\frac{1}{4}}$ **h** $3x$
3 a $x^{-\frac{1}{2}}$ **b** $x^{\frac{1}{3}}$ **c** $x^{-\frac{1}{4}}$
4 a $x = 8$ **b** $y = 100$

Practice 2

1 a 4   **b** 81   **c** 125
  d 16   **e** 25   **f** 49
2 a $x^{\frac{2}{3}}$ **b** $x^{\frac{3}{4}}$ **c** $x^{\frac{3}{2}}$ **d** $x^{\frac{5}{2}}$
3 a $x^{\frac{4}{3}}$ **b** $x^{\frac{13}{3}}$ **c** $x^{\frac{7}{6}}$ **d** $3^2 = 9$
  e $x^{\frac{4}{15}}$ **f** 16

Practice 3

1 a $\frac{1}{125}$ **b** $\frac{1}{8}$ **c** $\frac{64}{27}$ **d** 256
  e $\frac{125}{216}$ **f** $\frac{8}{27}$
2 a 5   **b** 2   **c** 2401   **d** 4
  e 81   **f** $\sqrt{3}$
3 Student's own answers.
4 a $\frac{1}{9}$ **b** 4   **c** 9   **d** 2

Practice 4

1 a $2^{\frac{7}{2}}$ **b** $2^{\frac{7}{3}}$ **c** $2^{\frac{7}{2}}$ **d** $2^{-\frac{5}{3}}$
2 a $7^{\frac{8}{15}}$ **b** $3^{\frac{3}{2}}$ **c** $2^{\frac{1}{4}}$ **d** $2^{\frac{1}{4}}$
3 a $2^{\frac{1}{2}} \times 3^{\frac{1}{2}}$ **b** $2^{\frac{1}{3}} \times 7^{\frac{1}{3}}$ **c** $3 \times 2^{\frac{1}{4}}$ **d** $2^{\frac{1}{2}} \times 5^{\frac{1}{2}}$
  e $2^{\frac{1}{4}} \times 5^{\frac{1}{2}}$ **f** $2^{\frac{1}{2}} \times 3^{\frac{1}{3}}$
4 a $2^{\frac{5}{6}}$ **b** $3^{\frac{7}{6}}$ **c** $7^{\frac{3}{10}}$ **d** $2^{\frac{1}{6}} 3^{\frac{1}{6}}$
  e $2^{\frac{4}{15}}$ **f** $2^{\left(\frac{1}{n} - \frac{1}{m}\right)} \times 5^{\left(\frac{1}{n} - \frac{1}{m}\right)} = 2^{\left(\frac{m-n}{mn}\right)} \times 5^{\left(\frac{m-n}{mn}\right)}$
  g $a^{\left(\frac{1}{n} + \frac{1}{m}\right)} = a^{\left(\frac{m+n}{mn}\right)}$ **h** $a^{\left(\frac{1}{n} - \frac{1}{m}\right)} = a^{\left(\frac{m-n}{mn}\right)}$

Practice 5

1 a 2   **b** 3   **c** $\sqrt[9]{5}$ **d** $\frac{1}{\sqrt[6]{11}}$ **e** 7
  f $\sqrt[6]{3^5}$ **g** $15\sqrt[6]{2^5}$ **h** $10\sqrt[4]{3^7}$ **i** $2\sqrt[9]{5}$ **j** $\frac{9}{2}$
2 a $\sqrt[3]{2^5}$ **b** $\sqrt[12]{3^7} \times \sqrt{2}$ **c** $\sqrt[3]{2^4} \times \sqrt[6]{3^7}$ **d** $\sqrt[6]{2^5}$ **e** $\frac{1}{\sqrt[6]{7^7}}$
  f $\sqrt[12]{5^5}$ **g** $4\sqrt{2}$ **h** $5\sqrt{5}$ **i** $\frac{3}{2}$ **j** 7
3 a $x = 3$ **b** $y = \sqrt[3]{3}$ **c** $z = \sqrt{3}$

Practice 6

1 a $\sqrt[3]{2^5}$ **b** 32   **c** 1000
  d $\sqrt[5]{3^3}$ **e** 8   **f** 1000
2 $32^{0.2} = (32)^{\frac{1}{5}} = (2^5)^{\frac{1}{5}} = 2$
3 a 2.4   **b** 1   **c** 4.6
4 $x = 1024$ **5 a** 3   **b** 2

Mixed practice

1. a $\frac{1}{3}$ b 3 c $\frac{1}{5}$ d 3
   e $\frac{3}{4}$ f 2 g $\frac{5}{2}$
2. a $7x^3$ b $\frac{1}{4x}$ c $\frac{3}{2x}$
3. a 1000 b 100 c $\frac{1}{9}$ d $\frac{1}{27}$
4. a 4 b $\frac{27}{8}$ c $\frac{25}{16}$ d $\frac{25}{27}$
5. a $2^{\frac{5}{2}}$ b $3^{\frac{5}{4}}$ c 2 d $6^{\frac{7}{3}}$
6. a 25 b $\frac{1}{2}$ c 7
7. a $\sqrt[3]{27}$ b $2\sqrt{6}$ c 2 d $\sqrt[3]{3}$
   e 2 f 8 g $\frac{1}{\sqrt[12]{12}}$ h $\frac{1}{\sqrt[12]{3}}$
8. Student's own answers
9. a $\sqrt[3]{27}$ b 12 c $\frac{1}{\sqrt{2}}$ d $\frac{\sqrt[12]{5}}{\sqrt[4]{2^5}}$
10. a $x = \frac{1}{3}$ b $x = 2$ c $x = 27$
11. a $\sqrt[5]{5^9}$ b 5 c 9 d 1 e $2^6 \times 3^3 = 1728$
12. Mathematics is as old as man.

E6.1

You should already know how to:

1. x^7
2. a 125 b $\frac{1}{2}$ c 4
3. a $6^{\frac{1}{2}}$ b $3^{\frac{2}{3}}$ c $5^{\frac{3}{2}}$
4. $2 \times 1.18^{52} = 10937$ (to the nearest whole number)

Practice 1

1. a $\log 500 = x = 2.70$ b $\log 150 = x = 2.18$
   c $\log 60 = x = 1.77$ d $\log_2 45 = x = 5.49$
   e $\log_2 6 = x = 2.58$
2. a $8^2 = 64$ b $8^{\frac{2}{3}} = 4$ c $10^{-1} = 0.1$
   d $5^5 = x$ e $b^4 = b$ f $x^0 = y$
3. a $\log_3 9 = 2$ b $\log_4 0.125 = -\frac{2}{3}$ c $\log_{1000} 10 = \frac{1}{3}$
4. a 4 b 5 c −2 d 5
   e −5 f −3 g 0 h −2
5. a $\frac{1}{2}$ b $\frac{3}{2}$ c 2 d 4
6. a 4.09 b 2.68 c 3.13 d 0.77
7. a $x = 3.36$ (3 s.f.) b $x = 2.86$ (3 s.f.)
   c $x = 3.20$ (3 s.f.) d $x = 6.49$ (3 s.f.)
8. a $\frac{1}{3}$ b 2 c $-\frac{1}{3}$ d 0 e $\frac{2}{5}$

Practice 2

1. 2.32 2. 0.661 3. 1.7 4. 0.232
5. 0.58 6. 2.58 7. 4.2 8. 4.3

Practice 3

1. a 303 900 and 307 850.7 ≈ 307 851
   b $P(t) = 300\,000 \times 1.013^t$
   c $350\,000 = 300\,000 \times 1.013^t$, year = 2012
2. 4.31 = 5 years
3.

Time	Time period t (hours)	Amount of caffeine (C mg)
08:00	$t = 0$	120
13.00	$t = 1$	60
18:00	$t = 2$	30
23:00	$t = 3$	15

   a $r = 0.5$
   b $C(t) = 120 \times 0.5^w$
   c $C(7) = 120 \times 0.5^7 = 0.9375$
   d $0.02 = 120 \times 0.5^w$, 13 hours (12.55)
4. $T = 72 \times 0.98^w$, $60 = 72 \times 0.98^w$, $w = 9.02$, thus after 10 weeks
5. a

year	population
0	100 000
1	90 000
2	81 000
3	72 900

   b $P(t) = 100\,000 \times 0.9^t$
   c $25\,000 = 100\,000 \times 0.9^t$, $t = 13.15$, thus after 14 years

Practice 4

1. 18 242 years
2. a $b = -0.866$ b $A = A_0 e^{-0.866t}$ c 0.265
3. a 1000 b 2013 c 5.58 or approximately 6 weeks
4. a 121 rabbits b 5 years
5. a 5 wombats b 1.73 years or approximately 21 months
6. a $r = 0.0199$ b 6224 c 34 years

Mixed practice

1. a $\log_7 23 = x$ b $\log 95 = x$ c $\log_8 6 = x$
   d $\log_4 47 = x$ e $\log_{12} 1200 = x$
2. a $5^3 = 125$ b $3^{-2} = \frac{1}{9}$ c $10^3 = 1000$
   d $7^4 = 2401$ e $a^n = m$
3. a 4 b 2 c −1 d 0
   e 1 f $\frac{1}{2}$ g $\frac{3}{2}$
4. a 2.81 b 1.79 c 3.57
   d 1.39 e 2.81 f 1.61
5. a 2.73 b 1.86 c 3.0
   d 1.46 e 0.91 f 0.65

6 a $-\dfrac{1}{2}$ **b** $\dfrac{7}{8}$ **c** $-\dfrac{1}{6}$

7 a $C(t) = 4 \times 1.055^t$
  b $C(26) = 4 \times 1.055^{26} = 16.09$
  c 30 years, or in the year 2020

8 a 100°C   **b** 54.8°C   **c** 5 minutes

Review in context

1 a 3.4
  b i 7.4   **ii** 5.27   **iii** 5.5   **iv** 5.9   **v** 5.9
  c 91 488 815 microns
  d 4280 microns

2 a 2.8   **b** 3.98×10^{-10} **c** 10^{-7}
3 a 4.2   **b** 0.005
4 a 110 dB, recommended   **b** 65 db   **c** 169 dB, necessary

E7.1

You should already know how to:

1 a $\dfrac{1}{2}$ **b** $\dfrac{\sqrt{2}}{2}$ **c** $\sqrt{3}$

2 a, b Suitable graphs of $\sin x$ and $\cos x$
3 a Amplitude: 2, frequency: 2
  b Vertical dilation of scale factor 3, vertical translation of 1 unit in the positive y direction
4 a Circumference = 24π = 75.4 cm
  b $AB = 4\pi = 12.6$ cm

Practice 1

1 a $\dfrac{\pi}{18}$ **b** $\dfrac{\pi}{12}$ **c** $\dfrac{\pi}{10}$ **d** $\dfrac{\pi}{5}$
  e $\dfrac{2\pi}{9}$ **f** $\dfrac{2\pi}{5}$ **g** $\dfrac{7\pi}{10}$ **h** $\dfrac{5\pi}{6}$
  i $\dfrac{5\pi}{4}$ **j** $\dfrac{5\pi}{3}$

2 a 30°   **b** 120°   **c** 135°   **d** 108°
  e 157.5°   **f** 160°   **g** 75°   **h** 240°
  i 300°   **j** 225°

3 $A = \dfrac{1}{2} r^2 \theta$

Practice 2

1 a 31.4 cm   **b** 70.7 cm   **c** 6 hours
2 a 39.0 m   **b** 0.292 km   **c** 1 591 500 rotations
3 12.00 m
4 a $\dfrac{61\pi}{2}$ **b** 143.7 m

Practice 3

1 a $\dfrac{\sqrt{3}}{2}$ **b** $\dfrac{\sqrt{2}}{2}$ **c** $-\sqrt{3}$
  d $-\dfrac{\sqrt{3}}{2}$ **e** 0   **f** 0
  g $-\dfrac{\sqrt{2}}{2}$ **h** $\dfrac{\sqrt{3}}{2}$ **i** -1

2 a $\dfrac{3\pi}{4}$ and $\dfrac{7\pi}{4}$ **b** $\dfrac{\pi}{3}$ and $\dfrac{5\pi}{3}$ **c** $\dfrac{5\pi}{4}$ and $\dfrac{7\pi}{4}$
  d $\dfrac{\pi}{2}$ and $\dfrac{3\pi}{2}$ **e** $\dfrac{\pi}{3}$ and $\dfrac{2\pi}{3}$

3 a i $\left(\dfrac{\sqrt{3}}{2}, \dfrac{1}{2}\right)$ **b i** $\left(-\dfrac{1}{2}, \dfrac{\sqrt{3}}{2}\right)$
  c i $\left(-\dfrac{\sqrt{2}}{2}, -\dfrac{\sqrt{2}}{2}\right)$ **d i** $\left(\dfrac{\sqrt{3}}{2}, -\dfrac{1}{2}\right)$

Practice 4

1 a $\dfrac{\sqrt{3}}{2}$ **b** 0   **c** $\dfrac{\sqrt{3}}{3}$
  d $-\dfrac{\sqrt{2}}{2}$ **e** 0   **f** $-\dfrac{1}{2}$
  g 1   **h** $\dfrac{\sqrt{2}}{2}$ **i** $\dfrac{1}{2}$

2 a For example $-\dfrac{\pi}{4}$ **b** For example $-\dfrac{2\pi}{3}$
  c For example $-\dfrac{\pi}{2}$ **d** For example $-\dfrac{3\pi}{2}$
  e For example $-\dfrac{\pi}{3}$

Practice 5

1 a Amplitude = 3; frequency = 4; period = $\dfrac{\pi}{2}$; horizontal shift = 0; vertical translation = 0
  b Amplitude = 1; frequency = 3; period = $\dfrac{2\pi}{3}$; horizontal shift = $\dfrac{\pi}{3}$; vertical translation = 2
  c Amplitude = 0.5; frequency = 2; period = π; horizontal shift = $\dfrac{\pi}{8}$; vertical translation = 7
  d Amplitude = 6; frequency = 3; period = $\dfrac{2\pi}{3}$; horizontal shift = $-\dfrac{\pi}{3}$; vertical translation = -3
  e Amplitude = 1; frequency = 3; period = $\dfrac{2\pi}{3}$; horizontal shift = $\dfrac{\pi}{3}$; vertical translation = 2

2 a
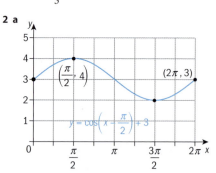

b
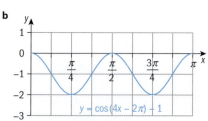

324 Answers

c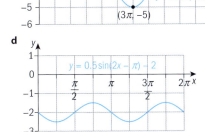

d

3 $y = 2\sin 2\left(x - \dfrac{\pi}{2}\right) - 1$

4 a i $y = 4\sin \dfrac{1}{3}\left(x + \dfrac{\pi}{2}\right)$ **ii** $y = 4\cos \dfrac{1}{3}(x - \pi)$

  b i $y = 3\sin\left(x - \dfrac{\pi}{4}\right) + 6$ **ii** $y = 3\cos\left(x - \dfrac{3\pi}{4}\right) + 6$

  c i $y = 6\sin 4\left(x - \dfrac{\pi}{2}\right) + 3$ **ii** $y = 6\cos 4\left(x - \dfrac{\pi}{8}\right) + 3$

  d i $y = 5\sin 3\left(x - \dfrac{5\pi}{6}\right) - 10$ **ii** $y = 5\cos 3\left(x - \dfrac{\pi}{3}\right) - 10$

Practice 6

1 a

b Amplitude = 1.5 m; period = 10 s; horizontal shift = 2.5 s; vertical translation = 0.
The amplitude is the maximum rise and fall in depth of the buoy. The period is the time taken for one rise and fall cycle of the buoy. The horizontal shift means that at 0 s the height of the buoy was above its mean height. The vertical translation is the mean height over the 10 s.

  c i 0.464 m **ii** −1.5 m

2 a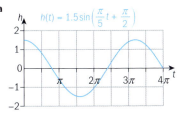

b Amplitude = 1.3 m, period = 12 hours, horizontal shift = 3, vertical shift = 0.67.
The amplitude is the maximum rise and fall in the heights of tides. The period is the time taken for a full cycle of high and low tide. The vertical shift is the mean depth over time, and the horizontal shift is the mean depth over time.

  c 1.97 m

  d Important for people whose livelihood is from the sea, e.g. fishermen, navigation, coastal engineering, etc.

3 a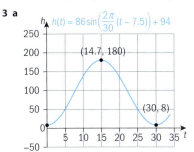

b Amplitude = 86 m (the radius of the wheel); period = 30 minutes (the time to make one complete revolution; horizontal shift = 7.5 min ($\dfrac{1}{4}$ of the period); vertical shift = 94 m, the maximum height above the ground.

  c 137 m

4 a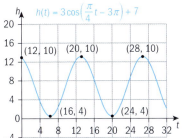

b Amplitude = 3 m (The range of the height of the car from the platform is 6 m). Period = 8 s, the time it takes to complete one complete cycle on the track; horizontal shift = 12 s (it takes 12 s to reach the maximum height); vertical shift = 7 m (the height of the platform).

5 b Amplitude = 22.5; period = 12; horizontal shift = 0; vertical shift = 54.5.
The amplitude is half the difference of the maximum and minimum temperatures in one year. The period is 12 months, the time taken for a full cycle of temperature highs and lows. The vertical shift is the mean temperature over the 12 months.

  c (Answers from GDCs will vary): $y = -22.5\left(\cos \dfrac{\pi}{6}x\right) + 54.5$

Mixed practice

1 a $\dfrac{\pi}{15}$ **b** $\dfrac{4\pi}{3}$ **c** $\dfrac{8\pi}{5}$ **d** $\dfrac{8\pi}{3}$

  e $\dfrac{25\pi}{36}$ **f** $-\dfrac{5\pi}{4}$ **g** $\dfrac{9\pi}{2}$ **h** $\dfrac{28\pi}{15}$

  i $-\dfrac{4\pi}{9}$ **j** $-\dfrac{35\pi}{9}$

2 a 15° **b** 72° **c** 270° **d** 96°
  e −75° **f** 252° **g** −99° **h** −45°
  i 138° **j** −51°

3 a 1 **b** $\dfrac{1}{2}$ **c** $\dfrac{\sqrt{2}}{2}$ **d** $\dfrac{-1}{\sqrt{3}}$ or $\dfrac{-\sqrt{3}}{3}$

  e $\dfrac{-\sqrt{3}}{2}$ **f** $\dfrac{\sqrt{3}}{3}$ **g** −1 **h** $\dfrac{\sqrt{2}}{2}$

  i 0 **j** $-\dfrac{1}{2}$

4 5.00 m

5 a $a = 2$; $f = \pi$; $p = 2$, $hs = 0$; $vs = 0$
  b $a = 5$; $f = 4$; $p = \dfrac{\pi}{2}$; $hs = -\dfrac{\pi}{2}$; $vs = 0$
  c $a = 0.2$; $f = 6$; $p = \dfrac{\pi}{3}$; $hs = -\dfrac{\pi}{3}$; $vs = -1$
  d $a = 3$; $f = \dfrac{\pi}{2}$; $p = 4$; $hs = 3$; $vs = 8$
  e $a = 1$; $f = 6$; $p = \dfrac{\pi}{3}$; $hs = 2$; $vs = -1$

6 a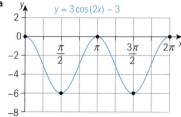

  b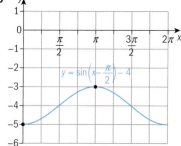

  c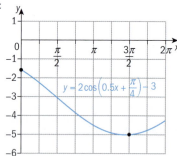

7 a $y = 2\sin\left(x - \dfrac{3\pi}{4}\right) + 1$ **b** $y = \sin\left(2x - \dfrac{\pi}{2}\right) - 3$

8 a The circumference of the tire depends on the diameter. The circumference is the distance travelled in one revolution of the tire.
  b i 1341 m   **ii** 1424 m   **iii** 1676 m
  c 1.2π
  d i 72 cm   **ii** 86 cm   **iii** 105 cm
  e i 2398.9 m   **ii** 2570.2 m   **iii** 3426.9 m

f The radian measure of the number of revolutions multiplied by the radius of the tire gives the total distance travelled by the tire.

9 b $y = -1000\cos\left(\dfrac{\pi}{6}x\right) + 4000$

  c 4866   **d** 3500
  e End of April/beginning of May, and end of August/beginning of September

Review in context

1 a $\dfrac{2\pi}{15}$ **b** $\dfrac{\pi}{3}$ **c** $\dfrac{7\pi}{30}$ **d** $\dfrac{5\pi}{12}$

2 a i $t(x) = a\sin(bx)$
   ii x represents the time after launch that the stone is in the air.
   iii a represents the maximum height of the stone during its flight. This is influenced by the catapult's size and power.
   iv b represents the frequency, or number of cycles per 2π. The more cycles there are, the closer the point of impact. The smaller the number of cycles, the further away the point of impact.
   v Domain: x is from 0 to $\dfrac{\pi}{b}$, range of $t(x)$ is from 0 to a.

  b c represents the horizontal shift, which is the position of the catapult from the origin. The catapult can be moved toward the target to hit it, for example.

  c Domain: x from $\dfrac{c}{b}$ to $\dfrac{\pi + c}{b}$, range: $t(x)$ from 0 to a (stays the same).

  d Instead of moving the catapult up, it can simply be moved backwards if the target is too close, for example. By moving the catapult up, the sine curve does not model the trajectory of the stone once the stone goes below the catapult's height – it would keep going down whereas the sine curve makes it advance forward.

  e i Domain: x from 2π to 6π, range: $t(x)$ from 0 to 8
   ii $t(x) = 8\sin\left(0.25x - \dfrac{\pi}{2}\right)$, $2\pi < x < 6\pi$

  f i $(2\pi, 0)$ **ii** $(4\pi, 0)$
   iii Max height 5, at the point $(3\pi, 5)$
   iv Domain x is from 2π to 4π, range $t(x)$ from 0 to 5.

E8.1

You should already know how to:

1 a $4n + 1$, 81
  b $-2.5n + 27.5$, -22.5
  c $1.2n + 41.5$, 65.5

2 a 480, 960; $15 \times 2^{n-1}$
  b 1.5625, $\dfrac{25}{64}$ or 0.390625; $100 \times (0.25)^{n-1}$
  c $\dfrac{2}{3}, -\dfrac{2}{9}$; $162 \times \left(-\dfrac{1}{3}\right)^{n-1}$

3 a 1, 3, 6, 10, 15   **b** $\dfrac{1}{2}n(n-1)$

4 a 2, 5, 8, 11, 14   **b** 6, 12, 18, 36, 72

Practice 1

1 465   **2** 1131   **3** 2112   **4** 18396

5 a 930   **b** 900

6 a 2601   **b** −2550   **c** 51

7 80

8 a 91, so correct.   **b** 385   **c** 2870   **d** 2485

9 190

Practice 2

1 a 459   **b** 198   **c** 438

2 a 513   **b** 10 380

3 a $a = 15, d = 4$ **b** 47   **c** 5029

4 725   **5** 120

Practice 3

1 a 185   **b** 56.4   **c** 780   **d** 351

2 a 825   **b** 441   **c** 924   **d** 1904   **e** 3630

3 720   **4** 308   **5** 630   **6** 13.2

Practice 4

1 a 413 336   **b** 23 327.5   **c** 23.857 690 7   **d** 420.04

2 a 1 797 558   **b** 6075.04   **c** 5147.5

3 a $x_n = 2 \times 3^{n-1}$ **b** 6560

Practice 5

1 a 26.622 976   **b** 39.921 875   **c** 917 503.125

2 a 19 530   **b** 329 554 400   **c** 1266

3 8 388 600   **4** 485 (3 s.f.)   **5** 484 275 610

6 a $r = \pm 2$ **b** 1778 or 602

7 8.65 (3 s.f.)   **8** $n = 30$

Practice 6

1 b $u_n = 3 \times 4^{n-1}$

   c 65 535   **d** $S_{15} = 1\,073\,741\,823$

2 a 11   **b** 660

3 a 8   **b** 32.8

4 a 962 mm²   **b** 1 969 214 mm² ≈ 1.97 m²

5 1371.6 mm (54 inches)

6 a $u_6 = 20 + 5 \times 2 = 30$

   b $u_n = 20 + 2(n-1) = 50$
   $n = 16$; the 16th day

   c Day 25

7 456 minutes (7 hours 36 minutes)

Practice 7

1 a 2000   **b** 100   **c** $\frac{200}{7}$

2 54   **3** 156.25

4 a $\frac{2}{9}$ **b** $\frac{35}{99}$ **c** $\frac{23}{111}$ **d** $\frac{16}{27}$ **5** $\frac{5}{9}$

Practice 8

1 a Yes   **b** No   **c** Yes   **d** Yes
   e Yes   **f** Yes   **g** No   **h** No

2 a $4, \frac{4}{3}, \frac{4}{9}$ **b** 5.98 (3 s.f.)   **c** 6

3 $r = \frac{15}{18} = \frac{5}{6}$. Since $-1 < r < 1$, the sum is finite.

4 a $r^2 = \frac{1.5}{24}$ so $r = \pm\frac{1}{4}$

   b $a = \pm 96$

   If $a = 96$, $S_\infty = \frac{96}{1-\frac{1}{4}} = 128$

   If $a = -96$, $S_\infty = -76.8$

5 $r = 0.9$ **6** $x = 15$ **7** $x = -3$

Mixed practice

1 a 9801   **b** 208   **c** 1365   **d** 62 499.84

2 2464   **3** 210   **4** 48.2 (3 s.f.)   **5** 394.0625

6 1071   **7** 159 (3 s.f.)   **8** $\frac{28\,394}{27} \approx 1050$ (3 s.f.)

9 $d = 3$ **10 a** $\frac{2}{9}$ **b** $\frac{2}{11}$ **c** $\frac{41}{333}$

11 37 700   **12** 4   **13** $a = -56, d = 24$

14 96   **15** $r = \frac{1}{3}$ **16** $a = 36, d = 6$

17 $x = 5$ **18** $x = 10$ **19** 165

20 396   **21** 1240 cm² (3 s.f.)

Review in context

1 a $8 + 2n$

   Since each week capacity increases by a constant amount (the common difference).

   b 28   **c** 180

2 4.16 million. It is only an estimate because she has estimated the rate at which attendance will fall, and other factors, such as reviews, might influence attendance.

3 a $\frac{120}{100} = \frac{144}{120} = 1.2$. There is a common ratio.

   b 3960 (3 s.f.)

4 a Because each new machine connects to the server and $n - 1$ other machines.

   b Adding the connections together gives $1 + 2 + 3 + \ldots + n$.

   c 1275   **d** 64

5 a $\frac{120\,000}{150\,000} = \frac{96\,000}{120\,000} = 0.8$. The sequence has a common ratio.

   b 553 392 (to the nearest whole number of followers)

   c 750 000

6 a 111 000   **b** 19 000   **c** 111 000 + 76 000 = 187 000

7 a 3   **b** 15   **c** 20 depots

E9.1

You should already know how to:

1 a 5   **b** 25

2 a (5, 8)   **b** (0, 7.5)

3 $\sqrt{57}$ **4** 53°

Practice 1

1 a $A(0, 0, 1)$, $B(4, 9, 0)$, $C(4, 0, 0)$, $D(0, 9, 1)$

   b $E(-5, 7, 0)$, $F(-5, 0, 0)$, $G(0, 7, 0)$, $H(-5, 7, -2)$

2 a 7   **b** 6   **c** 13

Practice 2

1. **a** 25 cm **b** 9 m **c** 27 cm
2. **a** 56.648… ≈ 57 cm **b** 137.2 mm
   c 25.729 ≈ 26 inches
3. $5\sqrt{3}$
4. 16.6 cm

Practice 3

1. **a** 7 **b** 3 **c** 11
   d 9 **e** 37 **f** 22.5
2. **a** 14 **b** 23
3. $OA = 9$, $OB = 9$. So OAB is isosceles because it has two sides of equal length.
4. Area = 36.3 (3 s.f.)

Practice 4

1. 17°
2. **b** 14°
3. **a** 54.7° **b** 54.7°
4. **a** 10.2 cm **b** 11.4 cm **c** 10° **d** 11°
5. **a** $8\sqrt{2}$ **b** $4\sqrt{2}$ **c** $4\sqrt{6}$ **d** 55°
6. **a** 13.6 m **b** 4.8°
7. 12.3 m

Practice 5

1. **a** (2, 4, 8) **b** (13, 6, 4) **c** (−3, −1, 11)
   d (4.5, 5, 8) **e** (−7.5, 12.5, 8.5) **f** (2, 1.85, − 0.15)
2. (21, 7, −2)
3. $a = 1$, $b = 3$, $c = −18$
4. $ABCD$ is a parallelogram.
5. $P(1, 7, 2)$, $Q(3, 8, 4)$, $R(5, 9, 6)$

Practice 6

1. **a** (10, 15) **b** (−1, 4) **c** (7, 12)
2. **a** (9, −1, −2) **b** (12, 4, 1.5) **c** (6, −4, 15)
3. (16, 2, 28)
4. $m : n = 10 : 15$ so let $m : n$ be 2 : 3.
   $a = −4$, $b = 5$

Mixed practice

1. 57 cm (to the nearest centimeter)
2. Yes. The space diagonal is around 25.7 inches long, and drumsticks are usually quite thin.
3. **a** (2, −0.5, 6); 7 **b** (5.5, 3, −6); 9 **c** (11, 0, 7.5); 9
4. **a** $\sqrt{270} \approx 16.4$ m; Midpoint = (7.5, 4.5, 5)
   b It gains 6 meters in height.
5. **a** $\sqrt{61} \approx 7.8$ m **b** 22.6°
6. **a** 41.77° **b** 43.11°
   c No, since the angles are not the same.
7. (19, 4, 17)
8. $OA = \sqrt{131}$ and $OB = \sqrt{189}$. A is closer to O.
9. $z = 1, 15$
10. Area = $2\sqrt{7897} \approx 178$

Review in context

1. **a** (112.5, 90, 15)
   b The greatest distance possible in the room is the space diagonal, which has length 193 feet. This is greater than the maximum range, so it is possible that some people might not hit targets.
   c Assume firing from ground level, 98.1 feet. If the laser were to be held at around 4 feet above the ground, 97 feet is a more realistic maximum range.
2. **a** $EF \approx 193.32$ m
   $GH \approx 687$ m
   7.98 m ≈ 8 m difference
   b (214, 29, 18.5)
   c $M(118.5, 22.5, 17)$; $N(120.5, 24.5, 18)$
   $MN = 3$ m
3. **a** 40 m **b** 76 m **c** Volume ≈ 687 m³
   This is an estimate because numbers will only have been given to a degree of accuracy; the tunnel will not be a perfect cylinder.
   d (99, 29.8, 24.4)
4. **a** 415 m
   b She will probably ski left and right, not in a straight line; the slope will have steeper and shallower parts.
   c 20.7 s
   d (−155, 268, −59)
   e 134 m
   f 6.87°

E9.2

You should already know how to:

1. **a** 10.5 cm² **b** 42 cm²
2. **a** 3.76 cm **b** 45.6°
3. 36.1 cm²
4. **a** 21.25 cm² **b** 1.75 radians

Practice 1

1. △ABC △DEF

△LMN △UVW

Practice 2

1. **a** 41.6 cm² **b** 181 m² **c** 968 km²
2. **a** 1114 cm² **b** 3402 cm²
3.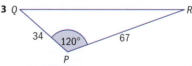

Area = 986 cm²

4 a 130 cm²
  b i It estimates the number of tiles needed to tile 10 m².
    ii It is an underestimate, as the tiles will probably not cover the space exactly, and will have to be cut leaving oddly shaped sections so extras will be needed.

5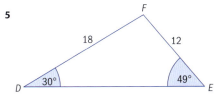
Area 106 cm²

6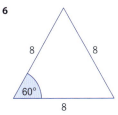
Area of one face = 27.7 m²
Total surface area = 111 m²
Thickness of paint is 1 mm, so volume = $111 \times 0.001 = 0.111$ m³ = 111 liters.
74 cans needed.

Practice 3

1 a Area = 0.0572 cm²
  b Area = 271 cm²
  c Area = 4.01 cm²
2 a 2.86 cm² **b** 0.112 cm² **c** 49.7 cm²
3 a 29.1 cm² **b** 90.0 cm² **c** 17.7 cm²
4 $2 = \frac{1}{2} \times 49(\theta - \sin\theta)$
  $\theta = 0.80$ (2 d.p.)

Practice 4

1 a $a = 22.9$ cm **b** $b = 16.1$ cm **c** $c = 6.14$ cm
2 a $A = 44°, B = 56°$ **b** $A = 82°, B = 28°$ **c** $B = 57°, C = 55°$
3 a $x = 59.5°$ **b** $x = 33.5°$ **c** $x = 39.3$ cm
4 $y = 12.4$ cm
5 a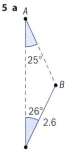
  b $\angle CBA = 129°, \angle CAB = 25°$
  c $c = 2.7$ km, distance = 5.3 km

6 a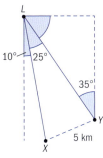
  b $L = 25°, X = 30°, Y = 125°$
  c 5.92 km

Practice 5

1 a $a = 6.3$ cm **b** $b = 6.0$ cm **c** $c = 61.1$ cm
2 a 91° **b** 103° **c** 42°
3 a $r = 19.0$ **b** $E = 104°$ **c** $y = 11.0$ **d** $D = 73.9°$
4 a N **b** 75° **c** 13.4 km

5 a 86.4° **b** 34.8°

Practice 6

1 a, b, e: the cosine rule; three sides and an angle are involved in the problem.
  c, d, f: the sine rule; each problem involves two sides and two angles.
2 a 11.7 cm **b** 95.0 cm **c** 132°
  d 74.6° **e** 112 cm **f** 26.8°
3 a $a = 21.3$ cm **b** $b = 8.27$ cm **c** $c = 24.1°$
  d $d = 70.6°$ **e** $e = 104°$ **f** $f = 10.1$ cm
4 Area = $\frac{15\sqrt{3}}{4} \approx 6.50$ units²
5 a 7.79 cm **b** 1.413 rad **c** 11.8 cm²

Practice 7

1 a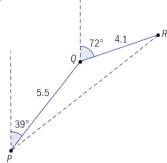
  b 147° **c** 9.21 km

2 a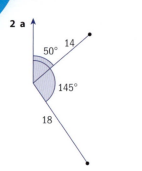

b 95°   **c** 23.7 km

3 a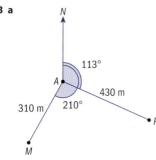

b 97°   **c** 560 m

4 a 50°   **b** 039°   **c** 112 m   **d** 211 m

Mixed practice

1 a 10.7 cm²   **b** 63.4 cm²   **c** 34.6 cm²
 d 47.6 cm²   **e** 37.6 cm²   **f** 36.3 cm²
 g 24.7 cm²

2 a 2.81 cm²   **b** 178 cm²

3 a 0.109 cm²   **b** 74.8 cm²

4 $a = 5.8$ cm, $b = 6.76$ cm, $c = 14.1$ cm, $d = 11.2$ cm, $e = 10.5$ cm, $f = 14.7$ cm

5 $a = 43.6°$, $b = 38.6°$, $c = 54.8°$, $d = 101°$

6 $a = 1.71$, $b = 0.547$, $c = 0.356$, $d = 0.364$

7 20.0°   **8** 42.6 m

9 $x = 62.7°$, perimeter = 66.1 cm

Review in context

1 b $\angle RST = 85°$, $\angle STR = 40°$, $\angle TRS = 55°$
 c $RT = 14.7$ km, $ST = 12.0$ km
 d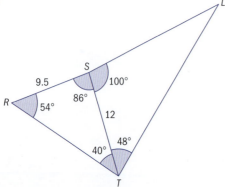
 e $LS = 16.8$ km, $LT = 22.2$ km

2 a 146°   **b** 18.1 km
3 a 77.6°   **b** 122°   **c** 241°
4 8.8 km

E10.1

You should already know how to:

1 $\log_5 y = x$

2 $x = 7^y$

3 a $y = \dfrac{x-2}{3}$ **b** $y = \pm\sqrt{x+1}$

4 Vertical dilation scale factor 2 and a vertical translation of one unit in the negative direction.

5 Algebraically: $f \circ f(x) = 5 \div \dfrac{5}{x} = x$
Graphically: $f(x)$ is a reflection of itself in the line $y = x$.

Practice 1

1 a $f^{-1}(x) = \log_4 x$ **b** $f^{-1}(x) = 6^x$

c $f^{-1}(x) = e^x$ **d** $f^{-1}(x) = 5^{\frac{x}{3}}$

e $f^{-1}(x) = e^{\frac{4x}{3}}$ **f** $f^{-1}(x) = \log_{11}\left(\dfrac{3x}{2}\right)$

2 a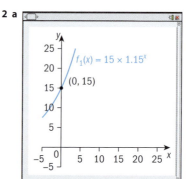

$P_0 = 15$ seagulls

b $f^{-1}(x) = \log_{1.15}\left(\dfrac{x}{15}\right)$

c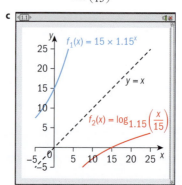

d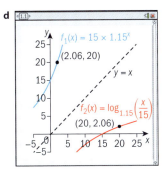

2.06 or approximately 2 months.

Practice 2

1 a B   **b** C   **c** A

2 a $a = 1.2$
  b $f^{-1}(x) = 1.2^x$ Domain is $\mathbb{R}$, range is $\mathbb{R}^+$.
  c The two functions are reflections of each other in the line $y = x$.

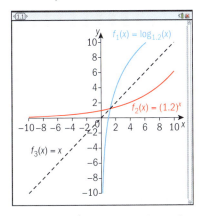

3 a Domain = $\mathbb{R}^+$; range = $\mathbb{R} - \{0\}$; $a = 3$
  b Domain = $\mathbb{R}^+$; range = $\mathbb{R} - \{0\}$; $a = 0.5$
  c Domain = $\mathbb{R}^+$; range = $\mathbb{R} - \{0\}$; $a = 5$

4 a i Domain is $\mathbb{R}$; range is $\mathbb{R}^+$ **ii** $f^{-1}(x) = \ln x - 2$
  iii The graphs are reflections in the line $y = x$.

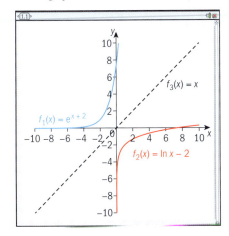

b i Domain is $x > -2$; range is $\mathbb{R}$ **ii** $f^{-1}(x) = 10^{\frac{x}{3}} - 2$
  iii The graphs are reflections in the line $y = x$.

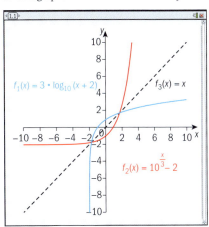

c i Domain is $x > -0.5$; range is $\mathbb{R}$.   **ii** $f^{-1}(x) = \dfrac{10^{\frac{x}{4}} - 1}{2}$
  iii The graphs are reflections in the line $y = x$.

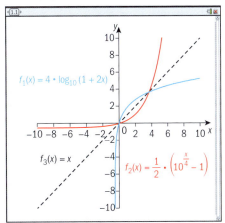

d i Domain is $x > \dfrac{2}{3}$; range is $\mathbb{R}$ **ii** $f^{-1}(x) = \dfrac{10^{2-x} + 2}{3}$
  iii The graphs are reflections in the line $y = x$.

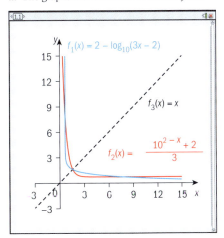

Answers 331

e i Domain is $x > 0$; range is $\mathbb{R}^+$.   **ii** $f^{-1}(x) = \dfrac{e^{x+4}}{2}$, $x > 0$

  iii The graphs are reflections in the line $y = x$.

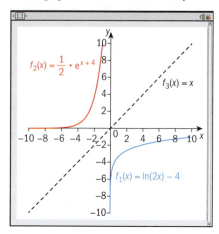

5 $y = \log_a x = \dfrac{\ln x}{\ln a}$, $b = \dfrac{1}{\ln a}$ hence $y = b \ln x$.

6 a $y = \log_2 x$ **b** $y = 5^{\frac{x-7}{2}}$ **c** $= \log_{0.2} x$

Practice 3

1 a Horizontal translation of 7 units in the positive x-direction, and vertical translation 1 unit in the positive y-direction.
  b Vertical dilation, scale factor 2 and vertical translation of 3 units in the negative y-direction.
  c Horizontal dilation scale factor $\dfrac{1}{6}$ and vertical translation 4 units in the negative y-direction.
  d Horizontal translation 2 units in the negative x-direction, vertical dilation scale factor 4, reflection in the x-axis, vertical translation 8 units in the positive y-direction.
  e Horizontal translation 5 units in the positive x-direction, vertical dilation scale factor 3, vertical translation 7 units in the positive y-direction.
  f Horizontal dilation scale factor $\dfrac{1}{3}$, horizontal translation 4 units in the negative x-direction, vertical dilation scale factor $\dfrac{2}{3}$, vertical translation 10 units in the negative y-direction.

2 a Horizontal translation 3 units in the positive x-direction, and vertical translation 2 units in the positive y-direction.
  b Vertical dilation scale factor of 0.5, and vertical translation 1 unit in the negative y-direction.
  c Horizontal dilation scale factor of $\dfrac{1}{2}$, and vertical translation 1 unit in the negative y-direction.
  d Vertical dilation scale factor 2, and vertical translation 1 unit in the positive y-direction.
  e Horizontal translation 1 unit in the negative x-direction, vertical dilation scale factor 3, and reflection in the x-axis.
  f Horizontal dilation scale factor 2 and a vertical translation 2 units in the positive y-direction.

  g Horizontal dilation scale factor $\dfrac{1}{5}$, vertical dilation scale factor 2, reflection in the x-axis, vertical translation 3 units in the negative y-direction.
  h Horizontal dilation scale factor $\dfrac{1}{2}$, horizontal translation 3 units in the positive x-direction, vertical dilation scale factor $\dfrac{1}{5}$, vertical translation 4 units in the positive y-direction.

3 a Horizontal translation of 5 units in the negative direction; $f(x+5)$.
  b Reflect the original in the y-axis, and then translate horizontally two units in the positive x-direction; $f(-x+2)$ or $f(-(x-2))$.
  c Reflect in the x-axis and translate vertically 2 units in the negative y-direction; $-f(x-2)$.

4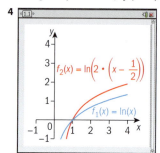

Translate $f(x)$ horizontally $\dfrac{1}{2}$ unit in the positive x-direction, and dilate horizontally by scale factor $\dfrac{1}{2}$.

5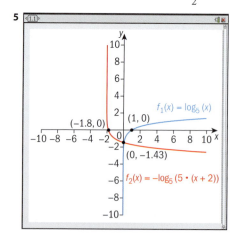

Translate $f(x)$ horizontally 2 units in the negative x-direction, dilate horizontally by a scale factor of $\dfrac{1}{5}$, and then reflect in the x-axis.

Mixed practice

1 a C   **b** A   **c** B

2 b $a = 2.2$
  c $w = f^{-1}(s) = 2.2^{s-14}$ tells what week you should be in for the speed you have achieved.

3 a i Domain is $x > 3$; range is $\mathbb{R}$ **ii** $f^{-1}(x) = e^x + 3$
   iii The graphs are reflections in the line $y = x$.

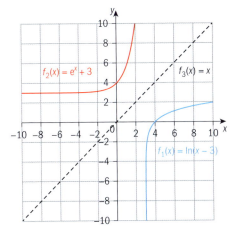

b i Domain is $\mathbb{R}$; range is $y > 2$.
   ii $f^{-1}(x) = 5\log_7\left(\dfrac{x-2}{5}\right); x > 2$
   iii The graphs are reflections in the line $y = x$.

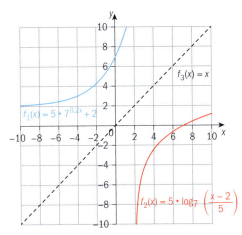

c i Domain is $x > 2$; range is $\mathbb{R}$.   **ii** $f^{-1}(x) = \dfrac{2^{2(x+5)} + 8}{4}$
   iii The graphs are reflections in the line $y = x$.

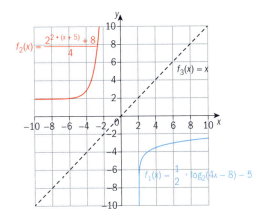

d i Domain is $x > -1$; range is $\mathbb{R}$ **ii** $f^{-1}(x) = 3^x - 1$
   iii The graphs are reflections in the line $y = x$.

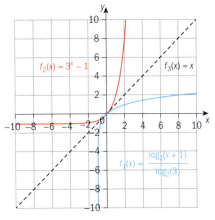

e i Domain is $\mathbb{R}$; range is $\mathbb{R}^+$.
   ii $f^{-1}(x) = 1 - \log_4 x\,; x > 0$
   iii The graphs are reflections in the line $y = x$.

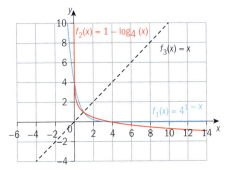

4 a Translate the graph 2 units vertically in the positive y-direction.

b Translate the graph 2 units horizontally in the negative x-direction.

c Translate the graph 2 units horizontally in the positive x-direction, then translate 3 units vertically in the negative y-direction.

d Reflect the graph in the y-axis and translate 2 units vertically in the positive y-direction.

e Dilate the graph horizontally by scale factor 4.

f Reflect the graph in the y-axis and translate 2 units horizontally in the positive x-direction.

g Reflect the graph in the y-axis, dilate vertically by scale factor 2, and reflect in the x axis.

h Translate the graph 2 units horizontally in the positive x-direction, dilate vertically by scale factor 2 and translate 1 unit vertically in the negative y-direction.

i Reflect in the y-axis, translate 1 unit horizontally in the negative x-direction, dilate vertically by scale factor 2, reflect in the x-axis, then translate 2 units in the positive y-direction.

5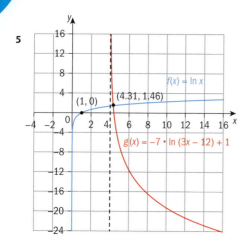

Dilate $f(x)$ horizontally by scale factor $\frac{1}{3}$; translate horizontally 4 units in the positive x-direction; dilate vertically by scale factor 7; reflect in the x-axis; translate vertically 1 unit in the positive y-direction.

6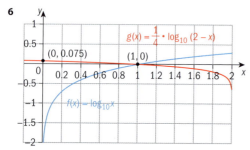

Reflect $f(x)$ in the y-axis; translate horizontally 2 units in the positive x-direction; dilate vertically by scale factor $\frac{1}{4}$.

7 a $y = \log_4(x-3)$
  b Students' own answers

Review in context

1 The standard model is reflected in the x-axis.
2 a The standard model is vertically dilated by scale factor 4.
  b The standard model would be vertically dilated by scale factor 2.
3 a i The graph of f is dilated vertically by scale factor 0.67.
    ii The graph of f is translated 7.6 units in the negative y-direction
  b Domain: $2.20 \times 10^{11} < E < 1.86 \times 10^{26}$
  c There is no solution for r when $E = 0$. For the model to have y-intercepts, it would have to include a horizontal translation in the negative x-direction.
4 Horizontal dilation by scale factor 2120; vertical dilation by scale factor 26 400; reflection in the x-axis.

E10.2

You should already know how to:

1 a Domain is $\mathbb{R}$, range is $\mathbb{R}$. b Domain is $\mathbb{R}$, range is $\mathbb{R}^+$.
  c Domain is $\mathbb{R}^+$, $x \neq 1$, range is $\mathbb{R}$.
2 a $y \propto x$, so $y = kx$ b $y \propto \frac{1}{x}$, so $y = \frac{k}{x}$
3 Horizontal translation 3 units in the negative x-direction, reflection in x-axis, vertical dilation scale factor 2, vertical translation 5 units in the positive y-direction.
4 a $y = \frac{4(x-2)}{3}$, which is a function
  b $f^{-1}(x) = 2 \pm \sqrt{x}$, which is not a function

Practice 1

1 a Domain: $x \neq 0, x \in \mathbb{R}$, range: $y \neq 0, y \in \mathbb{R}$

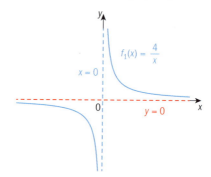

b Domain: $x \neq 0, x \in \mathbb{R}$, range: $y \neq 0, y \in \mathbb{R}$

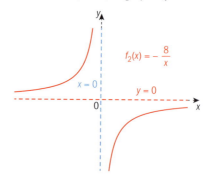

c Domain: $x \neq 0, x \in \mathbb{R}$, range: $y \neq 0, y \in \mathbb{R}$

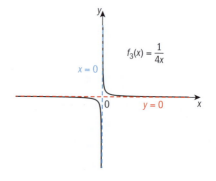

d Domain: $x \neq 0, x \in \mathbb{R}$, range: $y \neq 0, y \in \mathbb{R}$

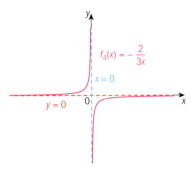

2 b $y = 200$ hours

3 a The resistance is inversely proportional to the current.
  b $k = 20$
  c $R(I) = \dfrac{20}{I}$ or $R(x) = \dfrac{20}{x}$

4 a 50 hours   **b** 10 hours   **c** 5 hours

5 a i Force and acceleration are directly proportional.
  ii Acceleration and mass are inversely proportional.
  b i 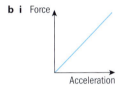   **ii**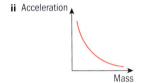

Practice 2

1 a $h = -3, k = -7$, asymptotes: $x = -3, y = -7$.
    Domain: $x \neq -3, x \in \mathbb{R}$, Range: $y \neq -7, y \in \mathbb{R}$.

  b $h = -2, k = 0$, asymptotes: $x = -2, y = 0$.
    Domain: $x \neq -2, x \in \mathbb{R}$, Range: $y \neq 0, y \in \mathbb{R}$.

  c $h = 5, k = 8$, asymptotes: $x = 5, y = 8$.
    Domain: $x \neq 5, x \in \mathbb{R}$, Range: $y \neq 8, y \in \mathbb{R}$.

  d $h = 0, k = -11$, asymptotes: $x = 0, y = -11$.
    Domain: $x \neq 0, x \in \mathbb{R}$, Range: $y \neq -11, y \in \mathbb{R}$.

  e $h = 1, k = \pi$, asymptotes: $x = 1, y = \pi$.
    Domain: $x \neq 1, x \in \mathbb{R}$, Range: $y \neq \pi, y \in \mathbb{R}$.

2 a Asymptotes: $x = 0$ and $y = -4$.
    Domain: $x \neq 0, x \in \mathbb{R}$, Range: $y \neq -4, y \in \mathbb{R}$.

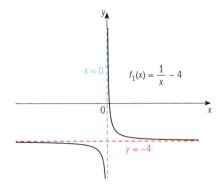

b Asymptotes: $x = 4$ and $y = 0$, Domain: $x \neq 4, x \in \mathbb{R}$,
  Range: $y \neq 0, y \in \mathbb{R}$.

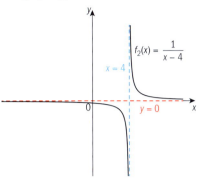

c Asymptotes: $x = 0$ and $y = 3$, Domain: $x \neq 0, x \in \mathbb{R}$,
  Range: $y \neq 3, y \in \mathbb{R}$.

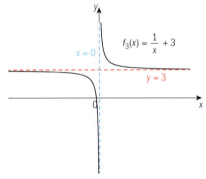

d Asymptotes: $x = 2$ and $y = 0$, Domain: $x \neq 2, x \in \mathbb{R}$,
  Range: $y \neq 0, y \in \mathbb{R}$.

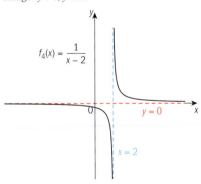

e Asymptotes: $x = -2$ and $y = 5$, Domain: $x \neq -2, y \in \mathbb{R}$,
  Range: $y \neq 5, y \in \mathbb{R}$.

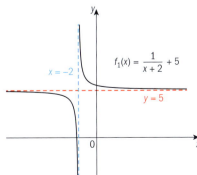

f Asymptotes: $x = 3$ and $y = 1$, Domain: $x \neq 3, x \in \mathbb{R}$, Range: $y \neq 1, y \in \mathbb{R}$.

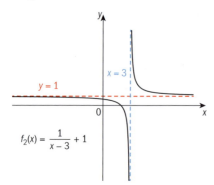

$f_2(x) = \dfrac{1}{x-3} + 1$

3 a $y = \dfrac{1}{x} - 5$ **b** $y = \dfrac{1}{x-2} - 7$

c $y = \dfrac{1}{x+4}$ **d** $= \dfrac{1}{x-3} + 2$

4 a $y = \dfrac{1}{x-3} + 4$ **b** $y = \dfrac{1}{x} + 2$

c $y = \dfrac{1}{x-2} - 1$ **d** $y = \dfrac{1}{x+\frac{3}{4}} + \dfrac{4}{3}$

5 a i $h > 0$ horizontal translation in the positive x-direction
  ii $h < 0$ horizontal translation in the negative x-direction
b i $k > 0$ vertical translation in the positive y-direction
  ii $k < 0$ vertical translation in the negative y-direction

Practice 3

1 a $f(x) = \dfrac{1}{2(x-4)} - 1 = \dfrac{1}{2x-8} - 1$

b $f(x) = \dfrac{1}{\frac{1}{3}(x-3)} - 2 = \dfrac{3}{x-3} - 2$

c $f(x) = \dfrac{-\frac{1}{4}}{x} + 8 = -\dfrac{1}{4x} + 8$

d $f(x) = \dfrac{\frac{3}{2}}{-(x+2)} + 4 = \dfrac{-3}{2x+4} + 4$

2 a Horizontal translation of 6 units in the negative x-direction, vertical dilation of scale factor 3 (or horizontal dilation of scale factor $\frac{1}{3}$), vertical translation of 5 units in the negative y-direction. Asymptotes: $y = 5, x = -6$

b Horizontal translation of 2 units in the positive x-direction, vertical dilation of scale factor 5 (or horizontal dilation of scale factor $\frac{1}{5}$), vertical translation of 3 units in the positive y-direction. Asymptotes: $y = 3, x = 2$

c Horizontal translation of $-\frac{1}{3}$ units in the negative x-direction, vertical dilation of scale factor 3 (or horizontal dilation of scale factor $\frac{1}{3}$). Asymptotes: $y = 0, x = -\frac{1}{3}$

d Horizontal translation of 4 units in the positive x-direction, reflection in the y-axis (or in the x-axis), vertical dilation of scale factor $\frac{5}{2}$ (or horizontal dilation of scale factor $\frac{2}{5}$), vertical translation of 2 units in the positive y-direction.

Asymptotes: $y = 2, x = 4$

e Horizontal translation of $\dfrac{1}{2}$ unit in the positive x-direction, reflection in the y-axis (or in the x-axis), vertical dilation of scale factor $\dfrac{5}{2}$ (or horizontal dilation of scale factor $\dfrac{2}{5}$), vertical translation of 8 units in the negative y-direction. Asymptotes: $y = -8, x = \dfrac{1}{2}$

f Reflection in the y-axis (or in the x-axis), vertical dilation of scale factor 13 (or horizontal dilation of scale factor $\dfrac{1}{13}$), vertical translation of 11 units in the negative y-direction. Asymptotes: $y = -11, x = 0$

Practice 4

1 a

b

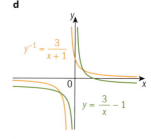

c

d

e

f

2 a $y = -\dfrac{5}{x} - 1$ **b** $y = \dfrac{7}{x-6} + 9$

c $y = -\dfrac{1}{x+8} + 12$ **d** $y = -\dfrac{6}{x+3} + 5$

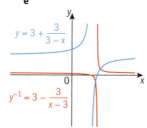

e $y = -\dfrac{10}{x-3} - 3 = \dfrac{3x+1}{3-x}$ **f** $y = -\dfrac{8}{x+2} + 4 = \dfrac{4x}{x+2}$

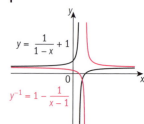

Mixed practice

1 a Domain: $x \neq 0, x \in \mathbb{R}$, range: $y \neq 0, y \in \mathbb{R}$, asymptotes: $x = 0, y = 0$.

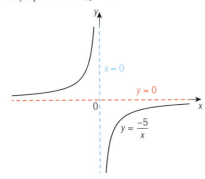

b Domain: $x \neq 0, x \in \mathbb{R}$, range: $y \neq 1, y \in \mathbb{R}$, asymptotes: $x = 0, y = 1$.

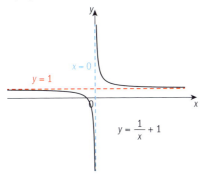

c Domain: $x \neq 2, x \in \mathbb{R}$, range: $y \neq -8, y \in \mathbb{R}$, asymptotes: $x = 2, y = -8$.

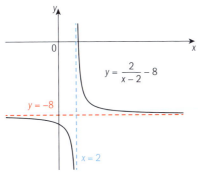

d Domain: $x \neq -3, x \in \mathbb{R}$, range: $y \neq 2, y \in \mathbb{R}$, asymptotes: $x = -3, y = 2$.

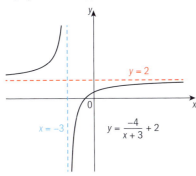

2 a $g(x) = \dfrac{3}{x-5} - 5$ **b** $g(x) = -\dfrac{1}{2x} + 1$

   c $g(x) = \dfrac{3-1}{x-3} - 2$ **d** $g(x) = -\dfrac{12}{x-4} - 1$

3 a Horizontal translation of 2 units in the positive x-direction, reflection in the y-axis (or in the x-axis), vertical translation of 9 units in the positive y-direction.
Asymptotes: $x = 2$ and $y = 9$.

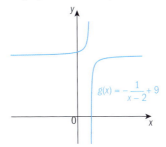

Inverse: $g^{-1}(x) = \dfrac{2x - 19}{x - 9}$

b Horizontal translation of 1 unit in the negative x-direction, vertical dilation of scale factor 3 (or horizontal dilation of scale factor $\dfrac{1}{3}$), vertical translation of 3 units in the negative y-direction.
Asymptotes: $x = -1$ and $y = -3$.

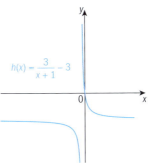

Inverse: $h^{-1}(x) = \dfrac{3}{x+3} - 1$

c Horizontal translation of 5 units in the positive x-direction, reflection in the y-axis (or in the x-axis), vertical dilation of scale factor 10 (or horizontal dilation of scale factor $\dfrac{1}{10}$)
Asymptotes: $x = 5$ and $y = 0$.

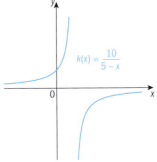

Inverse: $k^{-1}(x) = -\dfrac{10}{x} + 5$

d Horizontal translation of 1 unit in the positive x-direction, reflection in the y-axis (or in the x-axis), vertical dilation of scale factor 4 and horizontal dilation of scale factor 3, vertical translation of 1 unit in the positive y-direction.
OR
Horizontal translation of 1 unit in the positive x-direction, reflection in the y-axis (or in the x-axis), vertical dilation of scale factor $\frac{4}{3}$ (or horizontal dilation of scale factor $\frac{3}{4}$), vertical translation of 1 unit in the positive x-direction.
Asymptotes: $x = -\frac{2}{3}$ and $y = 1$.

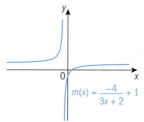

$m(x) = \frac{-4}{3x+2} + 1$

Inverse: $m^{-1}(x) = \frac{-4}{3x-3} - \frac{2}{3}$

e Horizontal translation of 2 units in the positive x-direction, reflection in the y-axis (or in the x-axis), vertical dilation of scale factor 5 and horizontal dilation of scale factor 4, vertical translation of 6 units in the positive y-direction.
OR
Horizontal translation of 2 units in the positive x-direction, reflection in the y-axis (or in the x-axis), vertical dilation of scale factor $\frac{5}{4}$ (or horizontal dilation of scale factor $\frac{4}{5}$), vertical translation 6 units in the positive y-direction.
Asymptotes: $x = 2$ and $y = 6$.

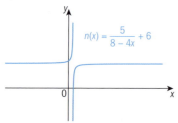

$n(x) = \frac{5}{8-4x} + 6$

Inverse: $n^{-1}(x) = \frac{-5}{4x-24} + 2$

f $p(x) = \frac{1}{4x-5}$ horizontal translation of $\frac{5}{4}$ units, vertical dilation of scale factor $\frac{1}{4}$ (or horizontal dilation of scale factor 4).
Asymptotes: $x = \frac{5}{4}$ and $y = 0$

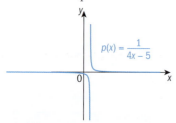

$p(x) = \frac{1}{4x-5}$

Inverse: $p^{-1}(x) = \frac{1}{4x} + \frac{5}{4}$

4 a 22.7 mg (3 s.f.)   **b** $C = -\frac{600}{x+12} + 50$

c Horizontal translation of 12 units in the negative x-direction, reflection in the x-axis vertical dilation of scale factor 600, vertical translation of 50 units in the positive y-direction.

Review in context

1 854.91 AUD

2 a slow pipe $= \frac{1}{3p}$ **b** $\frac{1}{p} + \frac{1}{3p}$

c $\frac{1}{p} + \frac{1}{3p} = \frac{1}{14}$, $p = 10.5$ hours.

Time to fill with slow pipe $= 31.5$ hours.

E11.1

You should already know how to:

1 a $\frac{2\pi}{3}$ **b** $\frac{3\pi}{4}$ **c** 30°   **d** 300°

2 a $\frac{1}{\sqrt{2}}$ or $\frac{\sqrt{2}}{2}$ **b** $-\frac{1}{2}$ **c** 1.07 (3 s.f.)

3 a 71.3° (3 s.f.)   **b** 30°

4 Horizontal dilation of scale factor $\frac{1}{2}$, vertical dilation of scale factor 3, vertical translation of 5 units in the positive y direction.

5 $x = -\frac{3}{2}$ or $x = 5$ **6 a** $\angle B = 25.7°$ (3 s.f.)   **b** 4.26

Practice 1

1 a 30°, 150°   **b** 104°, 256° (3 s.f.)
c 45°, 225°   **d** 225°, 315°
e 143°, 323°   **f** 150°, 210°

2 a $\frac{2\pi}{3}, \frac{4\pi}{3}$ **b** $\frac{\pi}{4}, \frac{3\pi}{4}$ **c** $\frac{4\pi}{3}, \frac{5\pi}{3}$
d 2.56, 3.73   **e** 1.43, 4.86   **f** 3.79, 5.64
g $\frac{3\pi}{4}, \frac{7\pi}{4}$ **h** $\frac{\pi}{6}, \frac{7\pi}{6}$ **i** $\frac{\pi}{2}, \frac{3\pi}{2}$
j $\frac{3\pi}{2}$ **k** $\frac{\pi}{3}, \frac{2\pi}{3}, \frac{4\pi}{3}, \frac{5\pi}{3}$
l 0.464, 2.68, 3.61, 5.82
m $\frac{\pi}{3}, \frac{5\pi}{3}, \pi$ **n** 1.05, 1.91, 4.37, 5.24
o $0, \frac{2\pi}{3}, \pi, \frac{4\pi}{3}, 2\pi$ **p** 1.11, 2.16, 4.25, 5.30   **q** $\frac{\pi}{4}, \frac{3\pi}{4}, \frac{3\pi}{2}$

3 a $\frac{\pi}{4}, \frac{5\pi}{4}$ **b** $0, \frac{3\pi}{4}, \pi, \frac{7\pi}{4}, 2\pi$

4 b 2.30, $\frac{7\pi}{6}$, 3.98, $\frac{11\pi}{6}$

Practice 2

1 a $\theta = \frac{3\pi}{4}$ **b** $\theta = \frac{4\pi}{3}$ **c** $\theta = \pi$
d $\theta = 0, \frac{2\pi}{3}, \frac{4\pi}{3}, 2\pi$
e $\theta = 0.068, 0.56, 1.32, 1.82, 2.58, 3.07$

f $\theta = \frac{3\pi}{8}, \frac{5\pi}{8}, \frac{11\pi}{8}, \frac{13\pi}{8}$ **g** $x = \frac{\pi}{18}, \frac{5\pi}{18}, \frac{13\pi}{18}, \frac{17\pi}{18}$

h $x = \frac{5\pi}{8}, \frac{7\pi}{8}, \frac{13\pi}{8}, \frac{15\pi}{8}$ **i** $\theta = 0, \frac{\pi}{6}, \frac{\pi}{3}, \frac{2\pi}{3}$

j $\theta = \pi$ **k** $x = \frac{\pi}{6}, \frac{\pi}{18}, \frac{5\pi}{18}$

l $x = 0.955, 2.19, 4.10, 5.33$

Practice 3

1 a $0, \pi, 2\pi$ **b** $\frac{\pi}{3}, \pi, \frac{5\pi}{3}$ **c** $\frac{\pi}{3}, \pi, \frac{5\pi}{3}$

d $\frac{\pi}{3}, \frac{2\pi}{3}, \frac{4\pi}{3}, \frac{5\pi}{3}$ **e** $0, \frac{\pi}{3}, \frac{5\pi}{3}, 2\pi$

2 a $0, 180°, 360°$ **b** $90°, 210°, 330°$
c $0°, 75.5°, 180°, 284°, 360°$
d $41.4°, 180°, 319°$ **e** $30°, 150°, 210°, 330°$

3 It is not an identity, since it is true only when $\sin x = \cos x$, and thus not true for all values of x.

Practice 4

2 a $\left(\frac{1}{\cos x} + \tan x\right)(1 - \sin x) \equiv \cos x$

b $\left(\frac{\sin x + \tan x}{-\cos x - 1}\right)(\cos x) \equiv -\sin x$

Practice 5

1 a $\angle C = 75.3°$ or $\angle C = 104.6°$
b $\angle C = 41.4°$ ($\angle C = 139°$ isn't possible)
c $\angle C = 12.8°$ ($\angle C = 167°$ isn't possible)
d $\angle C = 70.7°$ or $\angle C = 109°$

2 a $\angle Q = 53.1°; \angle P = 96.9°; p = 9.93$ cm or $\angle Q = 127°$; $\angle P = 23.1°; p = 3.93$ cm
b $\angle Q = 43.9°, \angle P = 81.1°, p = 31.4$ cm
c $\angle Q = 34.2°, \angle P = 124°, p = 44.4$ m or $\angle Q = 146°$, $\angle P = 13.2°, p = 11.3$ m
d $\angle Q = 35.0°, \angle P = 127°, p = 18.1$ km or $\angle Q = 145°$, $\angle P = 17.0°, p = 6.63$ km

3 $70.0°$ or $110.1°$ **4** $16.2°$

5 39 m or 7 m

6 a 1 triangle **b** 1 triangle
  c 0 triangles **d** 2 triangles

7 a
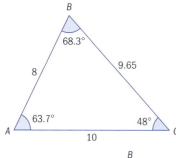

b 9.7 km or 3.7 km

8 a 2 **b** 10.9 m or 5.4 m

Mixed practice

1 a $90°, 270°$ **b** $150°, 330°$ **c** $158°, 338°$
  d $16.6°, 163.4°$ **e** No solution **f** $225°, 315°$

2 a $\frac{\pi}{3}, \frac{2\pi}{3}$ **b** $\frac{3\pi}{4}, \frac{5\pi}{4}$ **c** $\frac{\pi}{3}, \frac{4\pi}{3}$

  d $\frac{3\pi}{2}$ **e** $\frac{\pi}{3}, \frac{5\pi}{3}$ **f** $\frac{\pi}{6}, \frac{7\pi}{6}$

3 a $\frac{4\pi}{3}, \frac{5\pi}{3}$ **b** $53.1°$ or $306.9°$

  c $\frac{\pi}{12}, \frac{5\pi}{12}, \frac{13\pi}{12}, \frac{17\pi}{12}$ **d** $\frac{2\pi}{9}, \frac{4\pi}{9}, \frac{8\pi}{9}, \frac{10\pi}{9}, \frac{14\pi}{9}, \frac{16\pi}{9}$

4 a $225°, 315°$ **b** Quadrants I and IV
  c If $\sin\theta > 0$, $\tan\theta < 0$. If $\sin\theta = 0$, $\tan\theta = 0$. If $\sin\theta < 0$, $\tan\theta > 0$.

5 $-\sqrt{3}$ ($\theta = 120°$)

10 a $30°, 150°, 270°$ **b** $90°, 120°, 240°, 270°$
  c $0°, 180°, 360°$ **d** $109°, 180°, 251°$

11 a $\frac{\pi}{6}, \frac{\pi}{2}, \frac{11\pi}{6}, \frac{3\pi}{2}$ **b** $\frac{\pi}{4}, \frac{3\pi}{4}, \frac{5\pi}{4}, \frac{7\pi}{4}$ **c** $0, \pi, 2\pi$

  d $\frac{\pi}{6}, \frac{5\pi}{6}, \frac{\pi}{2}$ **e** $2.91, 3.38$ **f** $\frac{\pi}{4}, \frac{2\pi}{3}, \frac{4\pi}{3}, \frac{7\pi}{4}$

12 It is an identity because it is true for all values of θ.

13 a $B = 36.9°$ or $B = 143.1°$ **b** $B = 60.3°$ or $B = 119.7°$

Review in context

1 a $I = 74.6°$ or $105.4°$ **b** $T = 65.4°$ or $34.6°$

2 7.70 km or 2.08 km

3 a 228 m or 427 m
  b Because with the given information, there are two possible solutions for the position of B (because there are two possible solutions for the angle B).
  c 199 m

4 a
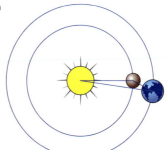

  b Either 235.2 million km or 44.8 million km.

E12.1

You should already know how to:

1 a 0.699 **b** 1.10 **c** 2.58

2 $x = 2.73$ **3** $x = -1$

4 $y = 3x - 4$
  Inverse function: $y = \frac{x+4}{3}$

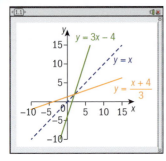

The functions are reflections in the line $y = x$.

5 $y^{-1} = 5^{\left(\frac{x+11}{2}\right)} - 3$

Practice 1

1 a $\log 12$ **b** $\log 15$ **c** $\log 200$

d $\log \dfrac{z^{\frac{1}{2}}x^3}{y}$ **e** $\ln \dfrac{x^9}{8}$ **f** $\log \dfrac{x^3}{96y^4}$

2 a $\ln a + \ln b$ **b** $\ln a - \ln b$ **c** $2\ln a + \ln b$

d $\dfrac{1}{2}\ln a$ **e** $-2\ln a$ **f** $\ln a + \dfrac{1}{2}\ln b$

g $3\ln a - \ln b$ **h** $3\ln a - 2\ln b$ **i** $\dfrac{1}{2}(\ln a - \ln b)$

3 a $x + y$ **b** $y - x$ **c** $x - y$ **d** $2x$
  e $2y$ **f** $3x$ **g** $y + 3x$

4 a $2x + 3y + 2$ **b** $5x - 4y - 2$

  c $\dfrac{1}{3}(6x + 8y + 3)$ **d** $-x - \dfrac{1}{2}y - 1$

Practice 2

1 a $f^{-1}(x) = \log_2 x + 3$; Domain of f is $\mathbb{R}$; range of f is $\mathbb{R}^+$. Domain of f^{-1} is $\mathbb{R}^+$; range of f^{-1} is $\mathbb{R}$.

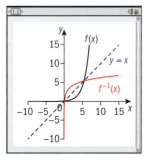

b $g^{-1}(x) = \dfrac{e^{x^2}}{2}$, $x > 0$; Domain of g is $x > \dfrac{1}{2}$; range of g is $\mathbb{R}^+$.
Domain of g^{-1} is $\mathbb{R}^+$ Range of g^{-1} is $y > 0$.

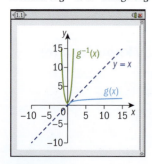

c $h^{-1}(x) = \dfrac{10^{\frac{x}{4}} - 1}{2}$

Domain of h is $x > -\dfrac{1}{2}$; range of h is $\mathbb{R}$. Domain of h^{-1} is $\mathbb{R}$; range of h^{-1} is $y > -\dfrac{1}{2}$.

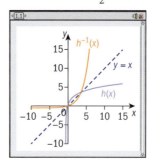

d $j^{-1}(x) = \dfrac{2 + e^x}{e^x - 1}$; Domain of j is $x > 1$; range of j is $\mathbb{R}^+$.
Domain of j^{-1} is $\mathbb{R}^+$; range of j^{-1} is $y > 1$.

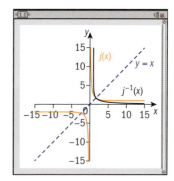

2 $g(x) = \log x^n = n \log x$. $g(x) = n \log x$ is a vertical dilation of $f(x)$ scale factor n.

3 $f(x) = -\log_3 x + 2$. Reflect the graph of $y = \log_3 x$ in the x-axis and translate it vertically 2 units in the positive y direction.

4 a i $\log_2\left(\dfrac{8}{x}\right) = \log_2 8 - \log_2 x = 3 - \log_2 x$.

  ii $\log_2(x+3)^2 = 2\log_2(x+3)$.

b i Reflect $y = \log_2 x$ in the x-axis then translate 3 units vertically in the positive y-direction.

  ii Translate 3 units horizontally in the negative x-direction, and dilate vertically by scale factor 2.

Practice 3

1 a $\log p^4 q^2$ **b** $\log \dfrac{p^n}{q^3}$ **c** $\ln \dfrac{(x-2)^3}{\sqrt{x}}$

d $\ln \dfrac{x}{(x-2)^2}$ **e** $\ln\left(\dfrac{e}{x}\right)$ **f** $\ln(x^4(x-1)^2)$

2 a $a - b$ **b** $a + b$ **c** $2a + c$

  d $\dfrac{1}{2}(b - a)$ **e** $\dfrac{1}{2}c + b$

3 a $\ln(x - 3) + \ln(x + 3)$
  b $\ln(x + 2) + \ln(x + 3)$
  c $\ln(x - 3) - \ln(x + 4)$

d $\ln(x+1) - \ln(x+2) - \ln(x-2)$
  e $\ln(x-6) + \ln(x+1) - \ln(x+2) - \ln(x-2)$
  f $2\ln x + \ln(x+2) + \ln(x-5) - \ln(5x-20)$

Practice 4

1 a 2.81 **b** 1.29 **c** 2.10
d 0.732 **e** 1 **f** 1.16

Practice 5

1 a $\dfrac{3}{5}\ln x$ **b** $\dfrac{2}{5}\ln x$ **c** $\dfrac{1}{10}\ln x$ **d** $\dfrac{1}{5}(\ln x - \ln 3)$

2 a $\ln x^5$ **b** $-\ln x$ **c** $\ln x^{\frac{7}{2}}$ **d** $\ln\left(\dfrac{x^{\frac{1}{2}}}{9}\right)$

Practice 6

1 a $x = 2.93$ **b** $x = 1.91$ **c** $x = 1.30$
2 a $x = 0.693$ **b** $x = 0.79$ **c** $x = 0.55$
3 a $x = 0.397$ **b** $x = 0.417$ **c** $x = 1.044$
4 a $x = 0$ or $x = 2.58$
  b $x = 1$ or $x = 1.29$
5 a $x = 1.2$ **b** $x = \dfrac{5}{34}$ **c** $x = -\dfrac{5}{8}$
  d $x = 2$ or $x = -7$. $x = -7$ is not possible, hence $x = 2$
  e $x = -4$ or $x = 5$. $x = -4$ is not possible, hence $x = 5$
6 b 10 years
7 a 15 is the initial number of video sharers.
  b 21 hours
  c 23 hours
  d Students' own answers
8 a $x = 0.746$ **b** $x = 1.35$
  c $x = 0.333$ **d** $x = 2$

Mixed practice

1 a $\log a + \log b + \log c$ **b** $\log a + \log c - \log b$
  c $3\log a + 2\log b + 4\log c$ **d** $\log a + \dfrac{1}{2}\log b$
  e $\log a + \dfrac{1}{2}\log b - \log c$

2 a $\log 12.5$ **b** $\ln 250$ **c** $\log_4 40$
  d $\log_a 4$ **e** $\ln\left(\dfrac{x^6}{(x-1)^2(2x+6)^3}\right)$ **f** $\log_4 54$
  g $\log_7 \dfrac{4x^{\frac{1}{3}}}{y^2}$

3 a $-y$ **b** $x + 2y$ **c** $-(x+y)$
  d $2x + 2y$ **e** $3x - 2y$
4 a $5x + y$ **b** $1 + \dfrac{1}{4}(x + 3y)$
  c $2(3x + 4y - 2)$ **d** $1 + 7x - \left(\dfrac{1}{2}\right)y$

5 a $f^{-1}(x) = \dfrac{\ln x - \ln 30}{2\ln 5}$; Domain of f is $\mathbb{R}$; range of f is $\mathbb{R}^+$.
  Domain of f^{-1} is $\mathbb{R}^+$; range of f^{-1} is $\mathbb{R}$.
  b $g^{-1}(x) = 10^{(x+2)^3}$; Domain of g is $\mathbb{R}^+$; range of g is $\mathbb{R}$;
  Domain of g^{-1} is $\mathbb{R}$; range of g^{-1} is $\mathbb{R}^+$.

  c $h^{-1}(x) = \dfrac{e^{-\frac{3}{2}x} + 1}{3}$
  Domain of h is $x > \dfrac{1}{3}$; range of h is $\mathbb{R}$.
  Domain of h^{-1} is $\mathbb{R}$; range of h^{-1} is $y > \dfrac{1}{3}$.

  d $k^{-1}(x) = \dfrac{5 \cdot 4^x}{7 - 4^x}$; Domain of k is $\mathbb{R}^+$; range of k is $\mathbb{R}$.
  Domain of k^{-1} is $\mathbb{R}$, $x \neq \dfrac{\ln 7}{\ln 4}$; range of k^{-1} is $\mathbb{R}^+$.

6 a $y = 3 + \log_3(x)$; Translate the graph of $y = \log_3 x$ by 3 units vertically in the positive y-direction.
  b $y = 4\log_3(x - 9)$; Dilate $y = \log_3 x$ vertically by scale factor 4 and translate horizontally 9 units in the positive x direction.

7 a $c - b$ **b** $a + b + c$ **c** $3a - 2c$
  d $\dfrac{1}{2}(2b - (a + c))$ **e** $\dfrac{1}{2}(3a + b)$

8 a $\ln(x - 5) + \ln(x + 5)$
  b $\ln(x - 7) + \ln(x + 4)$
  c $\ln(x + 1) - \ln(x - 8)$
  d $\ln(x - 5) + \ln(x + 1) - \ln(x + 4) - \ln(x + 2)$
  e $\ln(x - 12) + \ln(x + 2) - \ln x - \ln(x + 2) - \ln(x - 2)$;
  $\ln(x - 12 - \ln x - \ln(x - 2))$

9 a $x = \pm 1$ **b** $x = 2$
10 a $x = 15$ **b** $x = 112$ **c** $x = 3$
  d $x = 2$ **e** $x = 6$
11 a $\dfrac{x}{y}$ **b** $3x + y$ **c** $4x + 2y$
  d $y - x$ **e** $x - y$
12 a $x = 0.24$ **b** $x = -4.64$ **c** $x = 1.66$ **d** $x = 2.05$
13 a $\dfrac{y}{x}$ **b** $\dfrac{x}{y}$ **c** $\dfrac{2y}{x}$
14 a $\dfrac{3}{8}\ln x$ **b** $\dfrac{3}{8}\ln x - 16$
15 a 24 **b** 5
16 a 52 days **b** 2.8 million **17** $y = 15e^{0.05x}$

Review in context

1 a 8.90
  b Scale factor of 6.31 times bigger
  c Add 1.3
  d 40 times bigger
  e Eastern Sichuan was 40 times bigger.

2 a 6.8 acidic **b** 2.8 **c** 6.2
  d Yes **e** $1 \times 10^{-7} - 3.98 \times 10^{-8}$

3 a 137 dB **b** 6.3 times bigger

E13.1

You should already know how to:

1 Student diagrams in DGS

2 a Reflection in $x = -1$
  b Rotation 90° anticlockwise, center (3, 2)

c Translation 2 units left and 6 units down, or $\begin{pmatrix} -2 \\ -6 \end{pmatrix}$
d Reflection in $x = 0$ or the y-axis
e Reflection in the line $y = x + 1$
f Rotation 90° anticlockwise, center (0,1)

3 $a = 110°, b = 70°$ 4 $AC = 4.5$ cm, $DE = 6$ cm

Practice 1

1 A-F, B-E, C-D
2 I-K, G-L, H-J
3 M-R, N-Q, O-P
4 S-V, T-X, U-W
5 $\triangle ABC \cong \triangle NOM$
$\triangle DEF \cong \triangle PRQ$
$\triangle GHI \cong \triangle WXV$
$\triangle JKL \cong \triangle TSU$

Practice 2

1 a No b No c Yes (SSS)
  d Yes (AAS or ASA)
2 a $AC \cong CD$ (given)
   $\angle BAC \cong \angle DEC$ (given)
   $\angle BCA \cong \angle ECD$ (vertically opposite angles)
   Hence they are congruent. (ASA)
 b $AT \cong ET$ (given)
   $AD \cong ED$ (given)
   DT is common.
   Hence they are congruent. (SSS)
 c $AD \cong BD$ (definition of bisector)
   $\angle ADC = \angle BDC = 90°$ (definition of perpendicular)
   DC is common.
   Hence they are congruent. (SAS)
3 By rotating the pentagon by 144°, $\triangle CDE$ will line up with $\triangle ABC$.
4 $\triangle ABD, \triangle BFE, \triangle CDF, \triangle DEA, \triangle FAC, \triangle EFC, \triangle BAE, \triangle CBF,$
  $\triangle DCA, \triangle EDB, \triangle BCE, \triangle AFD.$

Practice 3

1 $AB \cong AC$ (given)
  $BM \cong MC$ (definition of midpoint)
  $AM \cong AM$ (reflexive property)
  Hence, the triangles $\triangle ABM$ and $\triangle ACM$ are congruent. (SSS)
2 $AB \cong CD$ (given)
  $BO \cong DO$ (definition of circle)
  $AO \cong OC$ (definition of circle)
  Hence, the triangles are congruent. (SSS)
3 $PQ \cong QR$ (given)
  $QX \cong QX$ (reflexive)
  $\angle QXR = \angle QXP = 90°$ (definition of altitude)
  Hence, the triangles are congruent. (RHS)
4 $\angle ABC \cong \angle ACB$ (given) implies $\triangle ABC$ isosceles
  $AB \cong AC$ ($\triangle ABC$ isosceles)
  $AO \cong OA$ (reflexive property)
  $OB \cong OC$ (definition of circle)
  Hence the triangles are congruent. (SSS)
5 $AB \cong AB$ (reflexive property)
  $AD \cong AE$ (definition of circle)
  $BE \cong BD$ (definition of circle)
  Hence the triangles are congruent. (SSS)

6 $HK \cong JK$ (definition of square)
  $HF \cong JF$ (definition of circle)
  $FK \cong FK$ (reflexive property)
  Hence the triangles are congruent. (SSS)
7 $\angle PQX \cong \angle SRX$ (alternate angles)
  $\angle PXQ \cong \angle SXR$ (vertically opposite angles)
  $PX \cong XS$ (definition of midpoint)
  Hence the triangles are congruent. (AAS)

Practice 4

1 $AB = CD, BC = DA$ from the definition of a rhombus.
  $AC = CA$ because it is a symmetric property.
  $\triangle ABC \cong \triangle CDA$ because of SSS congruence.
  $\angle DAC \cong \angle BCA$ because of CPCTC.
2 $BE = DA$ because it is given
  $\angle BCE = \angle DCA = 90°$ because it is given.
  $CE = CA$ because of the definition of an isosceles triangle.
  $\triangle BCE \cong \triangle DCA$ because of RHS congruency.
  $\angle CEB \cong \angle CAD$ because of CPCTC.
3 $AB = AC$ because the triangle is isosceles.
  $\angle BAX \cong \angle CAX$ because AX is the angle bisector.
  Therefore $\triangle ABX \cong \triangle ACX$ (SAS).
  Hence $\angle AXB \cong \angle AXC$ (CPCTC).
  $\angle AXB + \angle AXC = 180°$ because they lie on a straight line.
  $\Rightarrow 2 \times \angle AXC = 180°$
  $\Rightarrow \angle AXC = 90°$
4 $OA = OB$ because they are both radii.
  $AM = BM$ because M is the midpoint of AB.
  OM is common.
  Therefore $\triangle OAM \cong \triangle OBM$ (SSS).
  Hence $\angle AMO \cong \angle BMO$.
  $\angle AMO + \angle BMO = 180°$ because they lie on a straight line.
  $\Delta 2 \times \angle AMO = 180°$
  $\angle AMO = 90°$.

5 a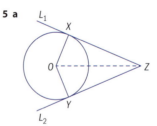

  b 1 $OX \cong OY$ (they are both radii)
    2 $\angle OXZ = \angle OYZ = 90°$ (the tangent to a circle in perpendicular to the radius)
    3 OZ is common
    4 Hence $\triangle OXZ \cong \triangle OYZ$ (RHS)
  c Hence $XZ \cong YZ$ (CPCTC)

10 $ABCDE$ has a line of symmetry perpendicular to DE through B (properties of a regular pentagon). Since K is equidistant to ED, it lies on the line of symmetry, and hence KC is a reflection of KA, they are equal length.

Mixed Practice

1 a No (only SA given)   b No (only SSA given)
  c Yes (AAS)   d No (only SSA given)

Review in context

1 a Since folding is equivalent to reflection, one rectangle is a mirror image of the other and hence they are congruent.
 b *EG* is the full width of the page. *EX* is half the width of the page.
 c 30°
 d Since *GDZ* is the empty space previously occupied by the triangle *GDE*, the two triangles are congruent.
 ∠*EGD* ≅ ∠*DGZ* = 30°
 e 30°
 f Since *GF* is half the width of the paper (the edge *GE*) and *EY* is half the width of the paper, they are the same length.
 g *GF* ≅ *EY* (by **f**, above)
 ∠*FGH* = ∠*YED* = 30° (by **d** and **e**, above)
 ∠*GFH* = 90° (since *ABFG* is a rectangle)
 ∠*EYD* = 90° (since the fold is parallel to *ABC*)
 Hence ∠*EYD* ≅ ∠*GFH*
 Hence ∠*GFH* = ∠*EYD* (ASA)
 h *FH* ≅ *DY* (CPCTC)

E14.1

You should already know how to:

1 $x < 7$

2 a

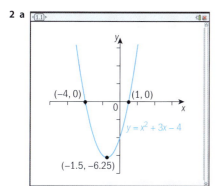

b

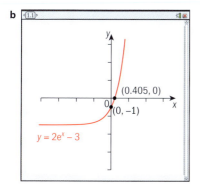

c

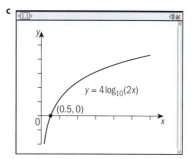

d

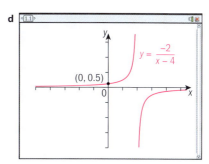

e

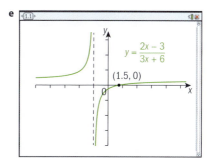

f

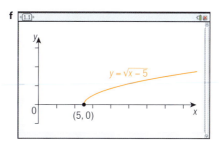

3 $x = -3 \pm \sqrt{6}$

Practice 1

1 a $x < -2$ or $x > 6$ **b** $-1.5 \le x \le 2$
 c $x \le -0.4$ or $x \ge 5$ **d** $-4 < x < 1.5$
 e $-\dfrac{3}{2} < x < \dfrac{2}{3}$

2 a $h(t) > 9600$ m
 b About 23 seconds

3 $x(x+6) > 216$, $x > 12$ cm, $(x+6) > 18$ cm

4 $x(12-x) \le 24$; 0 m $< x \le 2.54$ m or 9.46 m $\le x < 12$ m

5 $y = 0.005x^2 - 0.23x + 22 > 25$; $x > 56$ years

6 a $h(t) > 0$ **b** 3.56 seconds

7 $x > 2.73$ cm

8 4.47 cm × 4.47 cm × 2 cm

Practice 2

1 a $x > 1$ or $-1 < x < 0$ **b** $0 < x \leq \dfrac{1}{4}$

c $-1 < x < 4$ **d** $x > 5.09$
e $-1.5 < x < -1$ or $x > 1.57$ **f** $-1.5 < x < -1$ or $x < -5$
g $-3 < x < 0$ **h** $-4 < x < -2$ or $1 < x < 6$

2 a $R = \dfrac{2R_1}{2+R_1} \Rightarrow \dfrac{2R_1}{2+R_1} > 1, R_1 > 2$

b $R_2 > 7.48$

Practice 3

1 a

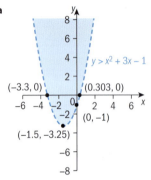

b

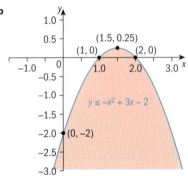

c

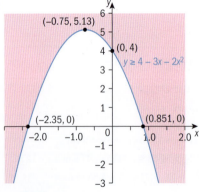

d

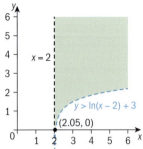

e

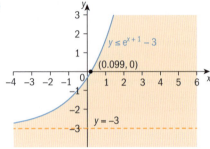

f

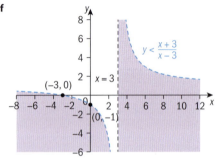

2 a $y < -x^2 - 4x + 5$ **b** $y \geq 0.5x^2 + 2x - 1$

c $y \leq e^{2x+1} - 4$ **d** $y > \dfrac{x}{x-2}$

3 a i $f(x) = x^2 - 4$ **ii** $y > f(x)$

b i $f(x) = -\dfrac{1}{3}(x-1)^2 + 3$ **ii** $y > f(x)$

c i $f(x) = x^2 - 6x + 5$ **ii** $y < f(x)$

4 a $f(x) = -\dfrac{3}{2025}(x - 45)^2 + 3$

b Domain $\{x \mid 0 \leq x \leq 90\}$, range $\{y \mid 0 \leq y \leq 3\}$

c $0 \leq y \leq f(x)$

Practice 4

1 a $-1 \leq x \leq 3$ **b** $x < 2$ or $x > 8$
c $-3 \leq x \leq 4$ **d** $-10 \leq x \leq 4$
e $-3 < x < \dfrac{1}{2}$ **f** $x < -\dfrac{1}{3}$ or $x > 1$
g $-\dfrac{1}{2} < x < \dfrac{2}{3}$

2 $0 < x \leq 1.70$ cm

3 Student's own answers

Practice 5

1. **a** $x < 0.209$; $x > 4.79$
   b $-6.32 < x < 0.317$
   c $-2.19 \leq x \leq 0.687$
   d $x < 0.333$ or $x > 2$
   e $-2.79 \leq x \leq 1.79$
   f $x < -0.414$ or $x > 2.41$
   g $-0.290 < x < 0.690$
   h $-9.12 \leq x$ or $x \geq -0.877$
   i $x < -1.18$ or $x > 0.425$
   j $0.209 < x < 4.79$

2. **a** $\Delta = 0$, hence only one root, $x = -1$. Since the graph lies entirely above the x-axis except for this one value, there is only solution, $f(x) = 0$ for $x = -1$.
   b $\Delta = 0$, hence only one root, $x = 2$. Since the graph lies entirely above or on the x-axis, the solution is $x \in \mathbb{R}$.
   c Since $\Delta < 0$, the quadratic has no roots. Since $a > 0$, its graph is always above the x-axis for all values of x. Hence $x \in \mathbb{R}$.
   d Since $\Delta < 0$, the quadratic has no roots. Since $a < 0$, its graph is always below the x-axis for all values of x. Hence $x \in \mathbb{R}$.

Practice 6

1. **a** $x < -1$ or $x > 1$ **b** $-1 < x \leq 4$
   c $x < -1$ or $x > 4$ **d** $x \geq 3$ or $x < 2$
   e $0 < x < \dfrac{1}{4}$ **f** $x > 5$ or $x \leq -\dfrac{2}{3}$

2. $-3 < x < -1$

3. About 17

Mixed practice

1. **a** $x < -3$ or $x > -1$ **b** $x \leq -8.58$ or $x \geq 0.583$
   c $x < -3.59$ or $x > 2.09$ **d** $x > 0$ or $x < -3$
   e $0 < x < 0.286$ **f** $0.333 < x < 5$

2. **a**

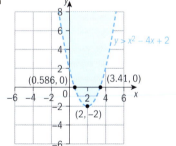

b

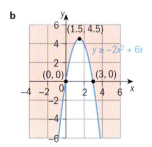

c

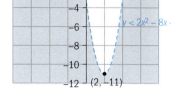

d

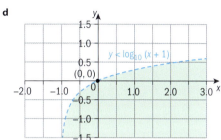

e

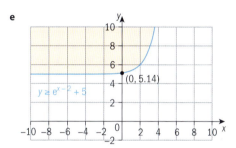

f

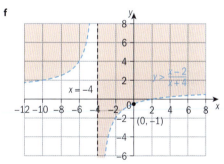

3. **a** $-6 < x < 1$ **b** $x < -2$ and $x > 12$
   c $-7 < x < 5$ **d** $-8 < x < 2$
   e $0 < x < 5.5$ **f** $x \leq \dfrac{3}{5}$ or $x \geq 2$
   g $x > 3$ or $x < -5$ **h** $-1 < x < 1$

4. 36.2 meters

5. Between 11 and 30 tourists

6. 4.6 cm < radius < 9.7 cm; 6.8 cm < height < 29.8 cm

7. 1.66 in < x < 6.16 in

Review in context

1, 2 Student's own answers.

E15.1

You should already know how to:

1 a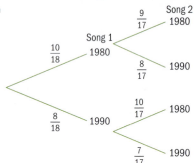

b $\frac{5}{17}$

2 a $\frac{32}{52}$ **b** $\frac{3}{169}$ **c** $\frac{4}{221}$

3 a $P(A) = \frac{15}{27}$ **b** $P(A \cup B) = 1$

4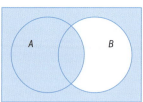

5 a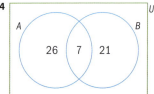

b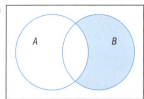

c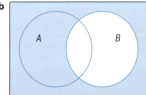

d

e

Practice 1

1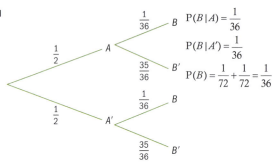

$P(B|A) = P(B|A') = P(B)$
∴ independent events

2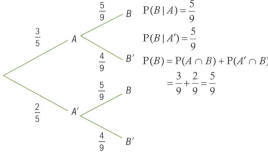

$P(B|A) = P(B|A') = P(B)$
∴ independent events

3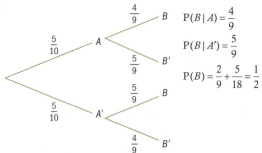

$P(B|A) \neq P(B|A') \neq P(B)$
The events are not independent.

4

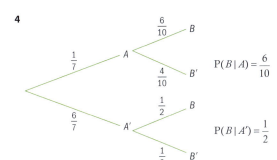

The events are not independent.

5 $P(B|A) = \frac{5}{9}$

$P(B|A') = \frac{6}{9}$, therefore not independent events.

6 $P(B|A) = \frac{5}{11}$

$P(B|A') = \frac{6}{11}$, therefore not independent events.

7 $P(B|A) = 90\%$

$P(B|A') = 65\%$, therefore not independent events.

8 $P(Y|X) = 0.8$

$P(Y|X') = 0.8$

$\therefore P(Y|X') = P(Y|X) = P(Y)$

$\therefore$ independent events

Practice 2

1 a $\frac{11}{26}$ **b** $\frac{11}{32}$

2 a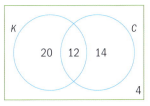

b i $P(K) = \frac{32}{50}$ **ii** $P(C|K) = \frac{12}{32}$

iii $P(C) = \frac{26}{50}$ **iv** $P(K|C) = \frac{12}{26}$

3 a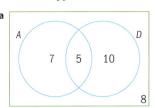

b i $P(A) = \frac{12}{30}$ **ii** $P(D|A) = \frac{5}{12}$

iii $P(D \cup A) = \frac{22}{30}$ **iv** $P(D \cap A | D \cup A) = \frac{5}{22}$

4 a $\frac{15}{25}$ **b** $\frac{9}{25}$ **c** $\frac{4}{9}$

5 a i $\frac{17}{35}$ **ii** $\frac{19}{35}$ **iii** $\frac{17}{35}$

b i $\frac{9}{19}$ **ii** $\frac{13}{17}$ **iii** $\frac{7}{18}$

c $\frac{7}{28}$

Practice 3

1 a

	Male (M)	Female (F)	Total
Professional (P)	5	12	17
Amateur (A)	5	6	11
Total	10	18	28

b i $P(M|P) = \frac{5}{17}$ **ii** $P(M|A) = \frac{5}{11}$ **iii** $P(M) = \frac{10}{28}$

c No they are not independent events

2

	soccer	No soccer	Total
Left	10	30	40
Right	15	45	60
Total	25	75	100

$P(L|S) = \frac{10}{25}$

$P(L|NS) = \frac{30}{75}$

$P(L) = \frac{2}{5}$

They are independent events.

3 Yes, because $P(F \text{ and } S) = P(F) \times P(S)$

Practice 4

1 a $P(LS | \text{operation}) = P(LS | \text{No operation}) = P(LS) = \frac{1}{9}$

The results are independent events the claim is not valid.

b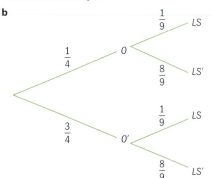

It is evident from the tree diagram $(LS | \text{operation}) = P(LS | \text{No operation})$

c Student's own answer with justification.

2 a

	T	T'	Total
G	10	15	25
G'	30	10	40
Total	40	25	65

b The events are not independent.
c Student's own answer.

3 a i $P(E) = 0.4$ **ii** $P(L|E) = 0.8$
iii $P(L|E') = 0.4$ **iv** $P(E \cap L) = 0.32$
v $P(E \cap L') = 0.08$ **vi** $P(E' \cap L) = 0.24$
vii $P(E' \cap L') = 0.36$ **viii** $P(L) = 0.56$
b No, because the conditional probabilities are different.

Practice 5

1 a

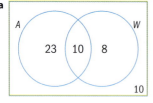

b i $P(A) = \dfrac{33}{51}$ **ii** $P(W) = \dfrac{18}{51}$
iii $P(A|W) = \dfrac{10}{18}$ **iv** $P(W|A) = \dfrac{10}{33}$
v $P(W|A') = \dfrac{8}{18}$

c

	A	A'	
W	10	8	18
W'	23	10	33
	33	18	51

2 a $\dfrac{12}{30}$ **b** $\dfrac{12}{70}$

c $P(M)$ does not equal $P(M|FT)$ does not equal $P(M|FT')$, therefore not independent.

3 a 88% **b** $\dfrac{4}{12}$

c No, from the tree diagram the second branches are not the same.

4 a $\dfrac{14}{30}$ **b** $\dfrac{14}{70}$

c Yes they are independent as
$P(M|L) = \dfrac{16}{30} = P(M|L') = \dfrac{64}{120}$

Mixed practice

1 a

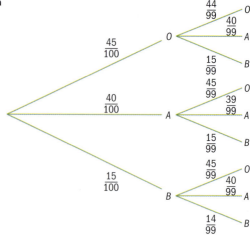

b $\dfrac{3750}{9900} = \dfrac{25}{66}$ **c** $\dfrac{40}{99}$ **d** $\dfrac{55}{99}$

2 a $\dfrac{6}{13}$ **b** $\dfrac{12}{13}$ **c** $\dfrac{2}{5}$ **d** $\dfrac{1}{3}$

e $-\dfrac{5}{13}$ **f** $\dfrac{1}{2}$ **g** $\dfrac{3}{4}$

3 a

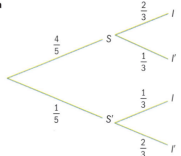

b $\dfrac{4}{5} \times \dfrac{2}{3} = \dfrac{8}{15}$

c $P(I|S) \neq P(I|S')$
∴ not independent

4 a

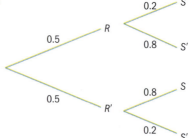

b Th 2: $P(R \cap S) = 0.1$
$P(R) = 0.5$
$P(S) = 0.1 + 0.4$
$= 0.5$
Th 3: $P(S|R) = 0.2$
$P(S|R') = 0.8$

$P(R \cap S) \neq P(R) \times P(S)$
∴ Not independent

∴ Not independent

5 a

	T	T'	
B	19	29	48
B'	11	11	22
	30	40	70

b $\frac{30}{70}$ **c** $\frac{29}{48}$ **d** Not independent, by Th 3

6 a

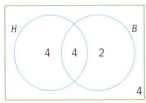

b i $\frac{8}{14}$ **ii** $\frac{6}{14}$ **iii** $\frac{10}{14}$

iv $\frac{4}{14}$ **v** $\frac{6}{14}$ **vi** $\frac{8}{14}$

vii $\frac{12}{14}$ **viii** $\frac{4}{14}$ **ix** $\frac{4}{14}$

c $P(H|B) = \frac{4}{6}$ $P(H|B') = \frac{4}{8}$

$P(B|H) = \frac{4}{8}$ $P(B|H') = \frac{2}{6}$

7 a

	Should have dress code	Should not have dress code	Total
Middle school	15	30	45
High school	40	80	120
Total	55	110	165

b Theorem 2
$P(MS \text{ and Dress code}) = P(MS) \times P(\text{dress code})$
$\frac{15}{165} = \frac{45}{165} \times \frac{55}{165}$

Theorem 3
$P(MS|DC) = P(MS|DC')$
$\frac{15}{55} = \frac{30}{110}$

Therefore they both work.

Review in context

1 a i $P(D) = \frac{45}{200} = \frac{9}{40}$ **ii** $P(N) = \frac{1}{2}$

iii $P(D|N) = \frac{1}{10}$ **iv** $P(D|N') = \frac{35}{100} = \frac{7}{20}$

b They are not independent from Theorem 3.

2 a

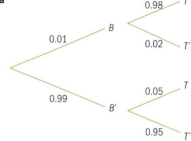

b $0.0098 + 0.9405 = 0.9503$

3 The events are independent.

4

	Smoker (A)	Non-smoker (A')	
Cancer (B)	1763	981	2744
(B')	73 237	124 019	197 256
	75 000	125 000	200 000

a $P(B|A) = \frac{1763}{75000}$

b $P(B|A') = \frac{981}{124\,019}$

c They are not independent from Theorem 2

d Students' own answer

5 a $P(B|A) = 0.69$ **b** $P(A) = 0.01$ **c** $P(B) = \frac{10}{330}$

b The events are not independent.

Index

A

A4 paper 203–4
additive systems 3
ambiguous case for the sine rule 243–5
amplitude of a sinusoidal function 135–7
Angle-Side-Angle Congruence (ASA/AAS) 269, 277
angular displacement 127
areas 183–4
   area of a parallelogram 188
   area of a segment 188
   area of a triangle 184–9
   area of a triangle formula 185, 199
   area of an equilateral triangle 188
   sine and cosine rules 189–94, 199
   sine and cosine rules, applying 195–9
argument of a logarithm 111
arithmetic series 146, 151, 164
asymptotes 218, 226, 229, 231

B

base of a logarithm 111
bases, number 3–17
bell curve 86
binary numbers 16–18

C

change of base formula 111
circles 124–5
   area of a segment 188
   radian measure and the unit
      circle 125–34, 142
column vectors 56–7, 73
common ratio, r 153–4
components, of vectors 59
composition of functions 29–32
compounded interest 117
conditional probabilities 299–300
   conditional probability axiom 302, 312
   making decisions with probability 309–12
   mathematical notation 301, 312
   representing conditional probabilities 305–9
   Theorem 1 302
   Theorem 2 302, 312
   Theorem 3 302, 312
   what is conditional probability? 301–4
congruence 264–5, 277
   conditions for congruence 269, 277
   CPCTC 273, 277
   establishing congruent triangles 265–73
   shapes 266
   sign 270
   using congruence to prove other
      results 273–6
   using similarity to prove other results 276–7

cosine rule 183–4, 189–94
   application 195–9
   formula 192, 199
CPCTC 273
cuboids 168–176

D

data 76–7
   dispersion 77–87
   mean 82
   normal distribution 86–7, 94
decay problems, growth and 114–116
decimal exponents 98, 105–6
decimal number system 5, 9–10
decomposing functions 32
diagonals 170, 171, 180
dilation 211, 213, 221–2, 229
   logarithmic functions 211, 213
   rational functions 221–2, 229
dimensions 167–8
   describing space 168–75
   distance formula 172, 180
   dividing space 175–9
   face diagonals 170
   finding the midpoint 176, 180
   higher dimensional space 179–80
   planes 169, 180
   section formula 178, 180
   space diagonals 170–1, 180
dispersion 77–87
distance formula 172, 180
dot product 68, 73

E

Euclid 273
exponential equations 109, 121
exponential functions 202, 203
   from exponential to logarithmic functions 203–5
   inverses 205
exponents
   decimal exponents 98, 105–6
   fractional exponents 97–8, 98–102, 106

F

face diagonals 170
four dimensions 179–80
fractals 163
fractional exponents 97–102, 106
frequency of a sinusoidal
      function 135–7
function operations 22
functions 21–2, 35–6
   decomposing functions 32–3
   doing and undoing 50–1

exponential functions 202–5
function operations 22–9
inverse functions 42, 44–50, 51
logarithmic functions 202–12, 213
modelling using compositions of functions 29–32
one-to-one functions 38–44, 51
onto functions 37–8, 51
rational functions 216–31
reversible and irreversible changes 44–50
sinusoidal functions 134–41, 142
types of 37–43

G

Gallivan, Britney 204
Gauss's method 149
generalizations 250–4
   proving generalizations 255–8
geometric series 146, 154, 164
   infinite geometric series 162, 164
grad/gradian 125
graphs of logarithmic functions 206–7, 209, 213
graphs of sinusoidal functions 134–41
growth and decay problems 114–116

H

hexadecimal 17
hieroglyphic number system, 3–14
horizontal asymptote 218, 231
   equation 226, 229
horizontal dilations 254
horizontal line test 37, 51
horizontal shift 136, 142
horizontal translation of a sinusoidal function 135–7
horizontal translations 254
hyperbolas 219

I

indices 97–8, 102–5, 106
infinite geometric series 162, 164
internet search ranking 255
inverse functions 42, 44–50, 51
   exponential functions 205
   inverse of a rational function 230–1
isosceles triangle theorem 274

K

Koch snowflake 163

L

Leibniz, Gottfried 6
line of symmetry 42

logarithmic functions 202–3, 213
   dilation 211, 213
   from exponential to logarithmic functions 203–5
   graphs of logarithmic functions 206–7, 209, 213
   properties of logarithmic functions 205–10
   reflection 211, 213
   transformations of logarithmic functions 210–12, 213
   translation 211, 213
logarithms 108–9, 109–14
   argument 111, 121
   base 111, 121
   change of base formula 111, 121, 257, 261
   creating generalizations 250–4
   laws of logarithms 249–54, 261
   natural logarithms 117–21, 121
   power rule 251, 261
   product rule 251, 261
   proving generalizations 255–8
   quotient rule 251, 261
   unitary rule 251, 261
   using logarithms to solve equations 114–17, 258–61
   zero rule 251, 261

M

magnitude, vectors 60–61
mapping diagram 49
Mayan number system 11–12
mean 82
midpoint, finding 176, 180
modelling compositions of functions 29–32
multiple transformations, order of 229
multiplication tables, different bases 16

N

Napier, John 118, 251, 256
non-linear inequalities 282–3, 297
   algebraic solutions to non-linear inequalities 291–6
   non-linear inequalities in one variable 283–8
   non-linear inequalities in two variables 288–91
   solving a quadratic inequality algebraically 292, 297
   solving a quadratic inequality graphically 284, 297
   solving non-linear equalities in two variables graphically 289, 297
normal distribution 86–7, 94
number systems 2–3
   binary numbers 16–18
   numbers in different bases 3–12
   performing operations 12–16

O

one-to-one functions 38–44, 51
onto functions 37–8, 51
ordered pairs 41

P

paper folding formula 204
parallelograms, area of 188
perpendicular height 185
perpendicular vectors 69, 73
phase shift 135, 136, 142
place value system 4
planes 169, 180
populations 79–80
   population growth
      equation 117, 121
   sample versus population 91–4
position vectors 57, 73
positive numbers 111, 121
positive real numbers 99, 101, 106
power rule of logarithms 250–1
   proving 255
product rule of logarithms 250–1
   proving 255
properties of graphs of logarithmic functions 206
pyramid scheme 157
Pythagoras' theorem 170, 192
Pythagorean identity 239
Pythagorean triples 171
Pythagorean quadruples 171

Q

quadrants 128–9
quotient rule of logarithms 250–1
   proving 255

R

radians 125–34, 142
   angles 127
   radian measures 126
radicals 97–8, 102–5, 106
radicands 104, 106
Ratio identity 239
rational functions 216–24, 231
   dilation 221–2, 229
   inverse of a rational
      function 230–1
   parent function, domain and range 226
   transforming rational functions 224–9
   translation 229
reciprocal functions 217, 219
rectangular hyperbola 219
reflection
   logarithmic functions 211, 213
reversible and irreversible changes 44
Right Angle-Hypotenuse-Side Congruence
      (RHS) 269, 277

S

samples 79
   sample versus population 91–4

scalars 61, 73
   scalar multiplication of a vector 65, 73
section formula 177–8
sequences 146
series 145–55, 164
   arithmetic series 146, 151, 164
   geometric series 146, 154, 164
   infinite geometric series 162, 164
   series and fractions 159–63
   series in real-life 155–9
Side-Angle-Side Congruence (SAS) 269, 277
Side-Side-Side Congruence (SSS) 269, 277
Sierpinski Triangle 163
sine rule 183–4, 189–94
   application 195–9
   formula 189, 199
sinusoidal functions 134–41, 142
   horizontal shift 136, 142
   order of transformations 141, 142
   phase shift 135, 136, 142
SOHCAHTOA 132
space 167–8
   describing space 168–75
   distance formula 172, 180
   face diagonals 170
   finding the midpoint 176, 180
   higher dimensional space 179–80
   planes 169, 180
   section formula 178, 180
space diagonals 170, 171, 180
special triangles 132
standard deviation 76–7, 94–5
   calculating from summary statistics 89–90
   different formulae for different purposes 87–91
   sample versus population 91–4
   ultimate measure of dispersion 77–87
sum of a sequence 147

T

theodolite 198
transformation
   exponential functions 211–12
   logarithmic functions 210–13
   rational functions 224–9
   sinusoidal functions 141, 142
translation
   logarithmic functions 211, 213
   rational functions 229
   vectors 63, 73
tree diagrams 301–11
trial and improvement 110–111
triangles 183–4
   area 184–9
   area of a triangle formula 185, 199
   area of an equilateral
      triangle 188
   CPCTC 273, 277
   establishing congruent triangles 265–73

triangular numbers 147
trigonometry 233–4
  ambiguous case 242–6
  cosine rule 183–4, 189–99
  Pythagorean identity 239, 246
  ratio identity 239, 246
  sine rule 183–4, 189–99
  trigonometric equations 234–8
  trigonometric identities 235, 238–42, 246

U

unbiased estimate 92
unit circle 126–133
unitary rule of logarithms 250–1

V

vectors 54–60
  column vectors 56–7, 73
  dot product 68, 73
  equal components 59, 73
  magnitude 60–2, 73
  operations with vectors 62–70
  perpendicular vectors 69, 73
  position vectors 57, 73
  resultants 63, 73
  scalar multiplication of a vector 65, 73
  translations 63, 73
  vector addition law 63–64
  vector geometry 70–2
  zero vectors 57, 73
Venn diagrams 305–310
vertical asymptote 218, 231
  equation 226, 229
volume diagonal 171

X

x-y plane 169

Z

zero rule of logarithms 250–1
zero vectors 57, 73

OXFORD
UNIVERSITY PRESS

Great Clarendon Street, Oxford, OX2 6DP, United Kingdom

Oxford University Press is a department of the University of Oxford. It furthers the University's objective of excellence in research, scholarship, and education by publishing worldwide. Oxford is a registered trade mark of Oxford University Press in the UK and in certain other countries.

© Oxford University Press 2017

The moral rights of the authors have been asserted.

First published in 2017

All rights reserved. No part of this publication may be reproduced, stored in a retrieval system, or transmitted, in any form or by any means, without the prior permission in writing of Oxford University Press, or as expressly permitted by law, by licence or under terms agreed with the appropriate reprographics rights organization. Enquiries concerning reproduction outside the scope of the above should be sent to the Rights Department, Oxford University Press, at the address above.

You must not circulate this work in any other form and you must impose this same condition on any acquirer.

British Library Cataloguing in Publication Data
Data available

978-0-19-835619-6

1 3 5 7 9 10 8 6 4 2

Paper used in the production of this book is a natural, recyclable product made from wood grown in sustainable forests. The manufacturing process conforms to the environmental regulations of the country of origin.

Printed in Great Britain by Bell and Bain Ltd. Glasgow.

Acknowledgements

The publishers would like to thank the following for permissions to use their photographs/illustrations:

Cover image: NASA; p2: MarcelClemens / Shutterstock; p3: Eric Isselee / Shutterstock; p6: wanchai / Shutterstock; p8: Jon Mackay; p11: Everett Historical / Shutterstock; p12: Wollertz / Shutterstock; p14: Richard Peterson / Shutterstock; p16: asharkyu / Shutterstock; p17: Jon Mackay; p20: dedek / Shutterstock; p21: RTimages / Shutterstock; p29: Sergey Dubrov / Shutterstock; p30: George Dolgikh / Shutterstock; p32: jeff Metzger / Shutterstock; p35: BortN66 / Shutterstock; p48: Andresr / Shutterstock; p54: Scott David Patterson / Shutterstock; p71: perspectivestock / Shutterstock; p76: PEPPERSMINT / Shutterstock; p80: sisqopote / Shutterstock; p85: Jacek Chabraszewski / Shutterstock; p91: Dionisvera / Shutterstock; p92: Boris Sosnovyy / Shutterstock; p93: Eric Isselee / Shutterstock; p94: Monkey Business Images / Shutterstock; p95: Jonathan Lenz / Shutterstock; p96: LeventeGyori / Shutterstock; p98: Stephen Finn / Shutterstock; p108: Ales Liska / Shutterstock; p113: Eugene Sim / Shutterstock; p116: visceralimage / Shutterstock; p117: wanchai / Shutterstock; p119: Seaphotoart / Shutterstock; p120: injun / Shutterstock; p120: Marco Tomasini / Shutterstock; p123: Christian Delbert / Shutterstock; p124: bunnyphoto / Shutterstock; p127: ILYA AKINSHIN / Shutterstock; p127: photka / Shutterstock; p128: AVS-Images / Shutterstock; p140: EpicStockMedia / Shutterstock; p144: SKA Organisation; p145: Pinosub / Shutterstock; p157: Jonathan Noden-Wilkinson / Shutterstock; p162: greenland / Shutterstock; p166: Andrey Pavlov / Shutterstock; p167: edobric / Shutterstock; p168: Jon Mackay; p171: SmileStudio / Shutterstock; p175: Jon Mackay; p182: IM_photo / Shutterstock; p183: Gail Johnson / Shutterstock; p197: Jon Mackay; p198: Dmitry Kalinovsky / Shutterstock; p200: Antiqua Print Gallery / Alamy Stock Photo; p202: photopixel / Shutterstock; p203: Paul André Belle-Isle / Shutterstock; p204: nuttakit / Shutterstock; p205: hironai / Shutterstock; p207: Konstantin Yolshin / Shutterstock; p215: photopixel / Shutterstock; p216: szefei / Shutterstock; p217: Amra Pasic / Shutterstock; p233: Teresa Levite / Shutterstock; p248: 3Dsculptor / Shutterstock; p249: Dmitry Pichugin / Shutterstock; p264: Podfoto / Shutterstock; p279: © The Carlisle Kid; p279: Tyler Olson / Shutterstock; p281: mama_mia / Shutterstock; p281: Joel_420 / Shutterstock; p282: noolwlee / Shutterstock; p286: Maks Narodenko / Shutterstock; p296: Monkey Business Images / Shutterstock; p299: Daxiao Productions / Shutterstock p306: Tina Renceij/123rf.

Although we have made every effort to trace and contact all copyright holders before publication this has not been possible in all cases. If notified, the publisher will rectify any errors or omissions at the earliest opportunity.

Links to third party websites are provided by Oxford in good faith and for information only. Oxford disclaims any responsibility for the materials contained in any third party website referenced in this work.